Non-local Structural Mechanics

Series Editor
Noël Challamel

Non-local Structural Mechanics

Danilo Karličić
Tony Murmu
Sondipon Adhikari
Michael McCarthy

iSTE

WILEY

First published 2016 in Great Britain and the United States by ISTE Ltd and John Wiley & Sons, Inc.

ISTE Ltd
27-37 St George's Road
London SW19 4EU
UK

www.iste.co.uk

John Wiley & Sons, Inc.
111 River Street
Hoboken, NJ 07030
USA

www.wiley.com

Library of Congress Control Number: 2015955807

British Library Cataloguing-in-Publication Data
A CIP record for this book is available from the British Library
ISBN 978-1-84821-522-1

Contents

Preface

Nanoscale experiments demonstrate that the mechanical properties of nanodimensional materials are much influenced by “size effects” or “scale effects”. Over the past decade, the non-local elasticity theory (non-local mechanics) has emerged as one widely promising size-dependant continuum theory. Significant progress has been made in fundamental and applied computational research in this area. The robust conventional local elasticity theory underpins the bulk of application of continuum mechanics in applied science and engineering since its inception in the early 19th Century. The utilization of the local elasticity theory in the context of nanoscale objects (such as carbon nanotube and graphene structures, etc.) has been questioned repeatedly in various research articles over the past decade. Non-local elasticity theory, pioneered from 1970s, can be applied over all scales and is considered to be more suitable for analyzing popular nanoscale objects such as carbon nanotube and graphene sheets.

This book is an initial comprehensive text to cover non-local elasticity theory for static, dynamic and stability analysis of a wide-ranging nanostructures. The authors have drawn on their own research experience to write this book. The text is written from a mechanics standpoint, comprising fundamental and complex theories that are relevant across a wide range of nanomechanical systems. The book introduces the reader to the fundamentals, as well as more in-depth aspects, of non-local mechanics and the associated latest research applications. The book brings together the vast research work for non-local mechanics in the context of nanoscale structures such as nanotubes and graphene sheets. The aim of this book is to systematically present the latest developments in the modeling and analysis of popular nanostructures. The authors have chosen to focus on the mathematical and computational aspects. This book will be relevant to aerospace, mechanical and civil engineering disciplines and various subdisciplines within them. The intended readers of this book include senior undergraduate students and graduate students doing projects or doctoral research in the field of small-scale structures. Researchers, professors and practicing engineers working in the field of small-scale structures will find this book useful.

There are very few books which are dedicated to non-local continuum mechanics, one of which is the classic book by Eringen from 1980s. Eringen's book explains the fundamental and origin of non-local theory. The book by Gopalakrishnan and Narendar [GOP 13] is based on wave propagations in nanostructures where the non-local continuum mechanics is presented. The fundamentals of wave propagation in nanotubes and topics such as rotating nanotubes, coupled nanotubes and nanotubes are addressed in this book. A recent book by Elishakoff [ELI 12] discusses in some detail the fundamental aspects of non-local beam mechanics for nanostructures applications. These books represent the state-of-the art at the time of their publications. And the contents of non-local mechanics theory are limited. The aim of this present book is to devote all its chapters on applications of non-local mechanics to nanoscale structures. As a significant amount of work has recently gone into the research of non-local mechanics and many recent publications have been achieved, this book also covers some of these latest developments with an introduction to fundamentals in a concise way, focusing on theoretical and computational aspects, although some references to experimental works are given. This book aims to give science and engineering graduate students and researchers a detailed understanding of the methods of non-local analysis necessary for nanoscale structures.

Our book covers the essential fundamental applications and important references related to non-local mechanics theory. Chapter 1 gives an introduction to non-local elasticity mechanics. Vibration analysis of the simplest non-local elasticity theory which is the non-local rod theory is considered in Chapter 2. Chapter 3 considers non-local elastic beam theories in details. Important theories such as non-local Euler–Bernoulli beam theory, non-local Timoshenko beam theory and non-local Reddy beam theory are presented in the context of vibration and buckling. Chapter 4 gives an introduction of non-local mechanics to two-dimensional small-scale structures via non-local plate theories. Non-local mechanics applied to simple double-nanobeam system is considered in Chapter 5. Chapter 6 considers double-nanoplate-system based on non-local elasticity theory. Chapter 7 describes the applications of non-local mechanics to multiple nanostructures. Cases related to multiple-nanorod, multiple-nanobeam and multiple-nanoplates are addressed in the context of vibration and buckling. Chapter 8 takes up the topic of use of computational method such as finite element method under the umbrella of non-local mechanics. Finite element methods for dynamics of non-local systems are concisely addressed in this chapter. Examples such as axial vibration of nanorod, bending vibration of nanobeams and transverse vibration of nanoplates are presented. How the non-local finite element is applied to nanodimension structure such as single-walled carbon nanotube, double-walled carbon nanotube and single layer graphene sheets is illustrated. Chapter 9 gives a detailed description of dynamic finite element analysis of axially vibrating non-local rods. Cases of mechanical damping are addressed in this chapter. Chapter 10 describes an important application of non-local mechanics to graphene structures such as in the field of vibration-based mass nanosensors. As non-local mechanics theories are recently validated with the molecular dynamics simulations, in Chapter 11 we give an introduction to molecular dynamics for small-scale structures.

This book is a result of 7 years of research in the area of non-local mechanics theory. The book's initial chapters began taking shape when Professor Adhikari and Dr Murmu were working on project of "scale dependent theory for nanomechanical systems" in Civil and Computational Engineering Center, University of Swansea, Wales, UK. Later chapters originated from research works with numerous colleagues, students, collaborators and mentors. We are deeply indebted to all of them for numerous stimulating scientific discussions, exchanges of ideas and in many occasions' direct contributions toward the intellectual content of the book. The authors particularly like to thank Dr S. C. Pradhan (IIT Kharagpur), Professor P. Kozić (University of Niš), Professor M.I. Friswell (Swansea University), Dr Y. Lei (Chansha), Professor F. Scarpa (University of Bristol), Dr C. Wang (Swansea University), Professor W.A. Curtin (École Polytechnique Fédérale de Lausanne) and Dr M. Cajić (Serbian Academy of Sciences, Belgrade).

Besides the names mentioned here, I am thankful to many colleagues, fellow researchers and students working in this field of research around the world, whose names cannot be listed here. The lack of explicit mentions by no means implies that their contributions are any less important. The opinions presented in this book are entirely of the authors, and none of our colleagues, students, collaborators and mentors has any responsibility for any shortcomings.

Tony MURMU
Danilo KARLIČIĆ
Sondipon ADHIKARI
Michael MCCARTHY
October 2015

1

Introduction to Non-local Elasticity

Recently, interest in nanotechnology is growing rapidly. The inventions of carbon nanotubes (CNTs) by Iijima [IIJ 91, IIJ 93] and successful extraction of graphene sheets [GEI 07] have motivated this interest. Because of its novel potential applications, recently nanomaterials have gained considerable attention among experimental, computational and theoretical research communities. As compared to more conventional materials, these nanomaterials possess superior mechanical, thermal, electrical and electronic properties [MOO 11]. Now, it is possible to arrange atoms into nanostructures that are only a few nanometers in size. For utilization and engineering of these nanoelements, proper experimental, computational and continuum mechanics-based methodologies are needed for future analysis in nanoengineering. One of the updated continuum mechanics methods for analysis of nanostructures is the non-local elasticity theory. In this chapter, we introduce some fundamental aspects to illustrate why nanostructures and non-local elasticity theory are important.

1.1. Why the non-local elasticity method for nanostructures?

The understanding of the mechanical response of nanoscale structures (small-scale structures of nanometer dimension), such as bending, vibration and buckling, is indispensable for the development and accurate design of nanostructures such as carbon nanotubes (CNTs) and graphene-based nanodevices. Figure 1.1 shows a single-walled CNT and single-walled graphene sheets. The dots in the figure represent carbon atoms. So far, experimentation on the study of actions of structures at the nanoscale is achievable, but quite difficult. Handling each and every parameter at the nanometer scale is a complicated task. Furthermore, computer simulation methods such as molecular dynamics (MD) modeling and simulation of nanostructures is computationally very expensive and time-consuming for macroscale material systems. Also, from an engineer's perspective, we may not be expertise in MD as it involves more of a chemistry dealing with atoms, molecules, bonds and interatomic forces.

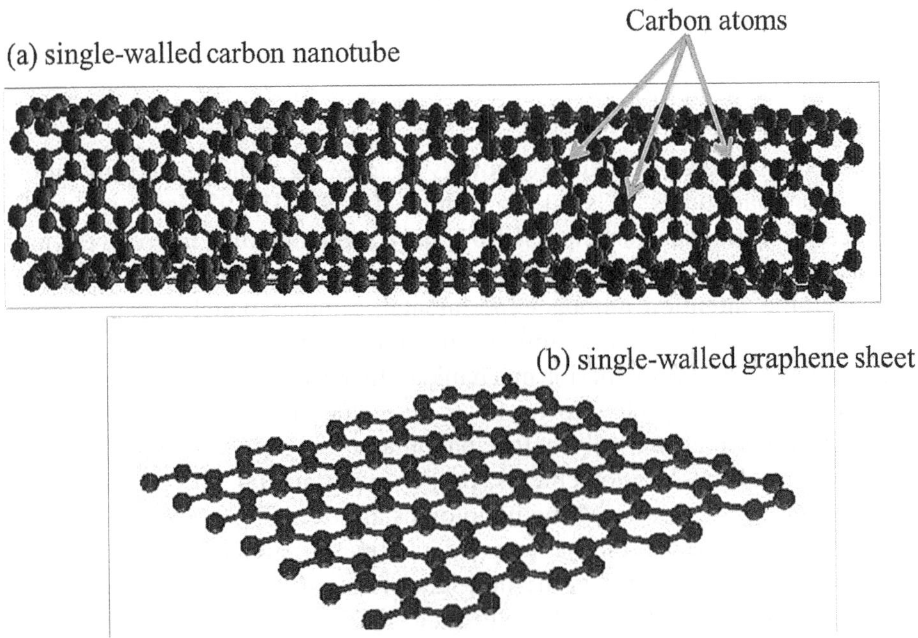

Figure 1.1. *Schematic diagrams of a) single-walled carbon nanotubes and b) single-walled graphene sheets. The mechanical behavior of these nanoscale structures can be analyzed by non-local elasticity theory along with molecular dynamics and experimental work. For a color version of the figure, see www.iste.co.uk/murmu/non-local.zip*

The experimentation and MD simulation for CNTs graphene and graphene-based systems are not always straightforward. So, how can these potential material nanostructures be effortlessly predicted in terms of bending, vibration, buckling and other studies for designing nanodevices (say in nanoelectromechanical systems, NEMS)? One approach is to utilize the enriched knowledge of available classical continuum mechanics. The continuum structural mechanics models continue to play an essential role in the mechanical study of CNT and graphene-based systems. Theories and design modules of macroscale structures, facilitated by engineers, are based on classical continuum models. The conventional local elasticity theory underpins the majority of application of continuum mechanics in applied science and engineering since its inception in the early 19th Century. However, the application of the local elasticity theory in the context of nanoscale objects has been repeatedly questioned in various research articles over the past decade. Classical continuum mechanics is a scale-effect-free theory and cannot be used in a nanoscale environment. Conventional continuum mechanics fails to predict size effects, which are present at small-length scales. At small scales, a material's microstructure becomes increasingly significant and its influence can no longer be ignored. The size effects are related to atoms and molecules and their interactions. Thus, updated size-dependent continuum-based methods are required in modeling graphene as they offer much faster solutions than MD simulations, while being capable of incorporating size effects due to the discontinuous and non-homogeneous nature of real materials. One popular size-dependent method frequently used to model bending, vibration and buckling behavior of CNTs and graphene sheets is the non-local elasticity theory. Local elasticity is based on the behavior of localness (point) irrespective of the surrounding, while non-local elasticity takes into account the influence of the surrounding. This effect is more prominent and intuitive at the atomic scale (nanoscale) where an atom is affected by other surrounding atoms. The beauty of the non-local method is that it can capture atomistic effects at the nanoscale and yet impart results for the whole body.

The new structural non-local method can bridge the gap between MD and scale-effect-free continuum mechanics to provide a viable means of studying such important nanoscale objects beyond CNTs and graphene.

1.2. General modeling of nanostructures

Modeling and simulation of nanostructures such as CNTs, buckyballs, graphene and nanoelectromechanical systems are important for an optimum design. It is the scientific and engineering work involved in the analysis and design of nanostructures that support or oppose loads. By loads, we mean the forces (atomic or non-atomic), deformations or accelerations applied to the structure or its components. Load on nanostructure elements can be static as well as dynamic and its understanding is crucial. Examples of elementary nanostructural components which build up the complex structural systems (nanorobots, nanomachines and nanoelectro mechanical, nanocolumns, nanoplates (graphene sheets), nanoshells (CNTs), etc. The reliable structural modeling of nanoscale models will depend on the application of physical laws (e.g. quantum mechanics), correct mechanics (e.g. non-local mechanics), theories of materials science (e.g. lattice dynamics) and applied mathematics. This structural model will then be able to predict how nanostructures would support and resist imposed loads. The structural model will help in understanding its reliable performance over time and failure criterion under practical loads.

1.3. Overview of popular nanostructures

A nanostructure is a small object of intermediate size between molecular and microscopic (micrometer-sized) structures. The remarkable properties of nanostructures are the cause of intense research around the world. Therefore, these days an increasing number of nanoscale structures are being fabricated worldwide and are being employed as the building blocks in the emerging field of nanotechnology. Some of the nanoscale structures include nanoparticles, nanowires, nanobeams, nanorings, nanoribbons, nanoplates, nanotubes (CNTs), and components of nanomachines:

– *Nanoparticles*: these are small nano-objects considered as a whole unit with respect to its transport and properties. These particles exhibit size-dependant properties and have dimensions in the range of 1–100 nm. These nanoparticles can be incorporated into parent material to form advanced nanocomposites.

– *Nanobeams and nanorods*: these small-scale structures are categorized as one-dimensional nanostructures. These have applications in microelectromechanical systems (MEMS) and nanoelectromechanical systems (NEMS). Figure 1.2(a) shows the typical atomic force image of a nanorod of nanometer dimensions [WON 97].

– *Nanowires*: these are one-dimensional nanostructures with diameters in the range of nanometers. These nanowires generally have an aspect ratio, i.e length-to-diameter of 1,000 or more. They can be used to build the next generation of computing devices, improve solar cell devices, etc. A typical image of nanowire is shown in Figure 1.2(b).

– *Nanoplates*: these are recognized as two-dimensional nanostructures. The nanoplates are a new subgroup of bottom-up grown nanostructures with a two-dimensional shape. Examples of thin nanoplates are graphene sheet, gold nanoplates [AH 05], etc. A typical image of nanoplate is shown in Figure 1.2(c). The two-dimensional nanostructures have potential application in information storage, catalyst, transducers, solar cells, MEMS/NEMS and components in nanomachines, etc.

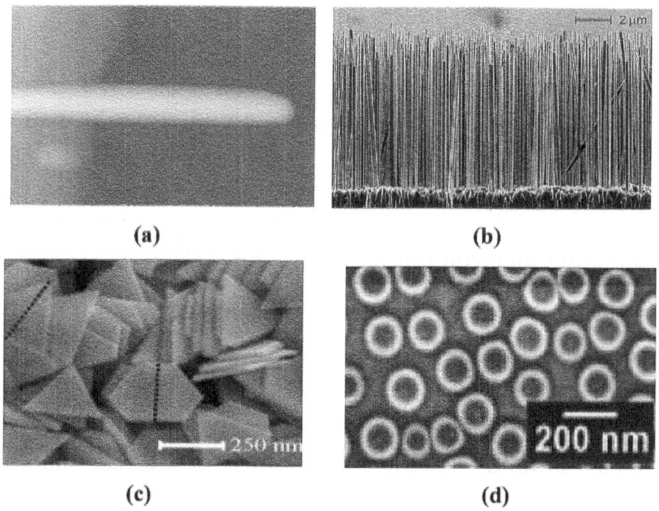

Figure 1.2. *a) An atomic force microscope image of nanorod with 35.3 nm diameter around 600 nm in length (courtesy of [WON 97]; b) nanowires (image from www.efocuss.com); c) high-magnification scanning electron microscope of single-crystalline gold nanoplates [AH 05]; d) scanning electron micrographs of nanorings made of 100 nm diameter (courtesy of [ZHU 04]). For a color version of the figure, see www.iste.co.uk/murmu/non-local.zip*

a) b)

Figure 1.3. *a) Scanning electron microscope of SWCNT grown on conical Si tip, b) MWCNT forest on glass substrate. Each rod-like element is the image of MWCNT with diameter of the order of tens of nanometers [DAI 02]*

– *Nanorings*: a nanoring is a small ring-formed crystal. The diameter is between 50 nm and 1 μm. The nanorings could serve as nanometer-scale sensors, resonators and transducers. These small-scale structures could provide a unique platform for studying piezoelectric effects and other phenomena at the small scale. Figure 1.2(d) shows an image of nanorings taken from scanning electron microscope [ZHU 04].

– *Nanoribbons*: these are thin strips of nanosheets or unrolled single-walled CNTs. Nanoribbons such as graphene nanoribbons may be a technological alternative to silicon semi-conductors due its semi-conductive properties.

– *Nanotubes*: among several nanoscale structures, nanotubes have aroused great interest in the scientific community because of their exceptional mechanical, electronic, electrochemical and electrical properties. Nanotubes are long and thin cylinders of macromolecules composed of carbon atoms in a periodic hexagonal arrangement. Generally, two varieties of these tubes have been distinguished, the single-walled CNT denoted as SWCNT (Figure 1.3(a)) and the multi-walled CNT denoted as MWCNT (Figure 1.3(b)), the latter consisting of a set of concentric single-walled tubes nested inside. A double-walled CNT is shown in Figure 1.4.

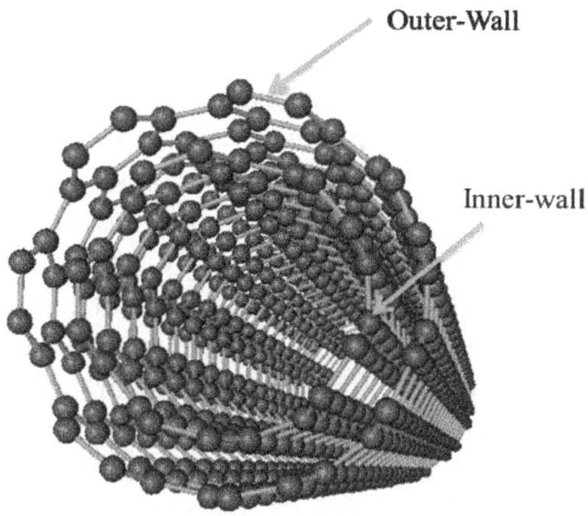

Figure 1.4. *Schematic diagram of double-wall carbon nanotubes. Study of bending, vibration and buckling of double-walled CNT is analyzed by using non-local elasticity as found in the scientific literature. In the later chapters, we show how the CNTs are modeled using non-local elasticity theory*

Carbon nanotubes (CNTs) hold exciting promise in useful potential applications, as electrodes in supercapacitors, as cable materials for space elevators, as structural elements in nanoscale devices and reinforcing element in superstrong and conducting nanocomposites, biomedical, bioelectrical, superfast microelectronics, solar cells, etc.

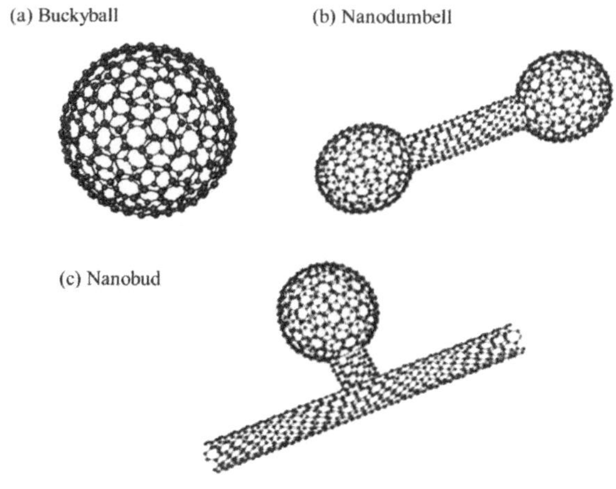

Figure 1.5. *Different types of carbon nanostructures*

– *Other complex nanostructures*: other nanostructures include hybrid complex nanostructures such as CNTs with attached buckyballs (spherical fullerenes) at the tip or at the span (nanobud). As cylindrical fullerenes, known as CNTs, spherical fullerenes are referred to as buckyballs. Buckyballs are cage-like fused-ring polycyclic systems of carbon atoms. As nanobuds [ARA 12] can be obtained by adjoining fullerene to CNT span, buckyballs can be incorporated into the tip of CNTs. The nanobuds are new structures where spherical fullerenes are covalently bonded to the outer sidewalls of the underlying nanotube. By similar method, buckyballs can be fixed to the ends of nanotubes. When buckyballs are added to the both ends of nanotube, we obtain a nanodumbbell. These CNT-buckyball systems can be utilized as state-of-the-art filler materials for strong tough nanocomposites. Figure 1.5 shows the different types of nanostructures.

1.4. Popular approaches for understanding nanostructures

The modeling and study of nanostructures such as graphene sheets can be done by various methods as depicted in Figure 1.6. Generally, the three popular methods by which the behavior of graphene is studied comprise experimental, MD simulation and continuum mechanics approach.

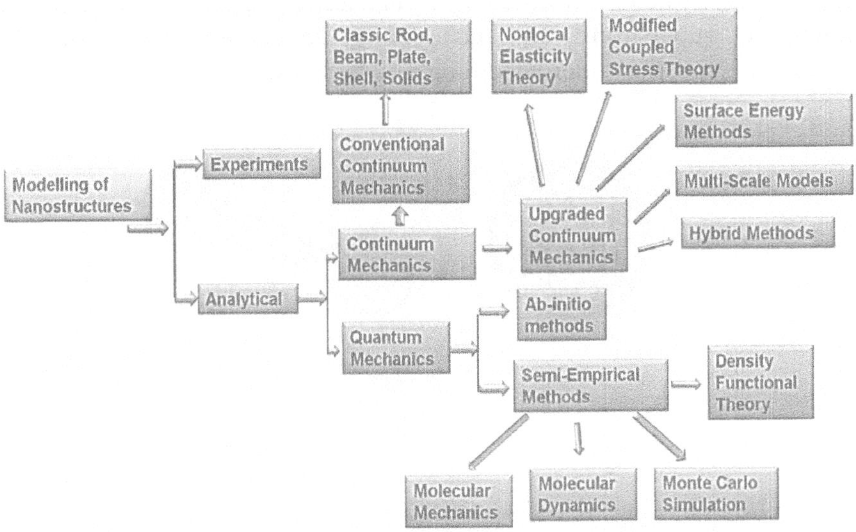

Figure 1.6. *General approaches used to model and study nanostructures*

1.5. Experimental methods

Some of the earlier and present day investigations on nanomaterials and nanostructures, such as CNTs and grapheme, are carried out by the experimental methods [HAN 11]. The details of experimental methods used are beyond the scope of this chapter. An experimental approach at the nanoscale is obviously a better way to analyze the behavior of grapheme as it is more realistic. However, in experimental study, controlling every parameter in nanoscale is a difficult task.

1.6. Molecular dynamics simulations

Due to the physical drawbacks and lack of scope for experimental method, many scientists and engineers resorted to atomistic-level simulation techniques. The behavior of graphene at atomistic levels can be simulated via the molecular dynamics (MD) [HAN 10, TAN 08]. MD simulation refers to expensive computer simulations depicting physical movements of atoms and molecules at the nanoscale. In an MD simulation, the motion of individual atoms within an assembly of N atoms or molecules is modeled on the basis of either a Newtonian deterministic dynamic or a Langevin-type stochastic dynamic, given the initial position coordinates and velocities of the atoms. Applying Newton's equations of motion, the trajectories of molecules and atoms are determined. Potential functions are defined according to which particles will interact. In MD simulations, the forces between the particles and potential energy are defined by molecular mechanics force fields. The molecular simulation methods, however, suffer from the disadvantage that these are sophisticated, require larger computational resources, require solving large number of equations and are highly expensive and time-consuming [MUR 12b].

1.7. Continuum mechanics approach

Because the experimental and atomistic computational approach, though realistic, suffers from the drawback that it is computationally expensive, time-consuming and requires greater expertise, one alternative is to utilize the available knowledge of classical continuum mechanics. Can classical continuum mechanics deal with structures of nanoscale dimensions? To answer this, classical continuum modeling (e.g. classical Kirchhoff's plate theories) of nanostructures such as graphene has thus received an increasing amount of attention. These continuum mechanics theories [AND 04] have thus started to play an important role in characterizing overall mechanical responses of nanoscale materials that are fundamental structural and functional building blocks in engineering nanostructures.

In continuum mechanics, the mechanical behavior of graphene is modeled as a continuous mass rather than as discrete particles and it is assumed that there is no empty space between particles (atoms). Theories of structures constructed on the foundation of continuum mechanics include Euler–Bernoulli beam theory, Timoshenko beam theory, Kirchoff's plate theory, Mindlin plate theory, and classical shell theory, etc. Using Euler–Bernoulli beam theory and Timoshenko beam theory, modeling and prediction of mechanical response of CNTs have been attempted. The continuum (local elasticity) theory is based on the constitutive relation that stresses a point which depends on the strain at that point only.

1.8. Failure of classical continuum mechanics

Though the elastic continuum models described earlier could provide quick and approximate predictions of the mechanical behavior of graphene, these classical elasticity models fall short of addressing important issues such as "surface effects" and "size effects" when dealing with nanostructures. The source of these discrepancies becomes clear when considering the physics of atomic-scale interactions and stress production at the atomic scale. A lot of experimental evidence shows the presence of size effects in nanostructures [BAU 11, KIA 98, TAN 09, TAN 08, XIA 06, ZIE 10] which we can call small-scale effects. This implies that the effects arising in the behavior of nanostructures due to their small size cannot be ignored if we need to create cutting-edge and accurate design. The classical theory of elasticity being the longwave limit of the atomic theory excludes the size effects at nanoscale.

According to continuum mechanics, graphene particles (carbon atoms) completely fill the space they occupy. Modeling objects in this way ignores the fact that matter is made up of atoms, and so is not continuous. However, on length scales much in the order of interatomic distances, such conventional continuum models are questionable. Thus, there is a need to upgrade the conventional continuum theory to account for discreteness or size effects in graphene sheets. A way to upgrade the conventional continuum theory to account for the small scale or size effects in graphene sheets is by introducing the concept of surface effects, coupled stress and non-localness.

Non-local elasticity theory [ERI 83] is one popular size-dependent method frequently used to model bending, vibration and buckling behavior in graphene. Local elasticity is based on behavior of localness (point) irrespective of the surrounding. While non-local elasticity emphasizes the effects of the surrounding (e.g. neighboring atoms), this effect is more prominent and intuitive at atomic scale (nanoscale) where an atom is affected by other surrounding atoms. The application of other size-dependent theories such as couple stress theory and modified couple stress theory (MCST) for analyzing microstructures, graphene and nanotubes is an area of open research, and will be discussed elsewhere. In this chapter, we look into the concept of non-local elasticity and how it is utilized to develop non-local plate theories to investigate the structural response of graphene sheets. Various reports devoted to non-local elastic theories for the bending, vibration and buckling analysis of graphene nanoplates are found in the scientific literature [AKS 11, ANS 11a, ANS 10, BAB 11, MAL 11, MUR 09a, MUR 09b, MUR 09c, PRA 10, SHE 10, SHI 11]. A good review on the work on non-local elasticity theory applied to CNTs and graphene sheets can be found in [ARA 12].

1.9. Size effects in properties of small-scale structures

Ruud *et al.* [RUU 94] reported that the mechanical properties of thin films decrease with smaller characteristic lengths. They conducted experiments on nanoindentation of multilayered thin films. The hardness and the elastic modulus were measured experimentally on Ag and Ni thin layers. The characteristic lengths of the nanoelements were used from 1.3 to 2.3 nm. Wong *et al.* [WON 97] showed that the moduli of small structures change depending on the diameter of the nanobeams. They presented a discussion on research in Young's modulus, strength and toughness of nanotubes and nanorods. Li *et al.* [LI 03] reported the size effects on Young's modulus of ultra-thin silicon in the range of 12–170 nm. Their study of nanocantilevers showed that Young's modulus decreases monotonously as the cantilevers become thinner. The phenomenon of size effects was shown to be in line with the atomistic simulation results. Furthermore, their results showed that there is a monotonous change of resonant frequency for a 38.5 nm thick nanocantilever with the increase of length. Sun and Zhang [SUN 03] used a semi-continuum model to study the size effects in plate-like nanomaterials. They observed that the mechanical properties of the nanoplates, such as the stiffness and the Poisson's ratio, are size-dependant at nanoscale. Cuenot *et al.* [CUE 04] investigated the effects of reduced size on the elastic properties nanomaterials using atomic force microscopy (AFM). The elastic modulus was measured on silver and lead nanowires and on polypyrrole nanotubes. Their research showed that the elastic properties of the nanomaterials are significantly affected by size. Smaller sizes of nanomaterials exhibited higher modulus values than that of larger ones. Furthermore, they interpreted that at nanometer scales, the surface effects become prominent and significantly modify the macroscopic properties. Gua and Zhao [GUA 05] showed that mechanical properties such as stiffness and Poisson's ratio are size-dependent (changes with atomic layers). For the investigation, a three-dimensional lattice model was used considering surface relaxation with size-dependent elastic constants of a nanofilm. Various other cross-references on the size-effects dependence are discussed in detail in the paper. Wang *et al.* [WAN 06] discussed the size dependency of properties at the nanoscale. They identified the intrinsic length scales of several physical properties at the nanoscale. Further, they showed that for nanostructures whose characteristic sizes are much larger than these scales, the properties obey a simple scaling law. Recently, Olsson *et al.* [OLS 07] carried out atomistic simulations of mechanical properties of iron nanobeams. Both tensile and bending stiffnesses have been determined employing molecular static simulations. From the molecular simulations, it was observed that there is strong size dependence in Young's modulus. The size dependence was attributed to the surfaces and edges deviating elastic properties, which can be stiffer and more compliant than the bulk.

1.10. Evolution of size-dependent continuum theories

To incorporate the size-dependence phenomenon, Miller and Shenoy [MIL 00] developed a continuum model for nanosize structural elements. The phenomenon of surface elasticity is accounted for in the continuum model. They found that the important length scales are the ratio of the surface elastic modulus to the bulk elastic modulus. He *et al.* [HE 04] extended the work of Miller and Shenoy [MIL 04] to develop a two-dimensional size-dependant continuum approach. The size-dependant continuum model was based on surface elasticity. The continuum model was developed for analyzing the static response of ultra-thin elastic films of nanoscale-thickness. They interpreted that the size-dependence is due to the dependence of surface stress on strain. Park and Gao [PAR 06] developed a continuum mechanics model based on MCST containing an internal length-scale parameter for capturing the size effects. A variational principle based on the principle of minimum potential energy is employed. The model was used for bending analysis of micro Euler–Bernoulli beam. A multi-scale continuum model based on micromorphic theory was developed by Vernerey *et al.* [VER 07]. The characteristic lengths of the microstructure were incorporated into the constitutive relation. The model has the capability of capturing the size effects arising from the microstructure and the behavior of the material across various scales. Kong *et al.* [KON 08] proposed MCST for dynamic analysis of microbeams. The governing equations were derived based on Euler–Bernoulli beam theory and Hamilton's principle. They found that the natural frequencies predicted by their new model are larger than that predicted by the classical beam model. They interpreted that due to intrinsic size dependence of materials, the stiffness increases and hence the frequency increases. The conclusions were presented for both simply-supported and cantilever microbeams. Furthermore, Kong *et al.* [KON 09] assessed the size effects on the microbeam bending response and its natural frequencies employing strain gradient elasticity theory. Chandraseker *et al.* [CHA 09] introduced an atomistic-continuum Cosserat rod model for the simulation of single-walled CNTs. A literature review on these continuum theories is highlighted in the paper. Recently, Tsiatas [TSI 09] proposed a modified Kirchoff's plate model for the static analysis of isotropic microplates based on MCST. The theory is able to capture the size effects. The size-dependent plate theory used for small-scale structures is the MCST. How well the theory can be applied to graphene sheets is not known, as the theory was originally used for microscale plates. However, recently, the theory is being applied to nanotubes and graphene sheets. It predicts that the frequencies and buckling load increase with the increase in scale parameter (equivalent to non-local parameter). Consequently, there is hardening of bending stiffness (contradictory to non-local elasticity). The difference arises due to the consideration of different extrinsic lengths, i.e. for the MCST it is thickness of the plate, whereas in non-local elasticity theory it is the length of the specimen (Figure 1.7).

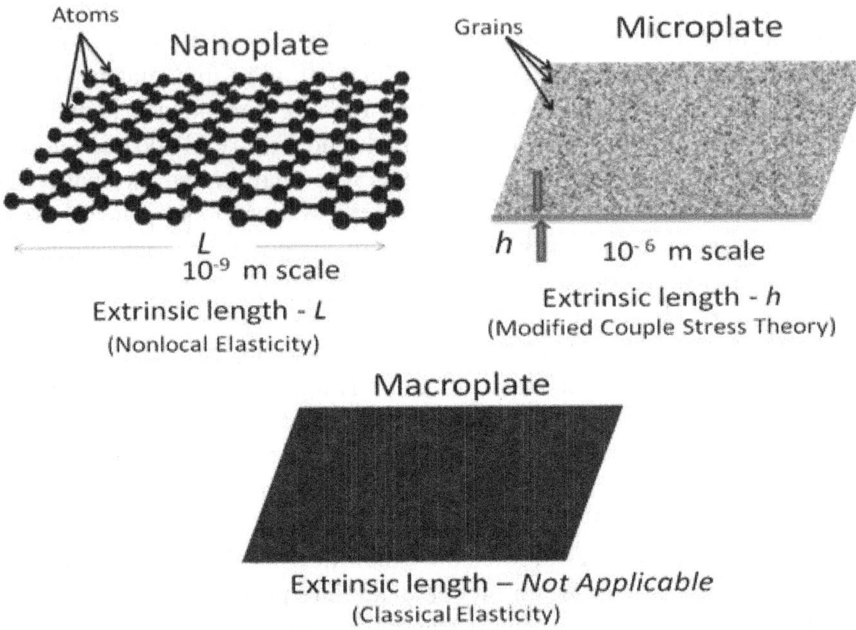

Figure 1.7. *Extrinsic lengths in plates at different scales*

1.11. Concept of non-local elasticity

It is well known that one of the basic principles of classical continuum mechanics is that it eliminates the influence of strains field of distant points on the reference point. Non-local continuum theory, or non-local elasticity theory, on the other hand, incorporates these distant effects known as non-local effects. The essence of the non-local elasticity theory is that the stress field at a reference point ξ in an elastic continuum depends not only on strain at that point but also on the strain at every other point ξ' in the domain [AMA 10, BES 11, ECE 07, ERI 83, HEI 10, RED 08]. This is in accordance with the atomic theory of lattice dynamics and experimental observations on phonon dispersion [ERI 83]. When the effects of strain at all other points are neglected, we obtain the classical elasticity theory. One of the advantages of non-local solutions is that it eliminates singularities predicted by the classical (local) theory of elasticity. It has been successfully applied on the problems of dispersion of waves, dislocations, fracture mechanics, etc. The most general form of the constitutive relation for non-local elasticity involves an integral over the whole body. The integral contains a kernel function that portrays the influences of strains at various locations on the stress at a given location. The kernel function is also referred to as influencing function or non-local modulus. In mathematical concept, classical elasticity is recovered when the non-local moduli become Dirac delta functions. By non-local elasticity, relations are established between the non-local constitutive moduli based on the continuum and atomic lattice dynamics.

The small-scale effects are accounted for in this non-local theory by considering internal size as a material parameter allowing working with non-local variables conceived. The interatomic forces and atomic length scales directly come to the constitutive equations as material parameters. The basic idea of the non-local elasticity is to establish a relationship between macroscopic mechanical quantities and nanoscale physical quantities within the framework of continuum mechanics. The general non-local elasticity continuum theory is generally depicted by a set of integropartial differential equations. The constitutive equations of non-local elasticity appear as integral equations, in terms of strain tensor, either the Fredholm equation of first kind or second kind. The non-local elasticity approach can be qualified as "integral" or "strongly" non-local when it expresses the stress at a point of a material domain as a weighted value of the entire strain field. It can be qualified as "gradient" or "weakly" non-local when the stress is expressed as a function of the strain and its gradients at the same points. For development of non-local models associated with different problems and applications, proper non-local kernels are to be selected.

In a nutshell, the non-local theory of Eringen [ERI 83] has the ability to predict behavior of the large nanosized structures, while avoiding solving the large number of equations. Non-local elasticity has great potential applications in closing the gap between the classical continuum limit and the atomic theory of lattices.

1.12. Mathematical formulation of non-local elasticity

1.12.1. *Integral form*

The following assumptions are considered for the non-local elasticity equation: elastic, linear, homogenous and isotropic solid. The basic concept of non-local elasticity without body forces is represented by the following set of equations:

$$\sigma_{ij,i} + \rho\left(f_j - \ddot{u}_j\right) = 0,$$

$$\sigma_{ij}(\mathbf{x}) = \int_V H(|\mathbf{x}' - \mathbf{x}|, \phi)\, \sigma_{ij}^{c}(\mathbf{x}')\mathrm{d}V(\mathbf{x}'),$$

$$\sigma_{ij}^{c}(\mathbf{x}') = \lambda\varepsilon_{rr}(\mathbf{x}')\delta_{ij} + 2\mu\varepsilon_{kl}(\mathbf{x}')$$

$$\varepsilon_{kl}(\mathbf{x}') = \frac{1}{2}\left(\frac{\partial u_i(\mathbf{x}')}{\partial x'_j} + \frac{\partial u_j(\mathbf{x}')}{\partial x'_i}\right) \qquad [1.1]$$

The terms σ_{ij}, σ_{ij}^{c}, ε_{kl}, C_{ijkl} are non-local stress, classical stress, classical strain and fourth-order elasticity tensors, respectively. The terms λ and μ are Lame's first and second parameters, respectively. The volume integral is over the region **V** occupied by the body. The kernel function $H(|\mathbf{x}' - \mathbf{x}|, \phi)$ is known as the non-local modulus or influencing kernel function. The non-local modulus acts as an attenuation function incorporating into the constitutive equations the non-local effects at the reference point **x** produced by local strain at the source $\mathbf{x}'$. The term $|\mathbf{x}' - \mathbf{x}|$ represents the distance in the Euclidean form and ϕ is a material constant that depends on the internal (e.g. lattice parameter, granular size and distance between the C-C bonds) and external (e.g. crack length and wave length) characteristic lengths. The non-local modulus reaches its maximum at $\mathbf{x} = \mathbf{x}'$ attenuating with $\mathbf{x}' - \mathbf{x}$. The non-local modulus has the dimension of $(\text{length})^{-3}$.

The single parameter ϕ based on non-local elasticity is expressed as:

$$\phi = \frac{e_0 l_c}{l_e} \qquad [1.2]$$

where e_0 is the constant appropriate to each material and l_c and l_e denote the internal characteristic length and the external characteristic length, respectively. Examples of the internal characteristic length are lattice parameter, granular size and distance between C-C bonds. While the external characteristic length includes crack length, wave length and sample size of the nanostructure, length of CNTs and graphene sheets (sample size) is generally used as external characteristic length. In other size-dependent continuum mechanics such as MCST, thickness of the structure is considered as external characteristic length (Figure 1.2). According to Eringen [ERI 83], the value of $e_0 = 0.39$, was obtained by matching the dispersion curves from Born–Karman model of lattice dynamics and non-local theory.

If the non-local term (parameter) approaches zero, i.e. $\phi \rightarrow 0$, H reverts to the Dirac delta measure. Consequently, classical elasticity is included in the limit of vanishing internal characteristic length. When $\phi \rightarrow 1$, non-local theory approximates atomic lattice dynamics.

1.12.2. *Non-local modulus*

For a given material, the non-local modulus H can be determined by matching the dispersion curves of plane waves with those of atomic lattice dynamics or experiments. By defining different forms of non-local modulus, various approximate models of non-local elasticity can be obtained. One such important example of two-dimensional non-local modulus is expressed as [ERI 83]:

$$H(|\mathbf{x}|, \phi) = \left(2\pi l_e{}^2 \phi^2\right)^{-1} K_0\left(\sqrt{\mathbf{x} \cdot \mathbf{x}}/l_e \phi\right) \quad [1.3]$$

where K_0 is the modified Bessel function. From equation [1.3], it can be seen that the integral of the equation over the domain yields unity. Further, one advantageous property of non-local modulus H is that when $\phi \rightarrow 0$, the non-local elasticity reduces to classical elasticity. The non-local modulus function of equation [1.3] is popularly used for the analysis of graphene sheets (and also for CNTs).

1.12.3. *Differential form equation of non-local elasticity*

Since the governing relation of non-local elasticity, equation [1.1], is an integro-partial differential equation, it is generally difficult to solve. To simplify the governing non-local relation, we assume that the non-local modulus $H(|\mathbf{x}' - \mathbf{x}|, \phi)$ as Green's function of linear differential operator:

$$L_o H(|\mathbf{x}' - \mathbf{x}|, \phi) = \delta(|\mathbf{x}' - \mathbf{x}|) \quad [1.4]$$

where L_o is the linear differential operator and δ is the Dirac-delta function. Green's function is chosen in conjunction with the properties of non-local modulus. By applying L_o to equation [1.1], we obtain:

$$L_o\sigma_{ij} = \sigma_{ij}^c \qquad [1.5]$$

The differential operator L_o has different forms for different expressions of non-local modulus. For the non-local modulus depicted in equation [1.4] by combining with equation [1.5] would yield:

$$L_o = 1 - {l_e}^2\phi^2\nabla^2 \qquad [1.6]$$

where ∇^2 is the Laplacian operator. Therefore, according to equations [1.5] and [1.6], the non-local constitutive relation [1.1] can be expressed in differential form as:

$$[1 - (e_0 l_c)^2\,\nabla^2]\sigma_{ij} = \lambda\varepsilon_{rr}\delta_{ij} + 2\mu\varepsilon_{kl} \qquad [1.7]$$

Equation [1.7] can be written in a simplified manner as [AYD 09, ERI 83, RED 08]:

$$[1 - (e_0 l_c)^2\nabla^2]\sigma_{ij} = C_{ijkl}\varepsilon_{kl} \qquad [1.8]$$

where C is the fourth-order elasticity tensor.

1.13. Non-local parameter

The material parameter e_0 is a constant appropriate to each material. The parameter e_0 is estimated such that the relations of the non-local elasticity model could provide satisfied approximation of atomic dispersion curves of plane waves with those of atomic lattice dynamics. Furthermore, the value of key parameter e_0 can be found by comparing non-local results with MD simulation results and atomistic-based techniques. The terms e_0, $e_0 l_c$ and $(e_0 l_c /l_e)$ are referred to as the non-local parameters or the scale coefficient. In many of the scientific literature, l_c and l_a are denoted by the symbol a and l, respectively. The common terms of nonlocal parameters are $e_0 a$ and $(e_0 a /l)$. Some different non-local parameter values proposed and used by some researchers are highlighted below.

Non-local parameters	Magnitudes	Researchers
a	0.142 nm	[SUD 03]
e_0	0.39	[ERI 83]
	0.288	[WAN 05]
	0–19	[DUA 07]
e_0a	0.7 nm	[WAN 08]
	0–2nm	[WAN 07]
	0–2 nm	[DUA 07]
	<2.1 nm	[WAN 05]
e_0a/l	0–0.8	[LU 06]

Table 1.1. *Non-local parameter proposed by various researchers*

1.14. Non-local elasticity theory versus molecular dynamics

Non-local elasticity theory has the potential to predict the vibration response and buckling phenomenon of graphene sheets under certain assumptions. The non-local elasticity theory has also been successfully applied to the prediction of CNTs. According to some recent works, non-local elasticity theory applied to CNTs has been compared with MD simulations, and a good agreement is found. However, this theory of non-local elasticity strongly depends on the optimized value of the non-local parameter. In this chapter, we mentioned the range of non-local parameter as 0–2 nm because the exact value of non-local parameter is not known. Details on the various values of non-local parameter as reported by various researchers are discussed in [NAR 11].

Various non-local beams and shell theories for studying free vibration response of single-walled and double-walled carbon nanotubes (SWCNTs, DWCNTs) were compared with MD simulation results [ANS 11b, ANS 12] and a good agreement was found. The two approaches match at some non-local parameter values. According to Hu *et al.* [HU 08], MD simulations indicate that the wave dispersion predicted by the non-local elastic cylindrical shell theory shows good agreement with that of the MD simulations in a wide frequency range up to the terahertz region.

Khademolhosseini *et al.* [KHA 12] demonstrated the superiority and accuracy of the non-local elasticity model over classical theories in predicting the size-dependent dynamic torsional response of SWCNTs by comparing their results with MD simulations. According to Murmu and Adhikari [MUR 11], the frequency results of a cantilever SWCNT from non-local theory match very well with the frequency from MD simulation for $e_0a \approx 1.0$ nm.

Furthermore, according to Murmu and Adhikari [MUR 12a], the frequency from non-local elasticity theory matches very well with MD simulations for CNT-based nanoscale biosensors for some optimized values of non-local parameter.

Similarly, the validity of non-local plate theory has been addressed by comparing with the MD simulation. MD simulations for the free vibration of various graphene sheets using non-local plate theory with different values of side length and chirality can be found to be equivalent to the non-local plate model [ANS 10]. Using some optimized non-local parameters, the non-local plate model can predict the resonant frequencies with great accuracy. Furthermore, it is also shown that the non-local plate models as compared with MD simulation can provide a remarkably accurate prediction of the graphene sheet behavior under nonlinear vibration in thermal environments [SHE 10]. Thus, non-local plate theory presented in this chapter can be a reliable theory to predict the mechanical behavior of graphene considering that optimized non-local parameter or scale coefficient is used.

In the following chapters, we introduce the non-local elasticity theory for the formulation and analysis of nanoscale structures. Explicit expressions of buckling and vibration of nanoscale structures are obtained.

2

Non-local Elastic Rod Theory

This chapter deals with the longitudinal vibration of one-dimensional nanostructures, such as single-walled carbon nanotubes (SWCNT) or ZnO (zinc-oxide) nanotubes, modeled as nanorods, using the non-local elasticity theory. In order to obtain accurate continuum models of nanorods, the classical rod theory needs to be reformulated according to the non-local elastic constitutive relation and for corresponding boundary conditions. Such modified continuum models take into account small-scale effects, which are the key points in the description of nanoscale structures. By using D'Alembert's principle, the governing equation of motion of a nanorod model is obtained for the non-local elastic constitutive relation. In addition, we analyze a nanorod system for two types of boundary conditions: clamped–clamped and clamped–free. The exact analytical solutions for natural frequencies of a nanorod are found by applying the method of separation of variables based on the Fourier infinite series. At the end of this chapter, several numerical examples are given in order to show the influence of non-local effects and boundary conditions on the vibration behavior of a nanorod.

2.1. Background

Recently, various one-dimensional nanostructures have been realized. They include nanodots, nanorods, nanowires, nanobelts, nanotubes, nanobridges, nanonails, nanowalls, nanohelixes and seamless nanorings. Among them, one-dimensional nanostructures such as nanotubes [WON 97], nanorods [NAN 11] and nanowires [WAN 01] are widely studied. The main reason for this is their simple material formation and device application [XIA 03].

Nanorods [WEN 03] are one-dimensional nanostructures whose lengths are in the range of 1 to 3,000 nm (Figure 2.1). These miniscule structures in the form of nanorods or nanowires can be grown using various methods. The popular methods include: (i) vapour phase synthesis [PAR 02], (ii) metal-organic chemical vapour deposition [GAL 06] and (iii) hydrothermal synthesis [LIU 03]. Nanorods have found applications in a variety of nanodevices. These include ultraviolet photodetectors, nanosensors, transistors, diodes and LED arrays [WAT 10].

The majority of the literature on non-local elasticity exists pertaining to the free transverse vibration of micro/nanobeams (CNTs), nanoplates (graphene sheets) and nanoshells (CNTS). This is due to the fact that the transverse modes are more prone to get excited by common types of external excitations. On the contrary, only a limited amount of work is devoted to the longitudinal vibration of nanobeams and nanoplates. In what follows, we will review some of the papers where the free or forced vibration of complex systems based on nanorods coupled with elastic, viscoelastic or other types of layers are examined for various boundary conditions, aspect ratios and small-scale parameter. Danesh *et al.* [DAN 12] accounted for the small-scale effect in the longitudinal vibration of tapered nanorod using the non-local elasticity theory. The authors used the differential quadrature method to solve motion equations for three different boundary conditions. A carbon nanotube embedded in an elastic medium was modeled by Aydogdu [AYD 12] as a nanorod surrounded with elastic layers by using Eringen's non-local elasticity theory. The author compared the longitudinal frequencies for the non-local and classical continuum models. Narendar and Gopalakrishnan [NAR 11] considered the non-local effects in the axial wave propagation within the system of two nanorods coupled with an elastic layer. The authors studied the influence of small-scale (non-local) parameter and stiffness of the layer on axial wave propagation. Hsu *et al.* [HSU 11] investigated the longitudinal frequencies of cracked nanobeams for different boundary conditions and using the theory of non-local elasticity. A wide study of the longitudinal, transversal and torsional vibration and instability was conducted by Kiani [KIA 13] for a system of SWCNTs. Şimşek [SIM 12] used a Galerkin approach to obtain the natural frequencies for the longitudinal vibration of axially functionally graded tapered nanorods. The author performed the analysis for nanorods with a variable cross-section, differently tapered ratios, material properties and boundary conditions. Longitudinal vibration of nanorods, which takes the non-local long-range interactions into account, was examined by Huang [HUA 12]. Chang

[CHA 12] considered the small-scale effects to investigate the axial vibration of elastic nanorods. The author used the differential quadrature method to solve the model equations. Filiz and Aydogdu [FIL 10] analyzed the longitudinal vibration of carbon nanotubes with heterojunctions using the non-local elasticity for different lengths, diameters and chirality of heterojunctions. Karličić *et al.* [KAR 15] performed a detailed analysis of the free longitudinal vibrational response of the system with coupled viscoelastic nanorods and investigated the influence of different physical parameters on complex natural frequencies. Murmu and Adhikari [MUR 10] analyzed the longitudinal vibration of double-nanorod system (DNRS) coupled with an elastic layer. They investigated the effects of the non-local parameter, layer stiffness and aspect ratio on the free longitudinal vibration of DNRS. Moreover, Karličić *et al.* [KAR 15] examined the longitudinal vibration of the multiple-nanorod system embedded in an elastic medium and subjected to a transversal magnetic field. The authors obtained the system of partial differential equations in terms of axial displacements based on D'Alamberts principle and non-local elastic constitutive relation. Exact analytical solutions for natural frequencies were also obtained by applying the method of separation of variables into the governing equation of the system.

Figure 2.1. *Multiple ZnO nanorods growing on central nanowire cores. The lengths of nanorods range from 100 to 3,000 nm, while the widths of nanowire cores are in the range of 10–20 nm [WEN 03]*

Like transverse vibration, axial vibration experiments can also be used to determine the Young modulus of CNTs [WEI 08]. Generally, the flexural modes occur at low frequencies. However, vibrating nanobeams (CNTs) may also have longitudinal modes at relatively high frequencies and can be very important for practical applications in high operating frequencies. When used as electromechanical resonators, nanorods can be externally excited and exhibit axial vibrations [ADH 13]. Furthermore, for a moving nanoparticle inside a single-walled carbon nanotube (SWCNT), the SWCNT generally vibrates both in the transverse and longitudinal directions. The longitudinal vibration is generated because of the existing friction between the outer surface of a moving nanoparticle and the inner surface of SWCNT. It is also reported [SAP 06] that transport measurements on suspended SWCNTs show signatures of phonon-assisted tunneling, influenced by the longitudinal vibration (stretching) modes.

In this chapter, we analyze the longitudinal vibration behavior of a single nanorod, based on the non-local elasticity theory. By using D'Alembert's principle, the governing equation of motion is derived and then solved by using the method of separation of variables. In this study, two types of boundary conditions are considered for the nanorod: clamped–clamped and clamped–free. The solutions for natural frequencies are obtained analytically in the exact form. At the end of this chapter, some conclusions are drawn according to the theoretical and numerical results obtained in the numerical study.

2.2. Governing equation of motion of the nanorod

Let us consider a nanorod of length L and cross-sectional area A, as shown in Figure 2.2(a)). Let us also assume that the cross-sectional areas are constant along the x coordinate, even though, in general, it could be arbitrary. We will further assume that the nanorod is elastic and homogeneous. In order to consider the free longitudinal vibration of a nanorod, we analyze the vibration in the direction of the x axis. Now, we can consider the equilibrium of an infinitesimal element of the length dx that is taken at a coordinate location x, as given in Figure 2.2(b). On the left side of the differential element, we take that the axial force N is the resultant of the normal stress σ_{xx} acting internally on the cross sectional area A. The stress resultant N varies along the length and it is a function of time $N = N(x, t)$. In addition, the continuously distributed axial load $f(x, t)$ that comes from externally applied sources is given. On the right end of the differential element, we have $+dN$.

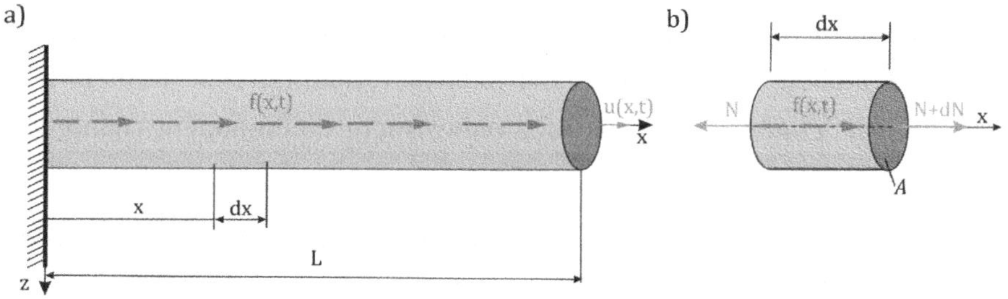

Figure 2.2. *a) Mechanical model of the longitudinal vibration of a nanorod; b) differential element with corresponding axial stress resultants*

By applying D'Alembert's principle, the sum of all forces with respect to x axes yields

$$-N + (N + dN) + f(x,t)dx = \rho A dx \frac{\partial^2 u}{\partial t^2} \qquad [2.1]$$

where $dN = \frac{\partial N}{\partial x}dx$ is the differential part of the axial stress resultant N,$u = u(x,t)$ is the axial displacement in the x-direction. After some transformations of the above equation, the equilibrium equations are given in the form (see [KAR 15])

$$\frac{\partial N}{\partial x} + f(x,t) = \rho A \frac{\partial^2 u}{\partial t^2} \qquad [2.2]$$

in which N is the stress resultant defined as

$$N(x,t) = \int_A \sigma_{xx} dA \qquad [2.3]$$

where the term σ_{xx} represents the normal stress in the infinitesimal element of a nanorod.

According to the non-local elasticity theory for one-dimensional case, which is presented in Chapter 1, we obtain the constitutive relation for a non-local elastic body in the differential form as [KAR 15]

$$\sigma_{xx} - (e_0 a)^2 \frac{\partial^2 \sigma_{xx}}{\partial x^2} = E\varepsilon_{xx} \qquad [2.4]$$

where the term (e_0a) in [2.1] denotes the non-local parameter also defined Chapter 1 and $\varepsilon_{xx} = \frac{\partial u}{\partial x}$ denotes the axial deformation. The term E is the conventional Young's modulus of a nanostructure component (nanorod or nanobeam).

Substituting [2.3] into [2.4], the axial stress resultant for the non-local theory is obtained as

$$N - (e_0a)^2 \frac{\partial^2 N}{\partial x^2} = EA \frac{\partial u}{\partial x} \qquad [2.5]$$

where A is the cross-section defined as

$$A = \int_A dA \qquad [2.6]$$

Now, the governing equation of motion can be expressed in terms of the axial displacement $u(x,t)$ for the non-local elastic constitutive relation. Introducing [2.2] into [2.5] we obtain the following equation of motion

$$\rho A \frac{\partial^2 u}{\partial t^2} - f(x,t) - EA \frac{\partial^2 u}{\partial x^2} = (e_0a)^2 \frac{\partial^2}{\partial x^2}\left(\rho A \frac{\partial^2 u}{\partial t^2} - f(x,t)\right) \qquad [2.7]$$

In what follows, we utilize [2.7] for the development of a mathematical model for the free longitudinal vibration of a non-localnanorod. The governing equation [2.7] can be simplified by assuming that that the external continuous distributed axial load $f(x,t)$ is equal to zero. Also, we consider two types of boundary conditions. Supposing that the ends of two nanorods are clamped–clamped (Figure 2.3(a)) and clamped–free (Figure 2.3(b)), the boundary conditions are given by:

– clamped–clamped:

$$u(0,t) = u(L,t) = 0, \qquad [2.8a]$$

– clamped–free:

$$u(0,t) = N(L,t) = 0, \qquad [2.8b]$$

where $N(L,t)$ is the non-local axial force at the end of a nanorod. From [2.2] and [2.5] we obtain

$$N(x,t) = (e_0 a)^2 \rho A \frac{\partial^3 u}{\partial x \partial t^2} + EA \frac{\partial u}{\partial x}. \qquad [2.9]$$

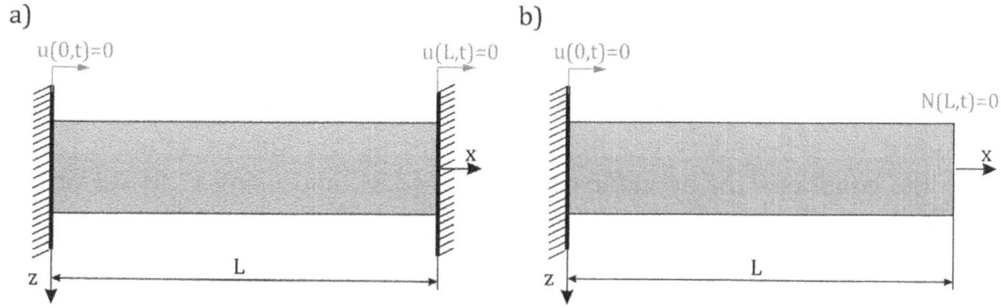

Figure 2.3. *Boundary conditions for the longitudinal vibration of a nanorod, a) clamped–clamped and b) clamped–free*

The exact closed form solution of the governing equation of motion with corresponding boundary conditions is obtained by using the method of separation of variables, which is based on the Fourier infinite series. By assuming the time harmonic motion and applying the method of separation of variables, the solution of governing equation of motion [2.7] takes the form

$$\theta(x,t) = \sum_{n=1}^{\infty} X_n(x) T_n(t), \qquad [2.10]$$

where $T_n(t) = e^{i\omega_n t}$ is the time function and $X_n(x)$ is the corresponding mode shape function, which depends on the boundary conditions of the system. Introducing the assumed solution [2.10] into [2.7] and neglecting the continuously distributed axial load $f(x,t)$, one obtains the following ordinary differential equation for the corresponding mode shape function $X_n(x)$ of the nanorod in the following form

$$\frac{d^2 X_n(x)}{dx^2} + \alpha_n^2 X_n(x) = 0, \qquad [2.11]$$

where α_n denotes the characteristic values determined from the corresponding boundary conditions as

$$\alpha_n^2 = \frac{\omega_n^2}{\frac{E}{\rho} - (e_0 a)^2 \omega_n^2}. \qquad [2.12]$$

We can note that the differential equation [2.11] is the well-known equation of the harmonic oscillator, where the solution is given as

$$X_n(x) = A_n sin\alpha_n x + B_n cos\alpha_n x, \tag{2.13}$$

for unknown constants A_n and B_n. These two unknown constants are obtained from boundary conditions. Introducing the known solution [2.13] into the equations for boundary conditions [2.8], and using [2.9] we obtain two transcendental equations from which we can find values of α_n.

For the clamped–clamped boundary conditions (Figure 2.3(a))

$$sin\alpha_n L = 0, \tag{2.14a}$$

the roots are

$$\alpha_n L = n\pi, \quad n = 1,2,\dots,\infty, \tag{2.14b}$$

with the corresponding mode shape function $X_n(x)$ as

$$X_n(x) = A_n sin\alpha_n x = A_n sin\frac{n\pi}{L}x. \tag{2.15}$$

For the clamped–free boundary conditions (Figure 2.3(b))

$$cos\alpha_n L = 0, \tag{2.16a}$$

the roots are

$$\alpha_n L = \frac{(2n-1)\pi}{2}, \quad n = 1,2,\dots,\infty, \tag{2.16b}$$

with the corresponding mode shape functions $X_n(x)$ as

$$X_n(x) = A_n sin\alpha_n x = A_n sin\frac{(2n-1)\pi}{2L}x. \tag{2.17}$$

Now, we can obtain non-local natural frequencies for the longitudinal vibration of a nanorod from expression [2.12] and corresponding boundary conditions as

$$\omega_n = \sqrt{\frac{E\alpha_n^2}{\rho[1+(e_0a)^2\alpha_n^2]}}, \tag{2.18}$$

where α_n is defined in [2.14a] for the clamped–clamped boundary conditions and in [2.16b] for the clamped–free boundary conditions.

2.3. Results and discussions

A non-local rod model for the longitudinal vibration of a nanorod is developed based on Eringen's non-local continuum mechanics. In this model, the small scale effects are explicitly taken into account. An exact analytical method is proposed to find non-local frequencies of an axially vibrating nanorod. Clamped–clamped and clamped–free boundary conditions are employed and their explicit relationships are derived. It is found that the longitudinal vibration frequencies of a nanorod are overestimated by the classical (local) nanorod model as it disregards the effect of the small-length scale (Table 2.1). The results are also compared with molecular dynamic simulations. Unlike the transverse vibration of nanobeams, the fundamental natural frequency of an axially vibrating nanorod has a decreasing nature for an increase in the non-local parameter. Similar to the transverse vibration of nanobeams, thenon-local effects for the longitudinal vibration of a nanorod are more pronounced in the higher modes of vibration.

In the following, we compare the longitudinal natural frequency obtained by the non-local elasticity theory and the MD simulation for one SWCNT. Thus, the observed fundamental frequency, i.e. the natural frequency of a nanorod model presented in this chapter, is compared with the frequency obtained for longitudinally vibrating SWCNT via the molecular dynamics simulation in [CAO 06]. For this purpose, we rewrite the natural frequency in the form $f = \omega_1/2\pi$, i.e. in a similar manner as given in [CAO 06]. The following dimensions and values of parameters are used to obtain the results given in Table 2.1: $\rho = 9517$ [kg/m^3], $E = 6.85$ [TPa], $L = 12.2$ [nm], $n = 1$ and $L = 12.2$ [nm]. The results are obtained for clamped–free boundary conditions and different values of $e_0 a$ and compared with the results for the natural resonant frequency of a single-walled carbon nanotube (SWCNT) (5, 5) obtained in [CAO 06] by the molecular dynamics simulation. In the local elasticity case when $e_0 a = 0$, the natural frequency is overpredicted and a better agreement is achieved when the non-local effect is considered by increasing the value of $e_0 a$ towards the unit value. Since the value of the internal length scale parameter $a$is fixed for C–C bonds in SWCNT, to fit the results for the natural frequency it is necessary to optimize the value of parameter e_0 for each problem of boundary conditions and geometrical properties of nanostructures. Since the molecular dynamics study is computationally prohibitive for nanoscale systems with a large number of atoms, the non-local theory should be used to obtain satisfying results for more complex systems as shown in Table 2.1.

Boundary conditions		Non-local parameter					MD simulation
C-F		$e_0a=$ 0 nm	$e_0a=0.5$ nm	$e_0a=$ 1 nm	$e_0a=1.5$ nm	$e_0a=$ 2 nm	[CAO 06]
[2.18]	f	0.549763	0.548627	0.545262	0.539788	0.532395	0.544

Table 2.1. *Comparison of the lower non-local natural frequency $f = \omega_1/2\pi$ of nanorod model given in (THz) with the results obtained with the MD simulation for the armchair SWCNT (5, 5)*

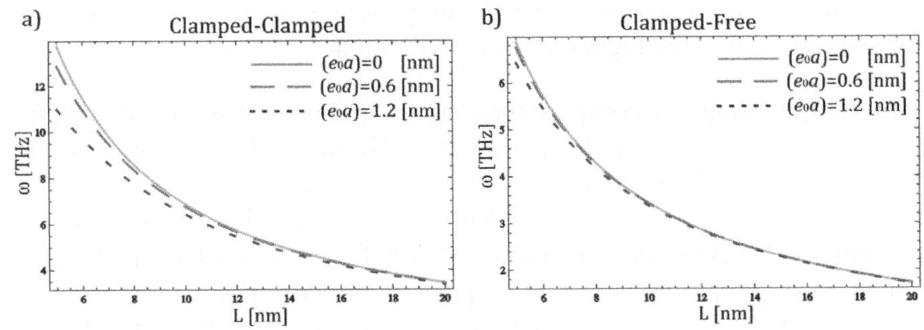

Figure 2.4. *Fundamental natural frequencies for the longitudinal vibration of a nanorod: a) clamped-clamped boundary conditions, and b) clamped-free boundary conditions*

Figure 2.4 shows the effects of the non-local parameter and the length of the nanorod on the natural frequency for the longitudinally vibrating nanorod for two types of boundary conditions, clamped–clamped and clamped–free. For this purpose, we consider the armchair (8, 8) SWCNT with thickness ($h = 0.34$ nm), Poisson's ratio ($\nu = 0.3$), mass density ($\rho = 2300$ kg/m^3) and Young's modulus ($E = 1.1$ TPa). In the present analysis, the values of the non-local parameter are adopted from Chapter 1. It should be noted that the fundamental natural frequency represents the lowest natural frequency of the nanorod for $n = 1$. From the figures, we note a significant decrease in the natural frequency for an increase in the length of the nanorod. Moreover, the natural frequency decreases for an increase in the non-local parameter but this effect is weakening for an increase in the length of the nanorod. From the physical point of view, we can conclude that the non-local parameter has a dampening effect on natural frequencies, as well as on the overall stiffness of the nanorod. Finally, the natural frequency of the clamped–clamped nanorod is significantly higher than the frequency of the clamped–free system. This is caused by the fact that the clamped–clamped nanorod is a more strongly constrained system than the clamped–free one.

3

Non-local Elastic Beam Theories

The aim of this chapter is to analyze static bending, vibration and buckling of one-dimensional nanostructures such as single-walled carbon nanotubes (SWCNTs) based on the non-local elasticity theory. In order to obtain an accurate continuum model of SWCNT, the Euler–Bernoulli, Timoshenko and Reddy beam theories are reformulated according to the non-local elastic constitutive relation with corresponding boundary conditions. Such modified continuum models take into account small-scale effects, which are playing an important role in description of a mechanical behavior of nanostructures. By using D'Alembert's principle, the governing equations of motion of a nanobeam model representing the SWCNT are obtained for non-local Euler–Bernoulli and Timoshenko beam theories. However, for higher beam theories such as the non-local Reddy beam theory, the Hamilton principle is more appropriate for derivation of governing equations of motion. The torsional vibrations of a nanobeam with two types of boundary conditions are also considered. The exact analytical solutions for static deflection, natural frequencies and buckling loads of a nanobeam are found by applying the method of separation of variables based on the Fourier infinite series. At the end of this chapter, several numerical examples are given to show the influence of non-local effects and beam size on the static bending, vibration and buckling of a nanobeam.

3.1. Background

In order to reach the full potential of nanomaterials such as SWCNTs, it is essential to understand their mechanical behavior. Recently, the bending, vibration and buckling studies of SWCNTs modeled as nanobeams have attracted much attention of the scientific community. Such one-dimensional nanostructures are being increasingly used in numerous nanodevices such as biosensors [WAN 05], mass sensors [CHI 08], resonators [LI 03] and nanoactuators [FRA 02]. Therefore, the issue of analyzing the mechanical behavior of nanobeam like nanostructures is important from the practical point of view and it has a wide application in nanoengineering design [AJA 01]. To perform and control experiments on the nanoscale level is a very difficult and expensive task. This is one of the reasons why many researchers focus their investigation on the development of theoretical models. Different atomistic methods are developed and split into the two main groups deterministic and stochastic methods observing the nanostructures as discrete systems. Some of the widely exploited methods are molecular dynamics, density function theory, etc. These theories are very important and useful in the analysis of simpler and smaller nanostructures. Nevertheless, they are very cumbersome and computational prohibitive when applied on larger or complex nanostructured systems. Therefore, theoretical models based on the continuum mechanics approach have attracted significant attention due to their simplicity and applicability. Since the mechanical behavior of a nanostructure strongly depends on the size effects, when the system is very small compared to the molecular distance, the classical continuum theories need to be modified in order to consider such effects. In the non-local elasticity theory, small-scale effects are introduced into the constitutive relation with a single material parameter and assuming the stress at a point is a function of strains at all other points of elastic body. As an extension of the classical beam theory based on the classical theory of elasticity, researchers have developed a non-local beam theory based on Eringen's constitutive equations, [ERI 72a] and [ERI 72b], which has been recently applied in a number of studies for the static and dynamic analysis of nanostructures [KAR 14a, KAR 14b, KAR 14c, KAR 15, KAR 14c, KAR 14d]. Reddy [RED 07] proposed reformulated equations of motion for Euler–Bernoulli, Timoshenko, Reddy and Levinson beam theories by using the Eringen's non-local theory and then evaluated static bending, vibrations and buckling response of beams for simply-supported boundary conditions. A comparison of the stress gradient (Eringen's non-local theory), and two strain gradient theories with molecular dynamics simulation, applied for the free vibration analysis of Euler–Bernoulli and Timoshenko beams, was carried out by Ansari *et al.*

[ANS 12]. Further, Aydogdu [AYD 09] used different beam theories including those of Euler–Bernoulli, Timoshenko, Levinson, Reddy and Aydogdu to analyze bending, buckling and vibration of nanobeams. The author applied different displacement functions and forms of non-local constitutive relations to derive motion equations and investigate the effects of non-locality and length of a nanobeam on displacement, natural frequency and critical buckling load. The results are also compared for all beam theories. By searching the literature, it was found that the problem of non-local torsional vibration is poorly researched. However, the torsional vibration has a great practical importance in design procedures of nanoshafts and their application in nanoelectromechanical systems (NEMS) devices. Lim *et al.* [LIM 12] developed new elastic non-local stress model for torsional vibration of circular nanotubes. The authors obtained the analytical solution and investigated the influence of some physical parameters on dynamical behavior of a nanotube. Murmu *et al.* [MUR 11] studied the problem of torsional vibration of SWCNT with a buckyball at the end of the nanotube. Based on the non-local continuum theory, the authors obtained torsional frequency equation and analytical solutions for natural frequencies of some special cases. In recent years, many papers on comparison of the results obtained from molecular dynamics simulations and non-local continuum mechanics have been published [DUA 07, ANS 11]. It can be concluded that the results obtained from both theories are in good agreement. More details on the molecular mechanics method will be given in Chapter 7.

This chapter is organized as follows. In section 3.2, we introduce the non-local Euler–Bernoulli, Timoshenko and Reddy beam theories static bending, transverse vibration and buckling of a nanobeam. Based on D'Alembert's and Hamilton's principle, the governing equations are derived to describe the motion of the system. Exact analytical expressions for static deflection, natural frequencies and buckling loads of simply-supported nanobeam are obtained in the form of trigonometric series. Torsional vibrations of a nanobeam for two types of boundary conditions are considered in section 3.3. Based on D'Alembert's principle and non-local constitutive relation, the governing equation of motion is derived and then solved by using the method of separation of variables. The solutions for natural frequencies are obtained analytically in the exact form. In section 3.4, some conclusions are drawn according to the theoretical and numerical results obtained in this chapter.

3.2. Non-local nanobeam model

Here, we will give some basic assumptions for mechanical modeling of SWCNTs as nanobeams. First, classical Euler–Bernoulli, Timoshenko and Reddy beam theories are reformulated and adopted for analysis of the bending, vibration and buckling of a nanobeam based on the non-local elasticity theory of Eringen. Afterward, a detailed analytical study of the free torsional vibration of nanobeam with two types of boundary conditions is considered.

In Figure 3.1, we show an SWCNT influenced by compressive axial forces with attached transverse load. It is assumed that an SWCNT is modeled as a slender simply-supported nanobeam, where geometrical and physical properties are defined as follows: A is the cross-sectional area, E is Young's modulus, ρ is the mass density and L is the length of a nanobeam. The x-coordinate is taken along the length of nanobeam, the z-coordinate is taken along the thickness and y-coordinate is taken along the width of a nanobeam. The transverse displacement over the nanobeam is denoted by $w(x,t)$ as shown in Figure 3.1(b).

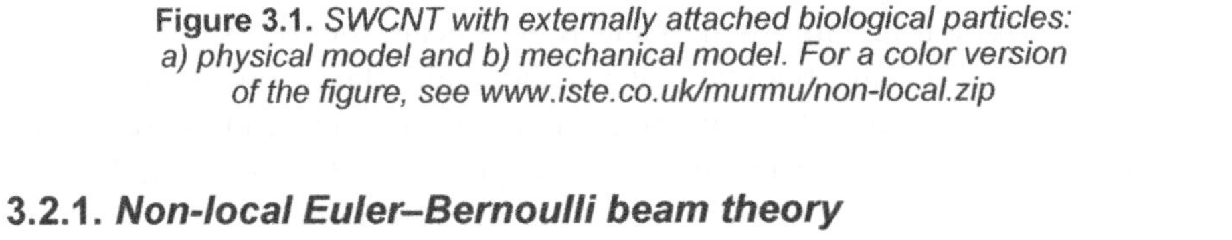

Figure 3.1. *SWCNT with externally attached biological particles: a) physical model and b) mechanical model. For a color version of the figure, see www.iste.co.uk/murmu/non-local.zip*

3.2.1. *Non-local Euler–Bernoulli beam theory*

Now, based on the non-local constitutive relation for one-dimensional case and Euler–Bernoulli beam theory, the partial differential equation of motion of the nanobeam can be derived in terms of transverse displacement. The beam model assumes that both the rotary inertia and shear deformation are negligible. According to the assumptions that the straight plane sections are perpendicular to the beam axis before deformation and after deformation, it remains (i) straight, (ii) perpendicular to the tangent line of the beam axis and (iii) inextensible [RED 02], the displacement field of Euler–Bernoulli beam theory can be written as:

$$\bar{u}_x = u(x,t) - z\frac{\partial w(x,t)}{\partial x}, \qquad \bar{v}_y = 0, \qquad \bar{w}_z = w(x,t), \qquad [3.1]$$

where $\bar{u}_x$ and $\bar{v}_y$ are the in-plane displacements of the nanobeam and $\bar{w}_z$ is the transverse displacement, i.e. displacement along x, y and z directions, respectively. By assuming the small deformations for given displacement fields, the only non-zero strains are:

$$\varepsilon_{xx} = \frac{\partial \bar{u}_x}{\partial x} = \frac{\partial u}{\partial x} - z\frac{\partial^2 w}{\partial x^2}, \qquad [3.2]$$

where ε_{xx} is the normal strain.

Let us consider an infinitesimal element of the nanobeam shown in Figure 3.2. The cross-sectional area is constant along the x coordinate. We can observe the shear force F_T , axial force N and bending moment M_f acting on the left face of the differential element. In the general case, all three quantities vary along the length of the nanobeam with time $F_T = F_T(x,t)$, $N = N(x,t)$ and $M_f = M_f(x,t)$, measured in the horizontal direction x. As shown in Figure 3.2, $q(x,t)$ is an external distributed force having a dimension of force per unit length and P is axial static load. On the right-hand side of the differential element, we have acting $F_T + dF_T$, $N + dN$ and $M_f + dM_f$, i.e. transverse force, axial force and bending moment, respectively. The terms α and $\alpha + d\alpha$ denote the angles of rotation of the cross-sectional area.

According to D'Alembert's principle applied on the infinitesimal element of the nanobeam (Figure 3.2), equilibrium equations can be obtained in the following form:

$$\frac{\partial F_T}{\partial x} + q + (N - P)\frac{\partial^2 w}{\partial x^2} = \rho A\frac{\partial^2 w}{\partial t^2}, \qquad [3.3a]$$

$$\frac{\partial N}{\partial x} = \rho A\frac{\partial^2 u}{\partial t^2}, \qquad [3.3b]$$

$$F_T = \frac{\partial M_f}{\partial x}, \qquad [3.3c]$$

where P is the axial static load, q denotes the transverse load and M_f and N are the moment and axial force stress resultants, respectively, defined as:

$$M_f = \int_0^A z\sigma_{xx}\, dA, \tag{3.4a}$$

$$N = \int_0^A \sigma_{xx}\, dA. \tag{3.4b}$$

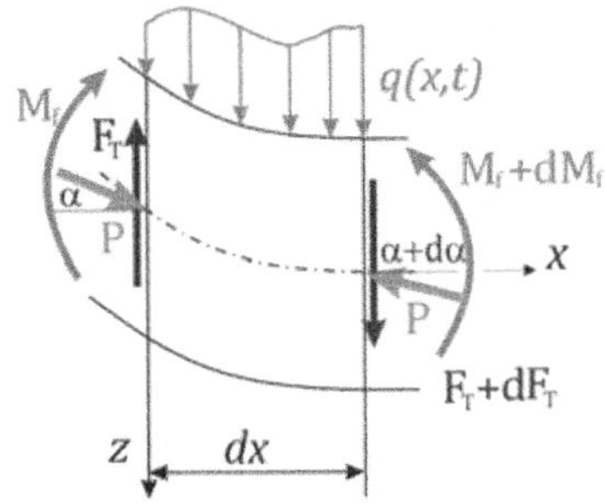

Figure 3.2. *Infinitesimal element of the nanobeam of length dx with stress resultants. For a color version of the figure, see www.iste.co.uk/murmu/non-local.zip*

By using the non-local constitutive relation derived in the previous chapter equation [2.1] with expressions [3.4a] and [3.4b], we can obtain the non-local axial force and the bending moment stress resultants as:

$$M_f - \mu \frac{\partial^2 M_f}{\partial x^2} = -EI \frac{\partial^2 w}{\partial x^2}, \tag{3.5a}$$

$$N - \mu \frac{\partial^2 N}{\partial x^2} = EA \frac{\partial u}{\partial x}, \tag{3.5b}$$

in which A denotes the cross-sectional area of the nanobeam and I denotes the cross-sectional moment of inertia. By assuming that the axial displacement u is equal to zero, we also have that the axial stress resultant N is equal to zero. Introducing equations [3.3] into [3.5], the equation of motion of the nanobeam in terms of transverse displacements $w(x,t)$ is obtained in the following form:

$$\rho A \frac{\partial^2 w}{\partial t^2} + P \frac{\partial^2 w}{\partial x^2} - q + EI \frac{\partial^4 w}{\partial x^4} = (e_0 a)^2 \frac{\partial^2}{\partial x^2}\left[\rho A \frac{\partial^2 w}{\partial t^2} + P \frac{\partial^2 w}{\partial x^2} - q\right]. \tag{3.6}$$

The boundary conditions for simply-supported non-local Euler–Bernoulli's nanobeam of the length L assume that the deflections and bending moments at the ends of the nanobeam are equal to zero, which is expressed as:

at $x = 0$

$$w(0,t) = 0,$$

$$M_f(0,t) = \left[(e_0 a)^2 \frac{\partial^2}{\partial x^2}\left(-q + P\frac{\partial^2 w}{\partial x^2} + \rho A\frac{\partial^2 w}{\partial t^2}\right) - EI\frac{\partial^2 w}{\partial x^2}\right]_{x=0} = 0, \quad [3.7a]$$

at $x = L$

$$w(L,t) = 0,$$

$$M_f(L,t) = \left[(e_0 a)^2 \frac{\partial^2}{\partial x^2}\left(-q + P\frac{\partial^2 w}{\partial x^2} + \rho A\frac{\partial^2 w}{\partial t^2}\right) - EI\frac{\partial^2 w}{\partial x^2}\right]_{x=L} = 0. \quad [3.7b]$$

It is interesting to note that the boundary conditions for simply-supported non-local nanobeam correspond to the boundary conditions from the classical beam theory as given by Lu *et al.* [LU 07]. In the continuation of this section, we will consider static bending ($P = 0$ and $\rho A\frac{\partial^2 w}{\partial t^2} = 0$), vibration ($P = 0$ and $q = 0$) and buckling ($q = 0$ and $\rho A\frac{\partial^2 w}{\partial t^2} = 0$) of the non-local Euler–Bernoulli nanobeam, where the static axial load $P = Const$ and transverse load $q = q(x)$ are constant with time.

3.2.1.1. *Bending of the non-local Euler–Bernoulli beam*

Now, we can consider an analytical solution for the static bending of a non-local Euler–Bernoulli nanobeam under the influence of the static transverse load $q = q(x)$, where the axial static load P and inertia term are neglected. Introducing $P = 0$ and $\rho A\frac{\partial^2 w}{\partial t^2} = 0$ into the governing equation of motion [3.6], we obtain partial differential equation for static bending of the nanobeam as:

$$EI\frac{\partial^4 w}{\partial x^4} = q - (e_0 a)^2\frac{\partial^2 q}{\partial x^2}. \quad [3.8]$$

Assuming the solution in the form of the trigonometric series known as the Fourier infinite series, we can solve partial differential equation [3.8] with corresponding boundary conditions [3.7]. Expressions for transverse displacements and the static transverse load can be written in the form:

$$w(x) = \sum_{n=1}^{\infty} W_n \sin \alpha_n x, \quad [3.9a]$$

$$q(x) = \sum_{n=1}^{\infty} Q_n \sin \alpha_n x, \quad [3.9b]$$

where $\alpha_n = \frac{n\pi}{L}$ $(n = 1,2,3,\ldots)$; W_n denotes the amplitude of static deflection which corresponds to the mode number and the coefficient Q_n is associated with the various types of static transverse load. By applying the inverse Fourier transformation (see [RED 02]), we obtain coefficient Q_n as:

$$Q_n = \frac{2}{L}\int_0^L q(x) \sin \alpha_n x, \quad [3.10]$$

where particular cases of transverse loads as illustrated in Figure 3.3, we can express as:

a) Continuous uniform transverse load (Figure 3.3(a))

$$q(x) = q_0 \quad \rightarrow \quad Q_n = \frac{4q_0}{n\pi}, \qquad n = 1,3,5,.. \quad [3.11a]$$

b) Continuous triangular transverse load (Figure 3.3(b))

$$q(x) = \frac{q_0}{L}x \quad \rightarrow \quad Q_n = (-1)^{n+1}\frac{2q_0}{n\pi}, n = 1,2,3,.. \quad [3.11b]$$

c) Concentrated transverse load (Figure 3.3(c))

$$q(x) = q_0\delta(x - x_0) \quad \rightarrow \quad Q_n = \frac{2q_0}{L}\sin\frac{\pi x_0}{L}, n = 1,2,3,.. \quad [3.11c]$$

d) Sinusoidal transverse load (Figure 3.3(d))

$$q(x) = q_0 \sin\frac{\pi x}{L} \quad \rightarrow \quad Q_1 = q_0,\ Q_n = q_0,\ n = 2,3,4\ldots \quad [3.11d]$$

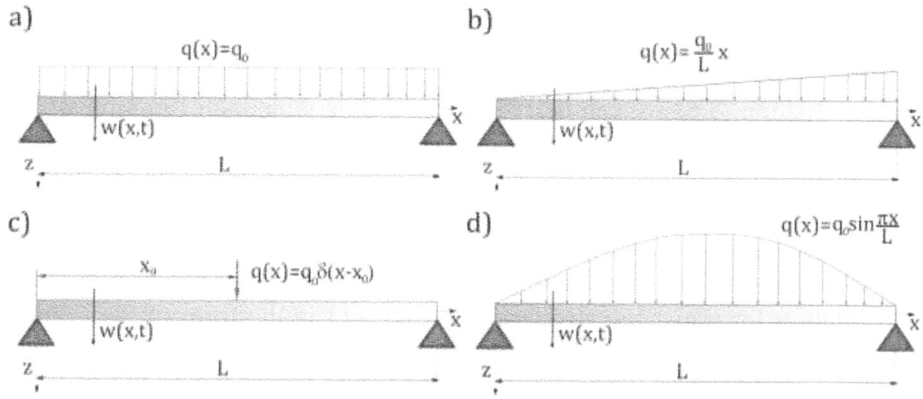

Figure 3.3. *Four different types of external transverse load $q(x)$ of the nanobeam: a) uniformly distributed load, b) triangular load, c) point load and d) sinusoidal distributed load*

Inserting the assumed solutions equation [3.9] into the partial differential equation for static bending equation [3.8], we can obtain the amplitude of static deflection for corresponding mode as a function of the static transverse load coefficients in the form:

$$W_{EBn} = \frac{1+(e_0 a)^2 \alpha_n^2}{EI\alpha_n^4} Q_n, \quad [3.12a]$$

where the total static deflection $w(x)$ is expressed as:

$$w_{EB}(x) = \sum_{n=1}^{\infty} W_{EBn} \sin \alpha_n x = \sum_{n=1}^{\infty} \frac{1+(e_0 a)^2 \alpha_n^2}{EI\alpha_n^4} Q_n \sin \alpha_n x. \quad [3.12b]$$

From equation [3.12b], we can see that for the corresponding transverse load given in equation [3.11], we must consider a larger number of terms in the series in order to achieve a better accuracy of the results for static deflection. It should be noted that by neglecting the parameter$(e_0 a)^2$ in the above expressions, results obtained by the non-local Euler–Bernoulli beam theory are reduced to the results from the classical Euler–Bernoulli beam theory.

3.2.1.2. *Vibration of the non-local Euler–Bernoulli beam*

By neglecting the static transverse load q and axial load P (Figure 3.4), from equation [3.6] we can derive the governing equation of motion which describes the free transverse vibration of nanobeam as:

$$\rho A\frac{\partial^2 w}{\partial t^2}+EI\frac{\partial^4 w}{\partial x^4}=(e_0 a)^2\rho A\frac{\partial^4 w}{\partial t^2\partial x^2}. \qquad [3.13]$$

Supposing the time harmonic motion of the nanobeam, we can assume the solution for transverse displacements $w(x,t)$ in the form of the Fourier trigonometric infinite series:

$$w(x,t)=\sum_{n=1}^{\infty}W_n\sin\alpha_n x\, e^{i\omega_n t}, \qquad [3.14]$$

where $i=\sqrt{-1}$, $\alpha_n=n\pi/L$, W_n are the amplitudes and ω_n is the natural frequency with n denoting the mode number.

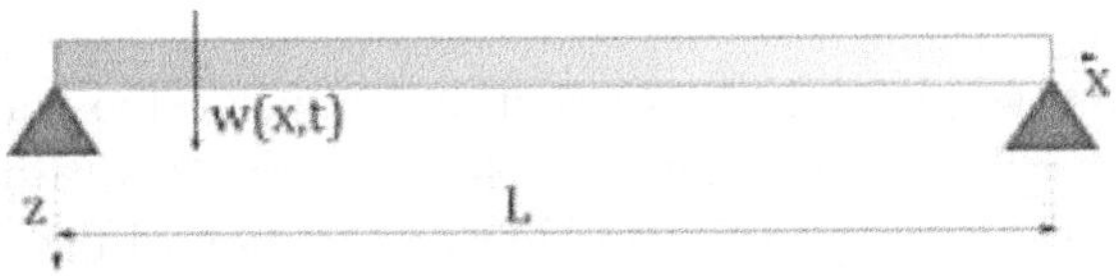

Figure 3.4. *The nanobeam without any external load*

Substituting the expansions for $w(x,t)$, equation [3.14], into the equation of motion of non-local Euler–Bernoulli nanobeam model, the closed form expression for natural frequencies is obtained as:

$$\omega_{EBn}=\sqrt{\frac{EI\alpha_n^4}{1+(e_0 a)^2\alpha_n^2}},\quad \text{for } n=1,2,3,\ldots \qquad [3.15]$$

3.2.1.3. *Buckling of the non-local Euler–Bernoulli beam*

At the end of this section, we discuss about the static stability, i.e. buckling of the non-local Euler–Bernoulli nanobeam subjected to the axial compression load P, as shown in Figure 3.5. By introducing $q=0$ and $\rho A\frac{\partial^2 w}{\partial t^2}=0$ into the partial differential equation of motion [3.6], we obtain the equation which describes the buckling state of the nanobeam as:

$$P\frac{\partial^2 w}{\partial x^2}+EI\frac{\partial^4 w}{\partial x^4}=(e_0 a)^2 P\frac{\partial^4 w}{\partial x^4}. \qquad [3.16]$$

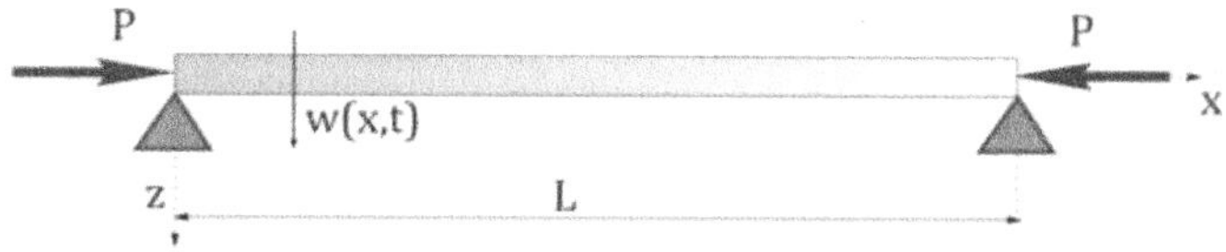

Figure 3.5. *The nanobeam subjected to the compressive static axial load*

Since the nanobeam is only under the influence of static axial load P, we introduce the assumed solution for transverse displacement $w(x)$ given in equation [3.9a] similar to the case when the nanobeam is subjected to transverse loads, which yields:

$$P_{EBn} = \frac{EI\alpha_n^2}{1+(e_0a)^2\alpha_n^2}, \quad \text{for n} = 1,2,3,\dots\ . \qquad [3.17]$$

Consequently, the smallest P_n is the critical buckling load which corresponds to the mode number n. In this case, the critical buckling load is obtained by introducing $n = 1$ into the expressions [3.17] that yields:

$$P_{EBcr} = P_{EB1} = \frac{EI\left(\frac{\pi}{L}\right)^2}{1+(e_0a)^2\left(\frac{\pi}{L}\right)^2}. \qquad [3.18]$$

By neglecting the non-local parameter $(e_0a)^2$ in the above expression for non-local critical buckling load, equation [3.18] is reduced to the equation for critical buckling load obtained from the classical beam theory.

3.2.2. *Non-local Timoshenko beam theory*

As already stated, in the Euler–Bernoulli beam theory plane sections remain perpendicular to the tangent line of a beam axis, which implies that the shear strain γ_{xz} is equal to zero. Moreover, this assumption leads to the fact that the shears stress and shear force are equal to zero as well. However, in the case of a short or thick beam and beams subject to high-frequency excitations when the wavelength approaches to the thickness of a beam, shear effects are having significant influence on the dynamic behavior of such systems. The Timoshenko beam theory was developed to take into account shear deformation and rotational inertia effects. This theory is based on the relaxed second assumption proposed in section 3.2, which enables plane section rotation to undergo the shear strain γ_{xz} and remain a plane.

Thus, in the Timoshenko beam theory, plane sections do not remain perpendicular to the beam axis after deformation. Based on this relaxed assumption and retaining the remaining two, the displacement field for the Timoshenko beam theory can be written as:

$$\bar{u}_x = u(x,t) + z\psi(x,t), \qquad \bar{v}_y = 0, \qquad \bar{w}_z = w(x,t), \tag{3.19}$$

where $\bar{u}_x$ and $\bar{v}_y$ are the in-plane displacements of a beam and $\bar{w}_z$ is the transverse displacement, i.e. displacement along x, y and z directions, respectively, and $\psi(x,t)$ is the rotation of a plane section around the y axis defined as:

$$\psi(x,t) = \gamma_{xz} - \frac{\partial w(x,t)}{\partial x}. \tag{3.20}$$

Considering only a small deformation of the Timoshenko beam, for a given displacement field [3.19], we obtain the following non-zero strains:

$$\varepsilon_{xx} = \frac{\partial \bar{u}_x}{\partial x} = \frac{\partial u}{\partial x} + z\frac{\partial \psi}{\partial x}, \tag{3.21a}$$

$$\gamma_{xz} = \frac{\partial \bar{u}_x}{\partial z} + \frac{\partial \bar{w}_z}{\partial x} = \psi(x,t) + \frac{\partial w(x,t)}{\partial x}, \tag{3.21b}$$

where ε_{xx} and γ_{xz} are the normal and shear strains.

Again, we consider an infinitesimal element of the nanobeam under the influence of the axial compressive force P and the transverse load q, with corresponding stress resultants shown in Figure 3.2. Applying the Timoshenko beam theory on a differential element, the rotary inertia of cross-section and shear effects would be incorporated into the equilibrium equations. By assuming the forces in the z-direction and moments about the y-axis with neglected axial displacement $u(x,t)$, we obtain the following expressions:

$$\frac{\partial F_T}{\partial x} + q - P\frac{\partial^2 w}{\partial x^2} = \rho A\frac{\partial^2 w}{\partial t^2}, \tag{3.22a}$$

$$\frac{\partial M_f}{\partial x} - F_T = \rho I\frac{\partial^2 \psi}{\partial t^2}, \tag{3.22b}$$

where P and q are defined in the section 3.2.1, the axial force stress resultant N vanishes by neglecting the axial displacement. The bending moment M_f and shear force F_T are defined as:

$$M_f = \int_0^A z\sigma_{xx}\, dA, \quad F_T = K_s \int_0^A \tau_{xz}\, dA, \qquad [3.23]$$

in which K_s is the term representing the shear correction factor. The shear correction factor K_s is introduced into the Timoshenko beam theory because of the assumption of constant shear stress across the cross-sections of a beam. Values of the shear correction factor K_s depend on the material and geometrical properties as well as on boundary conditions of a beam.

From equation [1.8], we can obtain constitutive relations for non-local elastic one-dimensional body in deferential form as:

$$\sigma_{xx} - (e_0 a)^2 \frac{d^2\sigma_{xx}}{dx^2} = E\varepsilon_{xx}, \qquad [3.24a]$$

$$\tau_{xz} - (e_0 a)^2 \frac{d^2\tau_{xz}}{dx^2} = G\gamma_{xz}, \qquad [3.24b]$$

where E and G are the elastic and shear modules of a nanobeam, $(e_0 a)^2$ is the non-local parameter, σ_{xx} andτ_{xz} are the normal and shear non-local stresses. By setting the non-local parameter $e_0 a$ equal to zero, the non-local elastic constitutive relation is reduced to the classical elasticity theory.

Applying the expressions for bending moment and transverse shear force (equation [3.23]) and the equilibrium equations [3.22] into the constitutive relations, we can obtain the non-local bending moment and non-local shear force in the following form:

$$M_f = (e_0 a)^2 \left(P\frac{\partial^2 w}{\partial x^2} - q + \rho A\frac{\partial^2 w}{\partial t^2} + \rho I\frac{\partial^3 \psi}{\partial x \partial t^2}\right) + EI\frac{\partial \psi}{\partial x}, \qquad [3.25a]$$

$$F_T = (e_0 a)^2 \left(P\frac{\partial^3 w}{\partial x^3} - \frac{\partial q}{\partial x} + \rho A\frac{\partial^3 w}{\partial x \partial t^2}\right) + GK_s A\left(\psi + \frac{\partial w}{\partial x}\right). \qquad [3.25b]$$

Now, introducing shear force expression [3.25b] into the first equilibrium equation [3.22a] and moment expression from [3.25a] into the second equilibrium equation [3.22b] and rearranging the terms, we obtain the system of partial differential equations in terms of generalized displacements as:

$$\rho A\frac{\partial^2 w}{\partial t^2}+P\frac{\partial^2 w}{\partial x^2}-q-GK_sA\left(\frac{\partial\psi}{\partial x}+\frac{\partial^2 w}{\partial x^2}\right)$$
$$=(e_0a)^2\frac{\partial^2}{\partial x^2}\left(\rho A\frac{\partial^2 w}{\partial t^2}+P\frac{\partial^2 w}{\partial x^2}-q\right), \quad [3.26a]$$

$$EI\frac{\partial^2\psi}{\partial x^2}-GK_sA\left(\psi+\frac{\partial w}{\partial x}\right)=\left(1-(e_0a)^2\frac{\partial^2}{\partial x^2}\right)\rho I\frac{\partial^2\psi}{\partial t^2}, \quad [3.26b]$$

For the non-local Timoshenko beam model, boundary conditions in the case of a simply-supported nanobeam of the length L are determined as in section 3.2.1. By assuming that the deflections and bending moments at the ends of the nanobeam are equal to zero, the mathematical expressions for boundary conditions are:

at $x=0$

$$w(0,t)=0,$$

$$M_f(0,t)=\left[(e_0a)^2\left(P\frac{\partial^2 w}{\partial x^2}-q+\rho A\frac{\partial^2 w}{\partial t^2}+\rho I\frac{\partial^3\psi}{\partial x\partial t^2}\right)+EI\frac{\partial\psi}{\partial x}\right]_{x=0}=0, \quad [3.27a]$$

at $x=L$

$$w(L,t)=0,$$

$$M_f(L,t)=\left[(e_0a)^2\left(P\frac{\partial^2 w}{\partial x^2}-q+\rho A\frac{\partial^2 w}{\partial t^2}+\rho I\frac{\partial^3\psi}{\partial x\partial t^2}\right)+EI\frac{\partial\psi}{\partial x}\right]_{x=L}=0. \quad [3.27b]$$

In the following, we will discuss the analytical solutions for static bending ($P=0$ and $\rho A\frac{\partial^2 w}{\partial t^2}=0$), vibration ($P=0$ and $q=0$) and buckling ($q=0$ and $\rho A\frac{\partial^2 w}{\partial t^2}=0$) of the non-local Timoshenko nanobeam, where the static axial load $P=Const$ and transverse load $q=q(x)$ are constant with time in a similar manner to section 3.2.1.

3.2.2.1. *Bending of non-local Timoshenko beam*

Here, we consider a solution for static bending of the non-local Timoshenko nanobeam subjected to the transverse static load $q=q(x)$ in an exact manner, by considering the four different cases as shown in Figure 3.3. By neglecting the axial static load ($P = 0$) and time derivatives $\left(\rho A \frac{\partial^2 w}{\partial t^2} = \rho I \frac{\partial^2 \psi}{\partial t^2} = 0\right)$, from the system of governing equations [3.26], we can obtain reduced partial differential equation for static bending of non-local Timoshenko beam as:

$$\left(1 - (e_0 a)^2 \frac{\partial^2}{\partial x^2}\right) q + GK_s A \left(\frac{\partial \psi}{\partial x} + \frac{\partial^2 w}{\partial x^2}\right) = 0, \quad [3.28a]$$

$$GK_s A \left(\psi + \frac{\partial w}{\partial x}\right) - EI \frac{\partial^2 \psi}{\partial x^2} = 0. \quad [3.28b]$$

Following the methodology developed in the sections 3.2.1, we assume the solution in the form of trigonometric series known as Fourier infinite series, which must satisfy the system of partial differential equation [3.28] with corresponding boundary conditions [3.27]. Assumed solutions are of the following form:

$$w(x) = \sum_{n=1}^{\infty} W_n \sin \alpha_n x, \quad [3.29a]$$

$$\psi(x) = \sum_{n=1}^{\infty} X_n \cos \alpha_n x, \quad [3.29b]$$

$$q(x) = \sum_{n=1}^{\infty} Q_n \sin \alpha_n x, \quad [3.29c]$$

where $\alpha_n = \frac{n\pi}{L}$ $(n = 1,2,3,\ldots)$; W_n denotes the amplitude of static deflection which corresponds to the mode number n, X_n is the coefficient of plane section rotation and Q_n denotes coefficients associated with the various types of static transverse loads, defined like in equations [3.10] and [3.11]. Introducing the assumed solutions expressed in equation [3.29] into the partial differential equation [3.28] yields the system of algebraic equations as:

$$GK_s A \alpha_n^2 W_n + GK_s A \alpha_n X_n = [1 + (e_0 a)^2 \alpha_n^2] Q_n, \quad [3.30a]$$

$$GK_s A \alpha_n W_n + (EI \alpha_n^2 + GK_s A) X_n = 0, \quad [3.30b]$$

or in the matrix form:

$$\begin{bmatrix} GK_s A \alpha_n^2 & GK_s A \alpha_n \\ GK_s A \alpha_n & EI \alpha_n^2 + GK_s A \end{bmatrix} \begin{Bmatrix} W_n \\ X_n \end{Bmatrix} = \begin{Bmatrix} [1 + (e_0 a)^2 \alpha_n^2] Q_n \\ 0 \end{Bmatrix}, \quad [3.31]$$

By applying Cramer's rule on the system of algebraic equations [3.31], we can obtain coefficients W_n and X_n as:

$$W_{Tn} = \frac{1+(e_0a)^2\alpha_n^2}{EI\alpha_n^4}\left(1 + \frac{EI}{GK_sA}\alpha_n^2\right)Q_n, \qquad [3.32a]$$

$$X_{Tn} = -\frac{1+(e_0a)^2\alpha_n^2}{EI\alpha_n^3}Q_n, \qquad [3.32b]$$

Inserting the obtained solutions for W_n and X_n, expressions [3.32], into equation [3.29a] and equation [3.29b], static deflection and rotation are expressed as:

$$w_T(x) = \sum_{n=1}^{\infty}\left[\frac{1+(e_0a)^2\alpha_n^2}{EI\alpha_n^4}\left(1 + \frac{EI}{GK_sA}\alpha_n^2\right)Q_n\right] sin\, \alpha_n x, \qquad [3.33a]$$

$$\psi_T(x) = -\sum_{n=1}^{\infty}\left[\frac{1+(e_0a)^2\alpha_n^2}{EI\alpha_n^3}Q_n\right] cos\, \alpha_n x. \qquad [3.33b]$$

Considering the different types of static transverse loads given in the section 3.2.1, we can determine static deflection of the non-local Timoshenko nanobeam by using equation [3.33]. Taking into account a larger number of terms in series leads to greater accuracy of the results.

3.2.2.2. *Vibration of the non-local Timoshenko beam*

To analyze the oscillatory behavior of the non-local Timoshenko nanobeam, we assume that the static transverse load q and axial load P vanish in equation [3.26], which gives the governing equations for the transverse vibration of the system in the form:

$$\rho A\frac{\partial^2 w}{\partial t^2} - GK_sA\left(\frac{\partial\psi}{\partial x} + \frac{\partial^2 w}{\partial x^2}\right) = (e_0a)^2\rho A\frac{\partial^4 w}{\partial x^2\partial t^2}, \qquad [3.34a]$$

$$EI\frac{\partial^2\psi}{\partial x^2} - GK_sA\left(\psi + \frac{\partial w}{\partial x}\right) = \left(1 - (e_0a)^2\frac{\partial^2}{\partial x^2}\right)\rho I\frac{\partial^2\psi}{\partial t^2}. \qquad [3.34b]$$

For the free vibration analysis, the solutions for the transverse displacement $w(x,t)$ and plane section rotation $\psi(x,t)$ are assumed as follows:

$$w(x,t) = \sum_{n=1}^{\infty} W_n \sin\alpha_n x\, e^{i\omega_n t}, \psi(x,t) = \sum_{n=1}^{\infty} X_n \cos\alpha_n x\, e^{i\omega_n t}, \qquad [3.35]$$

where $i = \sqrt{-1}$, $\alpha_n = n\pi/L$, ω_n is the natural frequency and coefficients W_n and X_n are defined in the section 3.2.2.1. Introducing assumed transverse displacement and rotation equation [3.35] into the equations of motion [3.34] yields:

$$[(1 + (e_0a)^2\alpha_n^2)\rho A\omega_n^2 - GK_sA\alpha_n^2]W_n - GK_sA\alpha_nX_n = 0\,, \qquad [3.36a]$$

$$-GK_sA\alpha_nW_n + [(1 + (e_0a)^2\alpha_n^2)\rho I\omega_n^2 - EI\alpha_n^2 - GK_sA]X_n = 0, \qquad [3.36b]$$

or in the matrix form:

$$\begin{bmatrix} (1 + (e_0a)^2\alpha_n^2)\rho A\omega_n^2 - GK_sA\alpha_n^2 & -GK_sA\alpha_nX_n \\ -GK_sA\alpha_nX_n & (1 + (e_0a)^2\alpha_n^2)\rho I\omega_n^2 - EI\alpha_n^2 - GK_sA \end{bmatrix}$$

$$\begin{Bmatrix} W_n \\ X_n \end{Bmatrix} = \begin{Bmatrix} 0 \\ 0 \end{Bmatrix}. \qquad [3.37]$$

For the non-trivial solution of the system of algebraic equations [3.37], we can obtain the characteristic polynomial as:

$$\tilde{a}\omega_n^4 - \tilde{b}\omega_n^2 + \tilde{c} = 0, \qquad [3.38]$$

in which $\tilde{a}, \tilde{b}$ and $\tilde{c}$ are the constants defined as:

$$\tilde{a} = \rho A\rho I(1 + (e_0a)^2\alpha_n^2)^2, \qquad [3.39a]$$

$$\tilde{b} = [GK_sA(\rho A + \alpha_n^2\rho I) + \rho AEI\alpha_n^2](1 + (e_0a)^2\alpha_n^2), \qquad [3.39b]$$

$$\tilde{c} = \alpha_n^4GK_sAEI. \qquad [3.39c]$$

Solving the characteristic polynomial [3.38] as quadratic equation gives the solution as:

$$\omega_{Tn(1/2)}^2 = \frac{\tilde{b}}{2\tilde{a}} \mp \sqrt{\frac{\tilde{b}^2}{4\tilde{a}^2} - \frac{\tilde{c}}{\tilde{a}}}, \qquad [3.40a]$$

or in the expansion form:

$$\omega_{Tn(1/2)}^2 = \frac{[GK_sA(\rho A+\alpha_n^2\rho I)+\rho AEI\alpha_n^2](1+(e_0a)^2\alpha_n^2)}{2\rho A\rho I(1+(e_0a)^2\alpha_n^2)^2}$$

$$\mp\sqrt{\frac{[GK_sA(\rho A+\alpha_n^2\rho I)+\rho AEI\alpha_n^2]^2(1+(e_0a)^2\alpha_n^2)^2-4(\rho AI)^2(1+(e_0a)^2\alpha_n^2)^2\alpha_n^4GK_sE}{4[\rho^2AI(1+(e_0a)^2\alpha_n^2)^2]^2}},$$

where ω_{Tn1} denotes lower non-local natural frequency and ω_{Tn2} is the higher non-local natural frequency of the non-local Timoshenko nanobeam.

3.2.2.3. *Buckling of the non-local Timoshenko beam*

When the axial static load $P$is applied on the nanobeam (Figure 3.5), it reaches a certain critical value and nanobeam becomes unstable, i.e. it begins to buckle. By setting $q=0$ and time derivatives to zero $\rho A\frac{\partial^2 w}{\partial t^2}=\rho I\frac{\partial^2 \psi}{\partial t^2}=0$ into the partial differential equation of motion [3.26], we can consider a buckling state of the non-local Timoshenko nanobeam as:

$$P\frac{\partial^2 w}{\partial x^2}-GK_sA\left(\frac{\partial\psi}{\partial x}+\frac{\partial^2 w}{\partial x^2}\right)=(e_0a)^2P\frac{\partial^4 w}{\partial x^4}, \qquad [3.41a]$$

$$EI\frac{\partial^2\psi}{\partial x^2}-GK_sA\left(\psi+\frac{\partial w}{\partial x}\right)=0. \qquad [3.41b]$$

Introducing the assumed solution, equation [3.29], proposed in the section related to the static bending of non-local Euler–Bernoulli beam, into system of partial differential equations [3.41], we can obtain the system of algebraic equations:

$$[-P\alpha_n^2(1+(e_0a)^2\alpha_n^2)+GK_sA\alpha_n^2]W_n+GK_sA\alpha_nX_n=0\,, \qquad [3.42a]$$

$$GK_sA\alpha_nW_n+(EI\alpha_n^2+GK_sA)X_n=0, \qquad [3.42b]$$

or in the matrix form:

$$\begin{bmatrix}-P\alpha_n^2(1+(e_0a)^2\alpha_n^2)+GK_sA\alpha_n^2 & GK_sA\alpha_nX_n\\ GK_sA\alpha_nX_n & EI\alpha_n^2+GK_sA\end{bmatrix}\begin{Bmatrix}W_n\\X_n\end{Bmatrix}=\begin{Bmatrix}0\\0\end{Bmatrix}. \qquad [3.43]$$

By assuming the non-trivial solution for the system of algebraic equations [3.43], we can determine the buckling load of the non-local Timoshenko nanobeam as:

$$P_{Tn}=\frac{EI\alpha_n^2GK_sA}{(1+(e_0a)^2\alpha_n^2)(GK_sA+EI\alpha_n^2)},\quad \text{for } n=1,2,3,\ldots. \qquad [3.44]$$

The same as in the case of non-local Euler–Bernoulli nanobeam, the smallest P_n is the critical buckling load which corresponds to the mode number n. In this case, the critical buckling load is obtained by introducing $n = 1$ into the expression [3.44] that yields:

$$P_{Tcr} = P_{T1} = \frac{EI\left(\frac{\pi}{L}\right)^2 GK_SA}{\left[1+(e_0a)^2\left(\frac{\pi}{L}\right)^2\right]\left[GK_SA+EI\left(\frac{\pi}{L}\right)^2\right]}. \quad [3.45]$$

By setting non-local parameter $(e_0a)^2$ equal to zero in the expression for non-local critical buckling load equation [3.45], it reduces to the equation for the critical buckling load of the classical Timoshenko beam.

3.2.3. *Non-local Reddy beam theory*

The displacement field of the Reddy beam theory [RED 07] is based on the relaxed assumptions that the straight plane sections perpendicular to the beam axis before deformation do not remain straight and perpendicular to the tangent line of the beam axis after deformation. The third assumption related to the inextensibility of the beam after deformation proposed for the Euler–Bernoulli beam theory remains the same. Furthermore, these assumptions imply that the shear strain accommodation across a beam thickness varies parabolically and vanishes on the top and bottom planes of a beam. Therefore, there is no need to use a shear correction factor in the third-order beam theory. From the viewpoint of nanobeam structures, difference between basic Euler–Bernoulli beam theory and Reddy beam theory is better applicability of higher order theory to shortened SWCNT nanostructures. Thus, we get the next displacement field for an arbitrary point of a beam as:

$$\bar{u}_x(x,z,t) = u(x,t) + z\psi(x,t) - c_1 z^3\left(\psi(x,t) + \frac{\partial w(x,t)}{\partial x}\right),$$

$$\bar{v}_y(x,z,t) = 0,$$

$$\bar{w}_z(x,z,t) = w(x,t). \quad [3.46]$$

where $c_1 = 4/(3h^2)$ and h is the height of a beam, $w(x,t)$ and $\psi(x,t)$ are the transverse and rotation components of the displacement. The non-zero strains of the proposed beam theory are:

$$\varepsilon_{xx} = \frac{\partial u}{\partial x} + z\frac{\partial \psi}{\partial x} - c_1 z^3\left(\frac{\partial \psi}{\partial x} + \frac{\partial^2 w(x,t)}{\partial x^2}\right),$$

$$\gamma_{xz} = (1 - c_2 z^2)\left(\psi + \frac{\partial w}{\partial x}\right), \qquad [3.47]$$

where $c_2 = 3,\ c_1 = 4/h^2$. In this case, the component u of the axial displacement $\bar{u}_x(x,z,t)$ is neglected. Based on the Hamilton's principle, motion of an elastic structure during the time interval $0 < t < T$ is such that the time integral of the total potential is extremum, Reddy [RED 07, RED 02]. Thus, for the nanobeam structure such an extremum would be of the form:

$$\int_0^T (\delta U + \delta V - \delta K)\, dt = 0, \qquad [3.48]$$

where δU is the variation of the strain energy, δV is the virtual work of external forces and δK is the variation of the kinetic energy of nanobeam. The variation of the strain energy of the nanobeam is:

$$\delta U = \int_0^A \int_0^L (\sigma_{xx}\delta\varepsilon_{xx} + \sigma_{xz}\delta\gamma_{xz})\, dxdA$$

$$= \int_0^L \left[(M_{xx} - c_1 P_{xx})\,\delta\frac{\partial \psi}{\partial x} - c_1 P_{xx}\delta\frac{\partial^2 w}{\partial x^2} + (Q_x - c_2 R_x)\delta\left(\psi + \frac{\partial w}{\partial x}\right)\right] dx, \qquad [3.49]$$

where M_{xx}, P_{xx}, Q_x and R_x are the stress resultants defined as:

$$(M_{xx}, P_{xx}) = \int_0^A (z, z^3)\sigma_{xx}\, dA; \quad (Q_x, R_x) = \int_0^A (1, z^2)\sigma_{xz}\, dA. \qquad [3.50]$$

The variation of the potential energy of external forces can be expressed as:

$$\delta V = -\int_0^L \left[q(x)w + P\frac{\partial w}{\partial x}\delta\frac{\partial w}{\partial x}\right] dx, \qquad [3.51]$$

where $q(x)$ is the continual transverse load and P is the axial load.

The variation of the kinetic energy is obtained as:

$$\delta K = \int_0^A \int_0^L \rho(\dot{u}\delta\dot{u} + \dot{w}\delta\dot{w})\, dxdA$$

$$= \int_0^L \left[\hat{I}_2\dot{\psi}\delta\dot{\psi} - c_1\hat{I}_4\left(\dot{\psi} + \frac{\partial\dot{w}}{\partial x}\right)\delta\dot{\phi}_x - c_1\hat{I}_4\dot{\psi}\delta\left(\dot{\psi} + \frac{\partial\dot{w}}{\partial x}\right)\right.$$

$$\left. + c_1^2\hat{I}_6\left(\dot{\psi} + \frac{\partial\dot{w}}{\partial x}\right)\delta\left(\dot{\psi} + \frac{\partial\dot{w}}{\partial x}\right) + \hat{I}_0\dot{w}\delta\dot{w}\right] dx, \qquad [3.52]$$

where ρ is the mass density and $\frac{\partial}{\partial t} = (\dot{\ })$ is the differentiation with respect to the time variable t, and $\left(\hat{I}_0, \hat{I}_2, \hat{I}_4, \hat{I}_6\right)$ are the mass inertias defined as:

$$\left(\hat{I}_0, \hat{I}_2, \hat{I}_4, \hat{I}_6\right) = \int_0^A (1, z^2, z^4, z^6)\, \rho dA. \qquad [3.53]$$

To derive the equations of motion associated with the present model, we substitute the expressions for δU, δV and δK from equations [3.49], [3.51] and [3.52] into equation [3.48] and after integrating by parts and then collecting the coefficients δw and $\delta\phi_x$, the following equations of motion are obtained:

$$\delta w: \quad \frac{\partial\bar{Q}_x}{\partial x} + c_1\frac{\partial^2 P_{xx}}{\partial x^2} + P\frac{\partial^2 w}{\partial x^2} + q = \hat{I}_0\ddot{w} + \hat{I}_4\frac{\partial\ddot{\psi}}{\partial x} - c_1^2\hat{I}_6\left(\frac{\partial\ddot{\psi}}{\partial x} + \frac{\partial^2\ddot{w}}{\partial x^2}\right), \qquad [3.54a]$$

$$\delta\phi_x: \quad \frac{\partial\bar{M}_{xx}}{\partial x} - \bar{Q}_x = \ddot{\psi}\hat{I}_2 - c_1\hat{I}_4\left(2\ddot{\psi} + \frac{\partial\ddot{w}}{\partial x}\right) + c_1^2\hat{I}_6\left(\ddot{\psi} + \frac{\partial\ddot{w}}{\partial x}\right), \qquad [3.54b]$$

where

$$\bar{Q}_x = Q_x - c_2 R_x, \qquad \bar{M}_{xx} = M_{xx} - c_1 P_{xx}. \qquad [3.55]$$

The general boundary conditions of the model are:

$$w \text{ or } V_x = c_1\frac{\partial P_{xx}}{\partial x} + \bar{Q}_x - P\frac{\partial w}{\partial x} - c_1\hat{I}_4\ddot{\psi} + c_1^2\hat{I}_6\left(\ddot{\psi} + \frac{\partial\ddot{w}}{\partial x}\right), \qquad [3.56a]$$

$$\frac{\partial w}{\partial x} \text{or} P_{xx}, \qquad [3.56b]$$

$$\phi_x \text{or} \bar{M}_{xx} = M_{xx} - c_1 P_{xx}, \qquad [3.56c]$$

where V_x denotes the equivalent shear force. By substituting equation [3.47] into equation [3.24] and the subsequent results into equations [3.50] and [3.55], the stress resultants for the non-local Eringen theory are obtained as:

$$P_{xx} - (e_0 a)^2 \frac{d^2 P_{xx}}{dx^2} = E(I_4 - c_1 I_6)\frac{\partial \psi}{\partial x} - c_1 I_6 E \frac{\partial^2 w}{\partial x^2}, \quad [3.57a]$$

$$\bar{Q}_x - (e_0 a)^2 \frac{d^2 \bar{Q}_x}{dx^2} = G\hat{A}\left(\psi + \frac{\partial w}{\partial x}\right), \quad [3.57b]$$

$$\bar{M}_{xx} - (e_0 a)^2 \frac{d^2 \bar{M}_{xx}}{dx^2} = -Ec_1 I_4\left(2\frac{\partial \psi}{\partial x} + \frac{\partial^2 w}{\partial x^2}\right) + EI_2 \frac{\partial \psi}{\partial x} + Ec_1^2 I_6\left(\frac{\partial \psi}{\partial x} + \frac{\partial^2 w}{\partial x^2}\right), \quad [3.57c]$$

where

$$(A, I_2, I_4, I_6) = \int_0^A (1, z^2, z^4, z^6)\, dA, \qquad \hat{A} = (A - 2c_2 I_2 + c_2^2 I_4). \quad [3.58]$$

The equations of motion can be expressed in terms of the displacement (w, ψ) for non-local constitutive relations. By substituting equation [3.57] into equation [3.54], we get the following equations of motion:

$$P\frac{\partial^2 w}{\partial x^2} + \hat{I}_0 \ddot{w} - K_1 \frac{\partial \ddot{\psi}}{\partial x} - c_1^2 \hat{I}_6 \frac{\partial^2 \ddot{w}}{\partial x^2} - q - c_1\left[E(I_4 - c_1 I_6)\frac{\partial^3 \psi}{\partial x^3} - c_1 I_6 E \frac{\partial^4 w}{\partial x^4}\right]$$

$$= (e_0 a)^2\left(P\frac{\partial^4 w}{\partial x^4} - \frac{\partial^2 q}{\partial x^2} + \hat{I}_0 \frac{\partial^2 \ddot{w}}{\partial x^2} - K_1 \frac{\partial^3 \ddot{\psi}}{\partial x^3} - c_1^2 \hat{I}_6 \frac{\partial^4 \ddot{w}}{\partial x^4}\right) + G\left(\frac{\partial \psi}{\partial x} + \frac{\partial^2 w}{\partial x^2}\right)\hat{A}, \quad [3.59a]$$

$$G\left(\psi + \frac{\partial w}{\partial x}\right)\hat{A} + K_2 \ddot{\psi} + K_1 \frac{\partial \ddot{w}}{\partial x} - (e_0 a)^2\left(K_2 \frac{\partial^2 \ddot{\psi}}{\partial x^2} + K_1 \frac{\partial^3 \ddot{w}}{\partial x^3}\right)$$

$$= EI_2 \frac{\partial^2 \psi}{\partial x^2} - EI_4 c_1\left(2\frac{\partial^2 \psi}{\partial x^2} + \frac{\partial^3 w}{\partial x^3}\right) + c_1^2 EI_6\left(\frac{\partial^2 \psi}{\partial x^2} + \frac{\partial^3 w}{\partial x^3}\right), \quad [3.59b]$$

where $P \Rightarrow -P$ is the applied axial compressive force, and:

$$K_1 = c_1^2 \hat{I}_6 - c_1 \hat{I}_4, \quad K_2 = \hat{I}_2 - 2c_1 \hat{I}_4 + c_1^2 \hat{I}_6. \quad [3.60]$$

The equation of motion of local Reddy beam theory can be obtained from equation [3.59] by setting the non-local parameter (e_0a) equal to zero. In the case of non-local Reddy beam theory, mathematical expressions for simply-supported nanobeam of the length L are determined by assuming that the deflections and bending moments at the ends of nanobeam are equal to zero, where, from equation [3.56], we have:

at $x = 0$

$$w(0,t) = 0,$$

$$\bar{M}_{xx}(0,t) = \left[(e_0a)^2\left(-c_1\frac{\partial^2 P_{xx}}{\partial x^2} + P\frac{\partial^2 w}{\partial x^2} - q + \hat{I}_0\ddot{w} + \hat{I}_2\frac{\partial\ddot{\psi}}{\partial x} - c_1\hat{I}_4\frac{\partial}{\partial x}\left(\ddot{\psi} + \frac{\partial\ddot{w}}{\partial x}\right)\right)\right.$$

$$\left.-Ec_1I_4\left(2\frac{\partial\psi}{\partial x} + \frac{\partial^2 w}{\partial x^2}\right) + EI_2\frac{\partial\psi}{\partial x} + Ec_1^2I_6\left(\frac{\partial\psi}{\partial x} + \frac{\partial^2 w}{\partial x^2}\right)\right]_{x=0}, \qquad [3.61a]$$

at $x = L$

$$w(L,t) = 0,$$

$$\bar{M}_{xx}(L,t) = \left[(e_0a)^2\left(-c_1\frac{\partial^2 P_{xx}}{\partial x^2} + P\frac{\partial^2 w}{\partial x^2} - q + \hat{I}_0\ddot{w} + \hat{I}_2\frac{\partial\ddot{\psi}}{\partial x} -\right.\right.$$

$$\left.c_1\hat{I}_4\frac{\partial}{\partial x}\left(\ddot{\psi} + \frac{\partial\ddot{w}}{\partial x}\right)\right)$$

$$\left.-Ec_1I_4\left(2\frac{\partial\psi}{\partial x} + \frac{\partial^2 w}{\partial x^2}\right) + EI_2\frac{\partial\psi}{\partial x} + Ec_1^2I_6\left(\frac{\partial\psi}{\partial x} + \frac{\partial^2 w}{\partial x^2}\right)\right]_{x=L}, \qquad [3.61b]$$

3.2.3.1. *Bending of non-local Reddy beam*

Here, we analyze static bending of the non-local Reddy nanobeam under the influence of transverse static load $q = q(x)$ as shown in Figure 3.3. Like in the previously studied beam theories, we assume that the axial static load $(P = 0)$ and time derivatives $\left(\frac{\partial^2 w}{\partial t^2} = \frac{\partial^2 \psi}{\partial t^2} = 0\right)$ are equal to zero in partial differential equation [3.59], which yields:

$$q - (e_0a)^2\frac{\partial^2 q}{\partial x^2} + G\left(\frac{\partial\psi}{\partial x} + \frac{\partial^2 w}{\partial x^2}\right)\hat{A} + c_1\left[E(I_4 - c_1I_6)\frac{\partial^3\psi}{\partial x^3} - c_1I_6E\frac{\partial^4 w}{\partial x^4}\right] = 0, \quad [3.62a]$$

$$G\left(\psi+\frac{\partial w}{\partial x}\right)\hat{A}=EI_2\frac{\partial^2\psi}{\partial x^2}-EI_4c_1\left(2\frac{\partial^2\psi}{\partial x^2}+\frac{\partial^3 w}{\partial x^3}\right)+c_1^2EI_6\left(\frac{\partial^2\psi}{\partial x^2}+\frac{\partial^3 w}{\partial x^3}\right), \quad [3.62b]$$

Assuming the solutions for equation [3.62] and boundary conditions equation [3.61], as in the previous case for the Timoshenko beam theory, and introducing them into the partial differential equations [3.62] we obtain the system of algebraic equations in terms of coefficients W_n, X_n and Q_n as:

$$\begin{bmatrix} s_{11} & s_{12} \\ s_{21} & s_{22} \end{bmatrix}\begin{Bmatrix} W_n \\ X_n \end{Bmatrix}=\begin{Bmatrix} (1+(e_0a)^2\alpha_n^2)Q_n \\ 0 \end{Bmatrix}, \quad [3.63]$$

where coefficients are:

$$s_{11}=EI_6c_1^2\alpha_n^4+G\alpha_n^2\hat{A}\,,\qquad s_{12}=s_{21}=G\alpha_n\,\hat{A}-Ec_1(I_4-c_1I_6)\alpha_n^3,$$

$$s_{22}=G\hat{A}+E(I_2-c_1I_4)\alpha_n^2-Ec_1(I_4-c_1I_6)\alpha_n^2. \quad [3.64]$$

The coefficient Q_n proposed within the section about Euler–Bernoulli beam theory and expressed in equation [3.11] depends on the type of transverse load applied to the nanobeam. Applying the Cramer's rule on the system of algebraic equations [3.63] with coefficients [3.64], we can get coefficients W_n and X_n as:

$$W_{Rn}=\frac{1+(e_0a)^2\alpha_n^2}{s_{11}s_{22}-s_{12}^2}s_{22}Q_n, \quad [3.65a]$$

$$X_{Rn}=-\frac{1+(e_0a)^2\alpha_n^2}{s_{11}s_{22}-s_{12}^2}s_{12}Q_n. \quad [3.65b]$$

By introducing coefficients W_n and X_n, expressions [3.65] into assumed solution for transverse displacement $w_R(x)$ and rotation ψ_R, for the Reddy beam theory we obtain static deflection and rotation as:

$$w_R(x)=\sum_{n=1}^{\infty}\left[\frac{1+(e_0a)^2\alpha_n^2}{s_{11}s_{22}-s_{12}^2}s_{22}Q_n\right]\sin\alpha_n x \quad [3.66a]$$

$$\psi_R(x)=-\sum_{n=1}^{\infty}\left[\frac{1+(e_0a)^2\alpha_n^2}{s_{11}s_{22}-s_{12}^2}s_{12}Q_n\right]\cos\alpha_n x. \quad [3.66b]$$

However, for different types of static transverse load shown in Figure 3.3, we can obtain static deflection for non-local Reddy nanobeam by using equation [3.66] and equation [3.11].

3.2.3.2. *Vibration of the non-local Reddy beam*

In order to study harmonic motion of the non-local Reddy nanobeam, we neglect the influence of the static transverse load $q(x)$ and the axial compressive load P on dynamic behavior. Partial differential equations that describe the free transverse vibration of the non-local Reddy nanobeam are derived by introducing $q(x) = 0$ and $P = 0$ into the system of equations [3.59], which yields:

$$\hat{I}_0 \ddot{w} - K_1 \frac{\partial \ddot{\psi}}{\partial x} - c_1^2 \hat{I}_6 \frac{\partial^2 \ddot{w}}{\partial x^2} - c_1 \left[E(I_4 - c_1 I_6) \frac{\partial^3 \psi}{\partial x^3} - c_1 I_6 E \frac{\partial^4 w}{\partial x^4} \right]$$

$$= (e_0 a)^2 \left(\hat{I}_0 \frac{\partial^2 \ddot{w}}{\partial x^2} - K_1 \frac{\partial^3 \ddot{\psi}}{\partial x^3} - c_1^2 \hat{I}_6 \frac{\partial^4 \ddot{w}}{\partial x^4} \right) + G \left(\frac{\partial \psi}{\partial x} + \frac{\partial^2 w}{\partial x^2} \right) \hat{A}, \qquad [3.67a]$$

$$G \left(\psi + \frac{\partial w}{\partial x} \right) \hat{A} + K_2 \ddot{\psi} + K_1 \frac{\partial \ddot{w}}{\partial x} - (e_0 a)^2 \left(K_2 \frac{\partial^2 \ddot{\psi}}{\partial x^2} + K_1 \frac{\partial^3 \ddot{w}}{\partial x^3} \right)$$

$$= E I_2 \frac{\partial^2 \psi}{\partial x^2} - E I_4 c_1 \left(2 \frac{\partial^2 \psi}{\partial x^2} + \frac{\partial^3 w}{\partial x^3} \right) + c_1^2 E I_6 \left(\frac{\partial^2 \psi}{\partial x^2} + \frac{\partial^3 w}{\partial x^3} \right). \qquad [3.67b]$$

By using the solution proposed for the free transverse vibration of Timoshenko nanobeam equation [3.35], from partial differential equations [3.67] we can obtain the system of algebraic equations as:

$$\left(\begin{bmatrix} s_{11} & s_{12} \\ s_{21} & s_{22} \end{bmatrix} - \omega_n^2 \begin{bmatrix} m_{11} & m_{12} \\ m_{21} & m_{22} \end{bmatrix} \right) \begin{Bmatrix} W_n \\ X_n \end{Bmatrix} = \begin{Bmatrix} 0 \\ 0 \end{Bmatrix}, \qquad [3.68]$$

where the coefficients are:

$$s_{11} = E c_1^2 I_6 {\alpha_n}^4 + G {\alpha_n}^2 \hat{A}, \quad s_{12} = s_{21} = -E c_1 (I_4 - c_1 I_6) {\alpha_n}^3 + G \alpha \hat{A},$$

$$s_{22} = G \hat{A} + E(I_2 - c_1 I_4) {\alpha_n}^2 - E c_1 (I_4 - c_1 I_6) {\alpha_n}^2, \qquad [3.69]$$

$$m_{11} = \hat{I}_0 + {\alpha_n}^2 c_1^2 \hat{I}_6 + (e_0 a)^2 {\alpha_n}^2 \hat{I}_0 + (e_0 a)^2 {\alpha_n}^4 c_1^2 \hat{I}_6,$$

$$m_{12} = m_{21} = K_1 \alpha_n + (e_0 a)^2 K_1 {\alpha_n}^3, \quad m_{22} = K_2 + (e_0 a)^2 K_2 {\alpha_n}^2.$$

The natural frequency is obtained by assuming the non-trivial solution of equation [3.68], i.e. by solving the following determinant:

$$\begin{vmatrix} s_{11} - \omega_n^2 m_{11} & s_{12} - \omega_n^2 m_{12} \\ s_{21} - \omega_n^2 m_{12} & s_{22} - \omega_n^2 m_{22} \end{vmatrix} = 0, \qquad [3.70]$$

from which we obtain the characteristic polynomial as:

$$\tilde{a}\omega_n^4 - \tilde{b}\omega_n^2 + \tilde{c} = 0, \qquad [3.71]$$

in which $\tilde{a}, \tilde{b}$ and $\tilde{c}$ are the constants defined as:

$$\tilde{a} = m_{11}m_{22} - m_{12}^2, \qquad [3.72a]$$

$$\tilde{b} = m_{11}s_{22} + m_{22}s_{11} - 2s_{12}m_{12}, \qquad [3.72b]$$

$$\tilde{c} = s_{11}s_{22} - s_{12}^2. \qquad [3.72c]$$

Solving characteristic polynomial [3.71] as quadratic equation, we obtain the solution as being:

$$\omega_{Rn(1/2)}^2 = \frac{\tilde{b}}{2\tilde{a}} \mp \sqrt{\frac{\tilde{b}^2}{4\tilde{a}^2} - \frac{\tilde{c}}{\tilde{a}}}, \qquad [3.73]$$

or in the expansion form as:

$$\omega_{Rn(1/2)}^2 = \frac{m_{22}s_{11} - 2m_{12}s_{12} + m_{11}s_{22}}{2(m_{11}m_{22} - m_{12}^2)}$$

$$\mp \frac{\sqrt{(m_{22}s_{11} - 2m_{12}s_{12} + m_{11}s_{22})^2 - 4(m_{11}m_{22} - m_{12}^2)(s_{11}s_{22} - s_{12}^2)}}{2(m_{11}m_{22} - m_{12}^2)}, \qquad [3.74]$$

where ω_{Rn1} denotes the lower non-local natural frequency and ω_{Rn2} is the higher non-local natural frequency of the non-local Reddy nanobeam.

3.2.3.3. *Buckling of the non-local Reddy beam*

Figure 3.5 shows the nanobeam subjected to the compressive axial static load P. When the axial static load P reaches a certain critical value, the system becomes unstable and we can say that the nanobeam begins to buckle. In order to analyze the buckling state of the non-local Reddy nanobeam, we neglect the transverse static load q and time derivatives from partial differential equation of motion [3.59] and obtain the following equations:

$$P\frac{\partial^2 w}{\partial x^2} - c_1\left[E(I_4 - c_1 I_6)\frac{\partial^3\psi}{\partial x^3} - c_1 I_6 E\frac{\partial^4 w}{\partial x^4}\right] = \\ (e_0 a)^2 P\frac{\partial^4 w}{\partial x^4} + G\left(\frac{\partial\psi}{\partial x} + \frac{\partial^2 w}{\partial x^2}\right)\hat{A}, \qquad [3.75a]$$

$$G\left(\psi + \frac{\partial w}{\partial x}\right)\hat{A} = EI_2\frac{\partial^2\psi}{\partial x^2} - EI_4 c_1\left(2\frac{\partial^2\psi}{\partial x^2} + \frac{\partial^3 w}{\partial x^3}\right) \\ + c_1^2 EI_6\left(\frac{\partial^2\psi}{\partial x^2} + \frac{\partial^3 w}{\partial x^3}\right). \qquad [3.75b]$$

Introducing the assumed solution equation [3.29], proposed in the section related to the static bending of the Timoshenko nanobeam, into system of partial differential equations [3.75], we obtain the system of algebraic equations in the matrix form as:

$$\left(\begin{bmatrix} s_{11} & s_{12} \\ s_{21} & s_{22} \end{bmatrix} - P\begin{bmatrix} n_{11} & 0 \\ 0 & 0 \end{bmatrix}\right)\begin{Bmatrix} W_n \\ X_n \end{Bmatrix} = \begin{Bmatrix} 0 \\ 0 \end{Bmatrix}, \qquad [3.76]$$

where the coefficients are:

$$s_{11} = Ec_1^2 I_6 \alpha_n{}^4 + G\alpha_n{}^2\hat{A}, \; s_{12} = s_{21} = -Ec_1(I_4 - c_1 I_6)\alpha_n{}^3 + G\alpha_n\hat{A},$$

$$s_{22} = G\hat{A} + E(I_2 - c_1 I_4)\alpha_n{}^2 - Ec_1(I_4 - c_1 I_6)\alpha_n{}^2, \qquad [3.77]$$

$$n_{11} = \alpha_n{}^2 + (e_0 a)^2\alpha_n{}^4.$$

The buckling load of the non-local Reddy nanobeam is obtained as a non-trivial solution from the system of algebraic equations [3.76] as:

$$P_{Rn} = \frac{s_{11}s_{22} - s_{12}^2}{n_{11}s_{22}}, \quad \text{for } n = 1,2,3,.... \qquad [3.78]$$

The smallest buckling load is known as the critical buckling load, which is obtained by taking the mode number $n = 1$ in equation [3.78]. Such a critical load represents the minimum required load that is needed for loss of the nanobeam stability.

3.3. Torsional vibration of nanobeam

Here, we consider a nanorod of length L and cross-sectional area A, as shown in Figure 3.1. In this case, cross-sectional areas are constant along the x coordinate but in general, it could have arbitrary shapes along the x coordinate. We consider the nanobeam as elastic and homogeneous. In order to consider the free torsional vibration of nanobeam around the x–axis, an infinitesimal element of the length dx is taken at a coordinate location x, as given in Figure 3.6(b). On the left side of differential element, we take that the torsional moment M_t is resultant of the shear stress $\tau_{r\varphi}$ in polar-cylindrical coordinate system, acting internally on cross-section area A, where $\tau_{r\varphi}$ is assumed to be uniform over the cross-section. Stress resultant M_t varies along the length, and it is a function of time $M_t = M_t(x,t)$. In addition, the axially distributed torsional moment $m(x,t)$ is shown, which results from the externally applied sources. On the right end of the differential element, we have $M_t + dM_t$.

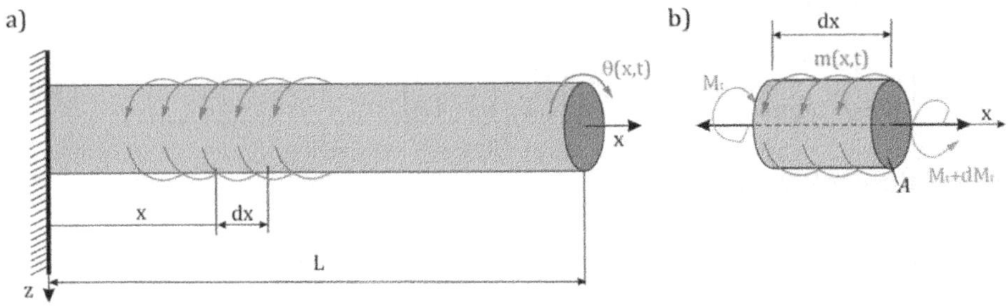

Figure 3.6. *a) Mechanical model of the torsional vibration of the nanobeam; b) differential element with corresponding moment stress resultants. For a color version of the figure, see www.iste.co.uk/murmu/non-local.zip*

Based on D'Alembert's principle, the sum of all moments with respect to x-axis yields:

$$-M_t + (M_t + dM_t) + m(x,t)dx = \rho I_o dx \frac{\partial^2 \theta}{\partial t^2}, \qquad [3.79]$$

where $dM_t = \frac{\partial M_t}{\partial x} dx$ is differential part of the torsional moment M_t, $\theta = \theta(x,t)$is the angular displacement around x-axis. After rearranging the above equation, the equilibrium equations are given in the form (see [MUR 11]):

$$\frac{\partial M_t}{\partial x} + m(x,t) = \rho I_o \frac{\partial^2 \theta}{\partial t^2}, \tag{3.80}$$

where M_t is the stress resultant defined as:

$$M_t(x,t) = \int_A \tau_{r\varphi} r \, dA, \text{ [3.81]} \tag{3.81}$$

in which term r represents the polar coordinate. From equation [1.8], we can obtain the constitutive relation for a non-local elastic body in deferential form as [MUR 11]:

$$\tau_{r\varphi} - (e_0 a)^2 \frac{d^2 \tau_{r\varphi}}{dx^2} = G\gamma_{r\varphi}, \tag{3.82}$$

where G is the shear modulus of the nanobeam, τ_{xz} is the shear non-local stresses in the polar-cylindrical system and $\gamma_{r\varphi} = r\frac{\partial \theta}{\partial x}$ is the shear deformation. Substituting equation [3.81] into equation [3.82], the stress torsional moment resultant for the non-local theory is obtained as:

$$M_t - (e_0 a)^2 \frac{d^2 M_t}{dx^2} = G I_O \frac{\partial \theta}{\partial x}, \tag{3.83}$$

where I_O is the polar moment of inertia defined as:

$$I_O = \int_A r \, dA. \tag{3.84}$$

The equation of motion can be expressed in terms of the angular displacement $\theta(x,t)$ for the non-local elastic constitutive relation. By substituting equation [3.80] into equation [3.83], we get the following equation of motion:

$$\rho I_o \frac{\partial^2 \theta}{\partial t^2} - m(x,t) - G I_O \frac{d^2 \theta}{dx^2} = (e_0 a)^2 \frac{d^2}{dx^2}\left(\rho I_o \frac{\partial^2 \theta}{\partial t^2} - m(x,t)\right). \tag{3.85}$$

Next, we utilize equation [3.85] for the development of a mathematical model for the free torsional vibration of non-local nanobeam. In the following, we neglect external continuous torsional moment $m(x,t)$ and consider two types of boundary conditions. Suppose that the ends of two nanorods are clamped-clamped (Figure 3.7(a)) and clamped-free (see Figure 3.7(b)), the boundary conditions are given by:

Clamped-Clamped:

$$\theta(0,t) = \theta(L,t) = 0, \qquad [3.86a]$$

Clamped-Free:

$$\theta(0,t) = M_t(L,t) = 0, \qquad [3.86b]$$

where M_t is the non-local torsional moment. From equation [3.80] and equation [3.83], we get:

$$M_t = (e_0 a)^2 \rho I_o \frac{\partial^3 \theta}{\partial x \partial t^2} + G I_O \frac{\partial \theta}{\partial x}. \qquad [3.87]$$

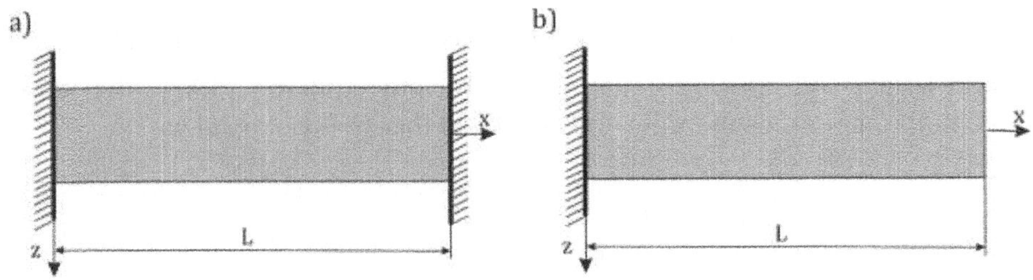

Figure 3.7. *Boundary conditions for the torsional vibration of nanobeam, a) clamped-clamped and b) clamped-free*

Analytical solution of the governing equation of motion with corresponding boundary conditions is obtained by using the Bernoulli–Fourier method, which is based on the Fourier infinite series. Assuming the time harmonic motion and using the separation of variables method, the solution of equation [3.85] taking into account boundary conditions equation [3.86] is proposed in the form:

$$\theta(x,t) = \sum_{n=1}^{\infty} X_n(x) T_n(t), \qquad [3.88]$$

where $T_n(t) = e^{i\omega_n t}$ is the time function and $X_n(x)$ is the corresponding mode shape function which depends on the boundary conditions of the system. By substituting the general solution equation [3.88] into equation [3.85] and neglecting the continuously distributed torsional moment $m(x,t)$, we get the following ordinary differential equation for corresponding mode shape function $X_n(x)$ of the nanobeam:

$$\frac{d^2X_n(x)}{dx^2} + \alpha_n^2 X_n(x) = 0, \qquad [3.89]$$

in which α_n denotes the characteristic values which are determined from the corresponding boundary conditions as:

$$\alpha_n^2 = \frac{\omega_n^2}{\frac{G}{\rho} - (e_0 a)^2 \omega_n^2}. \qquad [3.90]$$

The differential equation [3.89] is the well-known equation with the solution given in the form:

$$X_n(x) = A_n sin\alpha_n x + B_n cos\alpha_n x, \qquad [3.91]$$

where A_n and B_n are the unknown constants. Substituting the assumed solution equation [3.88] into the equations for boundary conditions equation [3.86], and using equation [3.91], we obtain two transcendental equations from which we can find values of α_n.

For the *Clamped-Clamped* boundary conditions (Figure 3.7(a)):

$$sin\alpha_n L = 0, \qquad [3.92a]$$

roots are

$$\alpha_n L = n\pi, \quad n = 1,2,\dots,\infty, \qquad [3.92b]$$

with corresponding mode shape function $X_n(x)$ as:

$$X_n(x) = A_n sin\alpha_n x = A_n sin\frac{n\pi}{L}x. \qquad [3.93]$$

For the *Clamped-Free* boundary conditions (Figure 3.7(b)):

$$cos\alpha_n L = 0, \qquad [3.94a]$$

roots are

$$\alpha_n L = \frac{(2n-1)\pi}{2}, \quad n = 1,2,\dots,\infty, \qquad [3.94b]$$

with corresponding mode shape functions $X_n(x)$ as:

$$X_n(x) = A_n sin\alpha_n x = A_n sin\frac{(2n-1)\pi}{2L}x. \qquad [3.95]$$

Now, we can obtain non-local natural frequencies for the torsional vibration of the nanobeam from expression [3.90] and corresponding boundary conditions as:

$$\omega_n^2 = \frac{G\alpha_n^2}{\rho[1+(e_0a)^2\alpha_n^2]}, \qquad [3.96]$$

where α_n is defined in equation [3.92b] for the *clamped–clamped* boundary conditions and in equation [3.94b] for the *clamped–free* boundary conditions.

3.4. Comparison of the non-local beam theories

In this section, we make a comparison of natural frequencies of nanobeams modeled by using the non-local Euler–Bernoulli, Timoshenko and Reddy beam theory: Table 3.1. In addition, we investigated the effects of non-local parameter (e_0a) and aspect ratio (L/D) on static banding, vibration and buckling of the non-local nanobeam modeled with the three mentioned theories, Figures 3.8–3.11. For this purpose, we consider the armchair (8, 8) SWCNT with thickness $(h = 0.34\,\text{nm})$, Poisson's ratio $(\nu = 0.3)$, mass density $(\rho = 2300\,\text{kg/m}^3)$ and Young's modulus $(E = 1.1\,\text{TPa})$. In the present analysis, the values of non-local parameter are adopted from Chapter 1. Also, it should be noted that the fundamental natural frequency represents the lowest natural frequency of the nanobeam for $n = 1$.

As shown in Table 3.1, the results for non-local fundamental natural frequencies of simply-supported SWCNT's obtained from different non-local beam theories are compared with the natural frequencies from local beam theories and validated with the molecular dynamic (MD) stimulation simulation results by [ANS 12]. In addition, the non-local parameter (e_0a) is adopted from [ANS 12]. From Table 3.1, it can be noticed that in general, fundamental natural frequencies of non-local models are lower compared to the local one. When the non-local parameter is sufficiently small or tends to the zero value, non-local natural frequencies are tending to the local one.

Furthermore, we can observe a decrease in the natural frequencies for an increase in the nanobeams aspect ratio for all three cases of beam theories. In addition, it can be observed that the results obtained from local and non-local Euler–Bernoulli, Timoshenko and Reddy beam models are the same for the aspect ratios higher than 17.3, which is attributed to the weakening of non-local effects for an increase in dimensions of the nanobeam. The numerical results show that natural frequencies of non-local nanobeam models are closer to the frequencies obtained by MD simulation in [ANS 12] compared to the frequencies obtained by local nanobeam models.

		Euler–Bernoulli beam theory [ANS 12]		Timoshenko beam theory [ANS 12]		Reddy beam theory [KAR 14d]	
L/D	MD simulation	Classical theory	Non-local elasticity theory	Classical theory	Non-local elasticity theory	Classical theory	Non-local elasticity theory
8.3	0.5299	0.5380	0.5376	0.5306	0.5302	0.5403	0.5393
10.1	0.3618	0.3640	0.3638	0.3606	0.3604	0.3669	0.3664
13.7	0.1931	0.1982	0.1982	0.1972	0.1971	0.2004	0.2003
17.3	0.1103	0.1244	0.1244	0.1240	0.1240	0.1260	0.1259
20.9	0.0724	0.0853	0.0853	0.0851	0.0851	0.0864	0.0864
24.5	0.0519	0.0621	0.0621	0.0620	0.0620	0.0629	0.0629
28.1	0.0425	0.0472	0.0472	0.0471	0.0471	0.0478	0.0478
31.6	0.0358	0.0373	0.0373	0.0373	0.0373	0.0378	0.0378
35.3	0.0287	0.0299	0.0299	0.0299	0.0299	0.0303	0.0303
39.1	0.0259	0.0244	0.0244	0.0244	0.0244	0.0247	0.0247

Table 3.1. *Comparison of the fundamental natural frequencies (THz) for different beam theories and (8, 8) armchair SWCNT's* $(R/(e_0a) = 0.6\,nm)$

In the following, we consider the Euler–Bernoulli, Timoshenko and Reddy beam theory to show the effects of non-local parameter (e_0a) and aspect ratio (L/D) on the natural frequency, buckling load and static deflection of the nanobeam. The parameter of the relative transverse displacement is defined as the ratio of the non-local and local static deflection solution:

$$\mathrm{RTD} = \frac{[w(x)]_{nonlocal}}{[w(x)]_{(e_0a)=0}}. \qquad [3.97]$$

Figure 3.8 shows the influence of non-local parameter on the relative transverse displacement of nanobeam for three different non-local beam theories. As it can be seen, the relative displacement increases for an increase in the non-local parameter. Also, the difference of relative displacements between three beam theories increases for an increase in the non-local parameter and decrease in the aspect ratio. As stated earlier, higher beam theories give better results for the structures with lower aspect ratios or short nanobeams than the Euler–Bernoulli beam theory, which are proved in Figure 3.8(b) where different values of relative displacement are obtained.

In Figure 3.9, we can observe the influence of non-local parameters on the fundamental natural frequency of the nanobeam for different beam theories. It can be noted that the natural frequency significantly decreases for an increase in non-local parameter, which is attributed to the softening of the system due to the presence of non-local effects. Further, differences between the curves of the Timoshenko and Reddy beam theory are minimal, while the difference compared to the Euler–Bernolli nanobeam is more pronounced. In addition, frequencies of the nanobeam with a lower aspect ratio are much higher.

Figure 3.10 shows the effect of the non-local parameter on the value of critical buckling load for different beam theories. We can observe that the value of critical buckling load decreases for an increase in non-local parameter. The values of critical buckling loads from Timoshenko and Reddy beam theory are lower with respect to value obtained from Euler–Bernoulli beam theory. However, critical buckling load is much higher for the lower aspect ratio than for the higher one (Figures 3.10(a) and (b)).

Finally, the effects of non-local parameter and the length of nanobeam on the torsional vibration of nanobeam are shown in Figure 3.11 for two types of boundary conditions, clamped-clamped and clamped-free. We can see a significant decrease in the natural frequency for an increase in the length of nanobeam. Also, the natural frequency decreases for an increase in non-local parameter but this effect is weakening for an increase in the length of nanobeam. At last, the natural frequency of the clamped–clamped nanobeam is significantly higher than the frequency of the clamped-free system.

a) b)

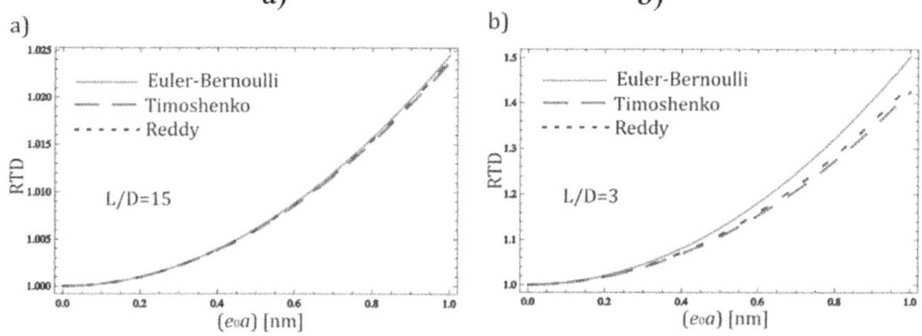

Figure 3.8. *Relative transverse deflection* $\left(RTD = \frac{[w(x)]_{nonlocal}}{[w(x)]_{(e_0a)=0}}\right)$ *of the nanobeam for Euler–Bernoulli, Timoshenko and Reddy beam theory. For a color version of the figure, see www.iste.co.uk/murmu/non-local.zip*

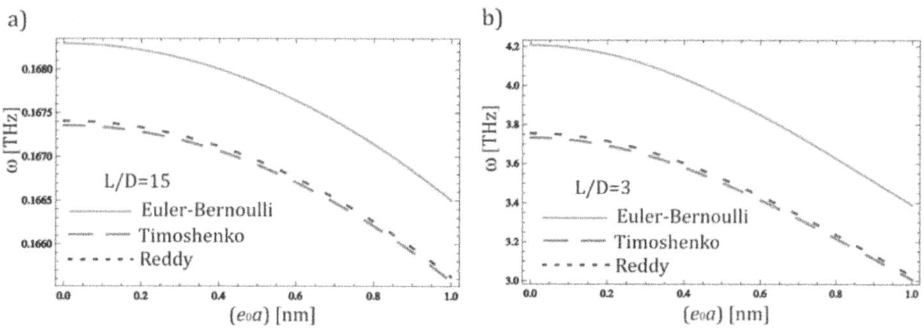

Figure 3.9. *Fundamental natural frequencies of the nanobeam for Euler–Bernoulli, Timoshenko and Reddy beam theory. For a color version of the figure, see www.iste.co.uk/murmu/non-local.zip*

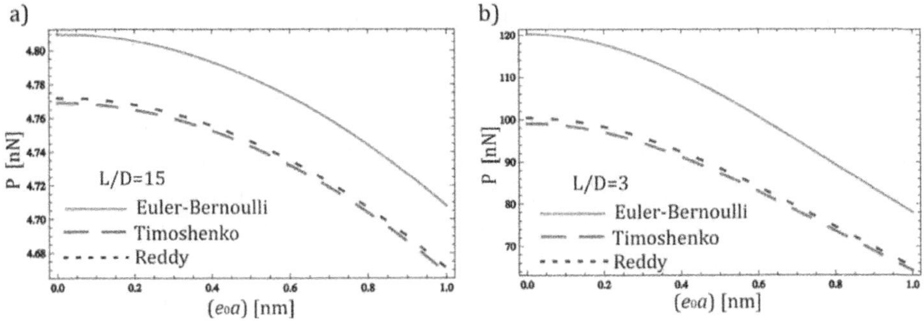

Figure 3.10. *Critical buckling load of the nanobeam for Euler–Bernoulli, Timoshenko and Reddy beam theory. For a color version of the figure, see www.iste.co.uk/murmu/non-local.zip*

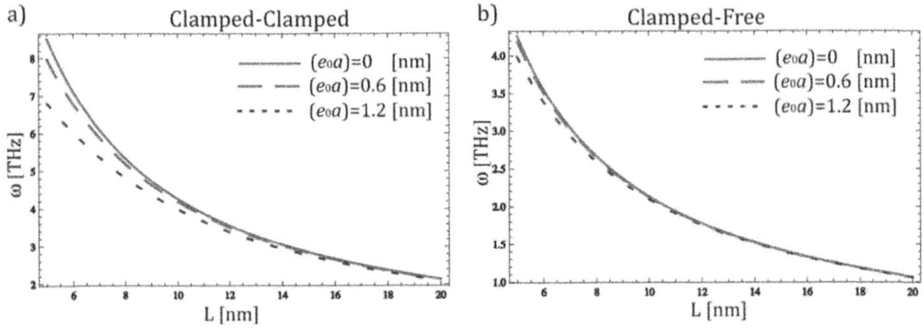

Figure 3.11. *Fundamental natural frequencies of torsional vibration of the nanobeam; a) clamped–clamped boundary conditions and b) clamped–free boundary conditions. For a color version of the figure, see www.iste.co.uk/murmu/non-local.zip*

4

Non-local Elastic Plate Theories

The development of an equation of motion requires non-local constitutive relations. Two-dimensional non-local constitutive relations are provided. Simple analytical relations are provided to determine the natural frequencies and buckling loads of idealized single-layer graphene sheets (SLGSs). An attempt has been made to emphasize the meaning of the effect of non-localness on the response of graphene sheets. The new structural non-local method can bridge the gap between molecular dynamics (MD) and scale-effect-free continuum mechanics to provide a viable means of studying such important nanoscale objects beyond graphenes. In this chapter, the theories for equation of motion for graphene sheets will be presented. The graphene sheets can be modeled as a non-local continuum nanoplate.

4.1. Non-local plate for graphene sheets

The schematic diagram of graphene sheets as a non-local plate is depicted in Figure 4.1. The carbon atoms of graphene at small scale are considered to be non-local in nature. This implies that the stress at a point not only depends on the strain at that point but also on the strains of all the points in the body. By using this concept, the discrete graphene sheets can be modeled as a continuum non-local elastic plate. The discrete nature of graphene is mapped into a non-local elastic continuum (Figure 4.1).

4.2. Non-local plate constitutive relations

Here, we present the non-local constitutive relations [PRA 09a] for the development of governing equations of graphene sheets. Consider the origin chosen at one corner of the rectangular SLGSs. The coordinate axes in the equivalent mapped non-local plate are shown in Figure 4.2. The x coordinate of the axis is taken along the length of the graphene sheets, the y coordinate is taken along the width and the z coordinate is taken along the thickness of the graphene (considering a constant atomic thickness, h).

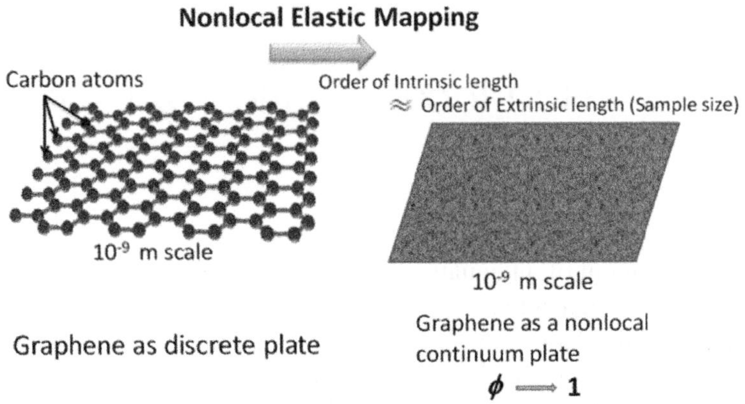

Figure 4.1. *Graphene sheet modeled as non-local continuum plate*

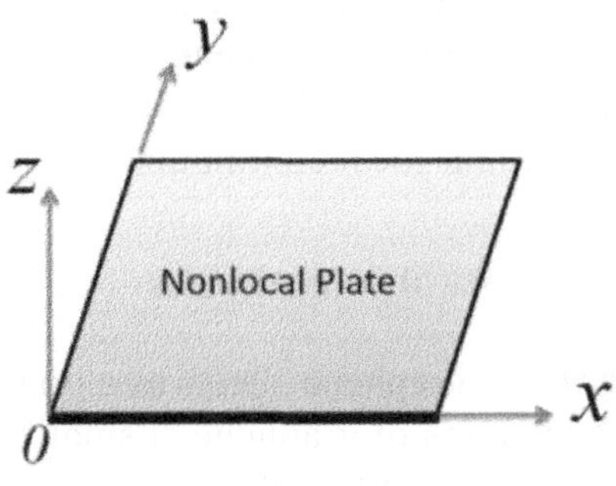

Figure 4.2. *Non-local plate model and coordinate system*

According to classical plate theory, the displacement field can be expressed as:

$$u_x = u(x, y, t) - z\frac{\partial w}{\partial x}; \quad u_y = v(x, y, t) - z\frac{\partial w}{\partial y}; \quad u_z = w(x, y, t) \tag{4.1}$$

where u, v and w are the displacement of the point $(x, y, 0)$ along the x, y and z directions, respectively. The corresponding strains ε are expressed as:

$$\varepsilon_{xx} = \frac{\partial u}{\partial x} - z\frac{\partial^2 w}{\partial x^2};\ \varepsilon_{yy} = \frac{\partial v}{\partial y} - z\frac{\partial^2 w}{\partial y^2};\ \varepsilon_{zz} = 0;\ \varepsilon_{xy} = \frac{1}{2}\left(\frac{\partial u}{\partial y} + \frac{\partial v}{\partial x} - 2z\frac{\partial^2 w}{\partial x \partial y}\right) \quad [4.2]$$
$$\varepsilon_{xz} = 0;\ \varepsilon_{yz} = 0$$

To use the present classical theory with nanoplates, the non-local effects due to small sizes are to be incorporated into the formulation. In non-local elasticity, it is assumed that the stress at a point depends on the strains at all other points of the continuum. In other words, according to non-local theory, strain at a point depends on both stress and spatial derivatives of the stress at that point. Using the concepts of non-local elasticity described earlier, the constitutive relations for two-dimensional nanostructures can now be developed. According to non-local elasticity, the two-dimensional non-local stress–strain relations can be expressed as:

$$\sigma_{xx} - (e_0 l_c)^2\left(\frac{\partial^2 \sigma_{xx}}{\partial x^2} + \frac{\partial^2 \sigma_{xx}}{\partial y^2}\right) = \frac{E}{(1-\nu^2)}(\varepsilon_{xx} + \nu\varepsilon_{yy}) \quad [4.3a]$$

$$\sigma_{yy} - (e_0 l_c)^2\left(\frac{\partial^2 \sigma_{yy}}{\partial x^2} + \frac{\partial^2 \sigma_{yy}}{\partial y^2}\right) = \frac{E}{(1-\nu^2)}(\varepsilon_{yy} + \nu\varepsilon_{xx}) \quad [4.3b]$$

$$\sigma_{xy} - (e_0 l_c)^2\left(\frac{\partial^2 \sigma_{xy}}{\partial x^2} + \frac{\partial^2 \sigma_{xy}}{\partial y^2}\right) = 2G\varepsilon_{xy} \quad [4.3c]$$

where $\sigma_{xx}, \sigma_{yy}, \sigma_{xy}$ and $\varepsilon_{xx}, \varepsilon_{yy}, \varepsilon_{xy}$ denote the non-local stress and strain tensors, respectively. E and G are the Young's and shear modulus of the graphene sheet, respectively. Here, we have assumed that there is constant and universal Young's modulus and shear modulus of graphene. For the development of the non-local constitutive relations, the following stress resultants are used:

$$N_{xx} = \int_{-h/2}^{h/2} \sigma_{xx} dz,\quad N_{yy} = \int_{-h/2}^{h/2} \sigma_{yy} dz,\quad N_{xy} = \int_{-h/2}^{h/2} \sigma_{xy} dz, \quad [4.4]$$

$$M_{xx} = \int_{-h/2}^{h/2} z\sigma_{xx} dz,\quad M_{yy} = \int_{-h/2}^{h/2} z\sigma_{yy} dz,\quad M_{xy} = \int_{-h/2}^{h/2} z\sigma_{xy} dz \quad [4.5]$$

Here, h denotes the thickness of the graphene sheets. The terms N_{xx}, N_{yy}, N_{xy} are the in-plane deformations, and M_{xx}, M_{yy}, M_{xy} are the bending and twisting moments, respectively.

Substitution of equations [4.4]–[4.5] and using the relations of classical plate theory, we can obtain the non-local constitutive relations as:

$$N_{xx} - (e_0 l_c)^2 \left(\frac{\partial^2 N_{xx}}{\partial x^2} + \frac{\partial^2 N_{xx}}{\partial y^2}\right) = C\left(\frac{\partial u}{\partial x} + v\frac{\partial v}{\partial y}\right) \quad [4.6a]$$

$$N_{yy} - (e_0 l_c)^2 \left(\frac{\partial^2 N_{yy}}{\partial x^2} + \frac{\partial^2 N_{yy}}{\partial y^2}\right) = C\left(\frac{\partial v}{\partial y} + v\frac{\partial u}{\partial x}\right) \quad [4.6b]$$

$$N_{xy} - (e_0 l_c)^2 \left(\frac{\partial^2 N_{xy}}{\partial x^2} + \frac{\partial^2 N_{xy}}{\partial y^2}\right) = C\frac{1-v}{2}\left(\frac{\partial u}{\partial y} + \frac{\partial v}{\partial x}\right) \quad [4.6c]$$

$$M_{xx} - (e_0 l_c)^2 \left(\frac{\partial^2 M_{xx}}{\partial x^2} + \frac{\partial^2 M_{xx}}{\partial y^2}\right) = -D\left(\frac{\partial^2 w}{\partial x^2} + v\frac{\partial^2 w}{\partial y^2}\right) \quad [4.6d]$$

$$M_{yy} - (e_0 l_c)^2 \left(\frac{\partial^2 M_{yy}}{\partial x^2} + \frac{\partial^2 M_{yy}}{\partial y^2}\right) = -D\left(\frac{\partial^2 w}{\partial y^2} + v\frac{\partial^2 w}{\partial xy^2}\right) \quad [4.6e]$$

$$M_{xy} - (e_0 l_c)^2 \left(\frac{\partial^2 M_{xy}}{\partial x^2} + \frac{\partial^2 M_{xy}}{\partial y^2}\right) = -D(1-v)\frac{\partial^2 w}{\partial x \partial y} \quad [4.6f]$$

where the terms C and D (bending rigidity) are expressed as:

$$C = \frac{Eh}{(1-v^2)}; \; D = \frac{Eh^3}{12(1-v^2)} \quad [4.7]$$

4.3. Free vibration of single-layer graphene sheets

Using fundamental relations [4.6] and classical equilibrium equations, an equation of motion of graphene sheets can be developed. Consider a rectangular SLGS which has dimension $(a \times b)$ in the x and y directions. The thickness of the SLGS h is assumed constant and generally considered to be one atomic dimension. In the analysis, it is assumed that the graphene sheet behaves as an ideal non-local elastic plate. Characteristics such as ripples properties and varying thickness of graphene sheets are neglected in the present theory. Furthermore, it is also assumed that there is no temperature effect on the graphene sheets.

4.3.1. *Transverse-free vibration*

Based on the non-local elasticity constitutive relations, the governing equations of free vibration of a rectangular graphene sheet at room temperature are given as [PRA 09a, MUR 09a, ARA 12]:

$$D\nabla^2.\nabla^2 w + \rho h[1-(e_0 l_c)^2\nabla^2]\frac{\partial^2 w}{\partial^2 t} = 0 \qquad [4.8]$$

where ρ is the density of the graphene sheets. Term w is the deflection in the z direction. D is known as the bending rigidity or stiffness and expressed as $D = Eh^3/12(1-\upsilon^2)$, where E and υ are the constant Young's modulus and Poisson's ratio of the SLGS. The term t is the conventional time. The term ∇^2 is the Laplacian operator and expressed as $\frac{\partial^2}{\partial x^2}+\frac{\partial^2}{\partial y^2}$. It should be noted that SLGS is considered isotropic in nature (though it may show orthotropic nature). In the same line, equations of motion for orthotropic graphene can also be developed.

You can see that equation [4.8] is not the conventional governing of an equation for free vibration of a graphene sheet but a modified and upgraded one which includes a scale parameter. Unlike in classical elasticity, in non-local elasticity, non-local boundary conditions are used. The boundary conditions for a simply-supported graphene sheet are considered as $w(0,y) = w(a,y) = w(x,0) = w(x,b) = 0$ and non-local moment $M(0,y) = M(a,y) = M(x,0) = M(x,b) = 0$. It should be noted that for simply-supported boundary conditions, the local and non-local boundary conditions are generally equivalent. However, for other arbitrary boundary conditions, local and non-local boundary conditions would be different. We assume the solution as:

$$w = \sum_{m=1}^{\infty}\sum_{n=1}^{\infty} W_{mn}\sin(m\pi/a)x\ \sin(n\pi/b)y\ e^{i\omega_{mn}t}) \qquad [4.9]$$

where ω is the circular frequency. Using equation [4.9] for a simply-supported graphene sheet with aspect ratio, $R = a/b$, the natural frequencies in radian per second are expressed as:

$$\omega = \frac{1}{a^2}\sqrt{\frac{D}{\rho h}}\sqrt{\frac{\psi^2}{1+\phi^2\psi}} \qquad [4.10]$$

where ψ is:

$$\psi = [(m\pi)^2 + R^2(n\pi)^2] \qquad [4.11]$$

In Hertz, the natural frequencies are $f = \omega/2\pi$. The term ϕ in the frequency expression is the non-local parameter (dimensionless) and is expressed as:

$$\phi = e_0\left(\frac{l_c}{a}\right) = e_0 l_R, \qquad [4.12]$$

where e_0 is a constant appropriate to each material. Here, the intrinsic l_c can be considered as the distance between two carbon atoms.

If a macroscopic (say thin graphite) plate is used, the expression applied to get the frequency would be:

$$\omega = \frac{1}{a^2}\sqrt{\frac{D}{\rho h}}\psi \qquad [4.13]$$

The expression is the cause of the effect that $\phi = 0$. In fact, ϕ is not exactly zero but $\phi \longrightarrow 0$, $(l_R \longrightarrow 0)$ and is very small and is neglected in classical elasticity theory.

Considering the SLGSs as an idealized non-local nanoplate, we assume the following properties. The Young's modulus of the SLGS is considered as $E = 1.06\,\text{TPa}$, the Poisson ratio $\upsilon = 0.25$, and the mass density as $\rho = 2250\,\text{kg/m}^3$. The thickness of the SLGS is taken as $h = 0.34$ nm [MUR 11]. If the natural frequencies are obtained based on the present non-local elasticity theory, the frequencies are influenced by size effects unlike those obtained by classical elasticity. In Chapters 6 and 7, we will see why the size effects influence the frequencies. According to equation [4.10], the non-local frequencies of graphene sheets are highly sensitive on the single small-scale coefficient or non-local parameter $e_o l_c$. Consider the non-local parameter $e_o l_c$ varying from 0 to 2 nm. In fact, it should be noted that to date we do not know the exact value of non-local parameter $e_o l_c$ (in fact, e_o). In the general case, the most common value of the non-local parameter which is used in the analysis of nano-engineering-structures is in the range 0-2 m. It should be noted that the range of a non-local parameter is not constant and universal. Researchers are attempting to find out the correct non-local parameter by comparing the non-local frequency results with MD simulations [ANS 10]. A description of various non-local parameters is discussed in [NAR 11].

The prediction of natural frequencies of graphene using non-local theory has been validated by many researchers using MD simulation and a good agreement has been found ([CHO 11c, CHO 11d, GUP 10]). In this section, we analyze the results predicted by the non-local elasticity theory (considered to be the correct theory at nanoscale) in comparison to classical elasticity theory.

The local frequency results from equation [4.13] and non-local frequency results from equation [4.10] are different, and the difference increases with the increasing value of non-local parameter $e_o l_c$. Another important observation that can be made is that the natural frequencies of the SLGS based on non-local elasticity theory are always smaller compared to solutions based on local elasticity theory [MUR 09a], i.e. is $\omega_{\mathrm{NL}} \leq \omega_L$. This can be attributed to the fact that for the same frequency with different non-local parameters $e_o l_c$, the bending stiffness of the graphene sheets considering a non-local model is lower than its local counterpart. Such a phenomenon arises due to small size effects. This observation will be explored in Chapters 6 and 7.

Consider the fact that the material properties of a square SLGS are kept fixed and the length is increased. It is observed that the increase in length reduces the non-local effects and the frequencies of graphene sheets converge with the classical theory results ($e_o l_c = 0$). For larger $e_o l_c$ values, the size independency length is larger. This implies that there is always a critical value of length of SLGS beyond which the non-local effects can be ignored. Within the critical value of length, the non-local governing equation for vibration should be used (equation [4.10]). Another consequence of using equation [4.10] is that frequencies of graphene for higher modes of vibration are subjected to higher non-local effects. Thus, non-local frequencies are generally much smaller than the frequencies via local elasticity.

4.3.2. *Graphene sheets embedded in an elastic medium*

In the previous section, the single-layer sheet was free and there was no effect of any surroundings. Suppose that a representative graphene sheet is embedded into an elastic medium. The embedding of graphene sheets in a polymer matrix is generally used for strengthening purposes. The polymer matrix for composites can be considered as an elastic medium. The elastic medium can be modeled both by the Winkler foundation model and the Pasternak foundation model (Figure 4.3).

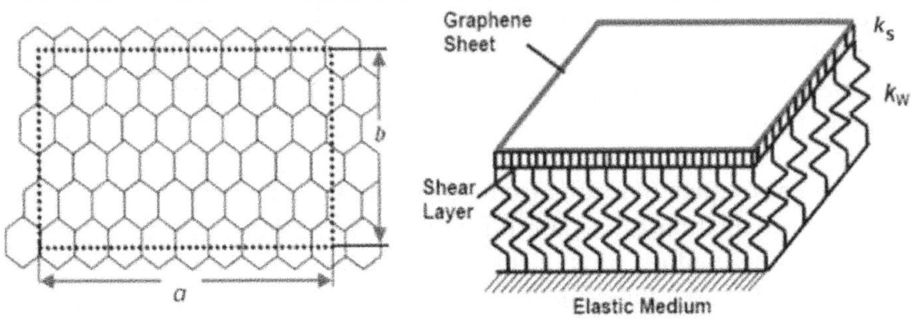

Figure 4.3. *Graphene sheets embedded in an elastic medium*

Winkler foundation model

The governing equation of motion of the graphene sheet using non-local elasticity theory by considering the Winkler model is expressed as:

$$D\nabla^2.\nabla^2 w + \rho h[1-(e_0 l_c)^2\nabla^2]\frac{\partial^2 w}{\partial^2 t} + [1-(e_0 l_c)^2\nabla^2]K_w w = 0 \qquad [4.14]$$

Pasternak foundation model

The governing equation of motion of the graphene sheet using non-local elasticity theory by considering the Pasternak model is expressed as:

$$D\nabla^2.\nabla^2 w + \rho h[1-(e_0 l_c)^2\nabla^2]\frac{\partial^2 w}{\partial^2 t} + [1-(e_0 l_c)^2\nabla^2]\left(K_w w - K_S\frac{\partial^2 w}{\partial^2 x}\right) = 0(23)$$

The terms K_w and K_s are the constants to each elastic medium, known as the Winkler modulus and the shear foundation model, respectively. In these foundation models, the elastic mediums are represented by virtual springs whose stiffnesses are K_w and K_S.

For a simply-supported graphene plate in an elastic medium with aspect ratio, the natural frequencies are found as follows.

Winkler foundation model

$$\omega_W = \frac{1}{a^2}\sqrt{\frac{D}{\rho h}}\sqrt{\frac{\psi^2+\frac{K_W a^4}{D}(1+\phi^2\psi)}{1+\phi^2\psi}} \quad [4.15]$$

Pasternak foundation model

$$\omega_P = \frac{1}{a^2}\sqrt{\frac{D}{\rho h}}\sqrt{\frac{\psi^2+\frac{K_W a^4}{D}(1+\phi^2\psi)+\frac{K_S a^2}{D}\psi(1+\phi^2\psi)}{1+\phi^2\psi}} \quad [4.16]$$

If the graphene sheet length is very large, the non-local effects will become diminished in the frequency response, and in this case the conventional classical equation can be used by employing ($\phi = 0$). The non-local natural frequencies are sensitive to the surrounding matrix. The surrounding matrix is modeled as a Winkler elastic medium. If the SLGS is embedded into a highly stiff elastic medium, the non-local frequencies will be a higher value (same as classical frequencies). As the stiffness of the elastic medium (surrounding matrix) increases, the frequencies of the SLGS will also increase. This increasing trend of frequency with surrounding matrix is noted to be influenced significantly by the non-local parameter ($e_o l_c$). For lower values of $e_o l_c$, the frequencies are higher while it is lower for high $e_o l_c$. In other words, this interprets that if the graphene sheets are embedded into a soft elastic medium, the fundamental frequency will be quite low for very small-size graphene sheets.

Consider the surrounding matrix is modeled as a Pasternak elastic medium. Applying the geometrical and material properties of graphene, the frequencies are obtained from equation [4.16]. Keeping the Winkler stiffness constant, if the value of the shear modulus increases, the fundamental frequencies of SLGS increase. The frequencies obtained from the Pasternak foundation model are relatively larger than those obtained from the Winkler foundation model. Similar non-local effects are seen here. For a higher $e_o l_c$ value, the frequencies with the Pasternak foundation model are lower. Unlike the Winkler foundation model, the increases in frequencies are linear.

4.4. Axially stressed nanoplate non-local theory

Consider the case when the SLGS is uniaxially stressed by a load N. The load may be external or developed due to residual stress during fabrication.

The equation for the vibration of SLG can be simply written as [MUR 09a]:

$$D\nabla^2.\nabla^2 w + \rho h[1-(e_0 l_c)^2\nabla^2]\frac{\partial^2 w}{\partial^2 t} + [1-(e_0 l_c)^2\nabla^2]N\frac{\partial^2 w}{\partial^2 x} = 0 \qquad [4.17]$$

and the natural frequency for the axially loaded graphene can be determined as:

$$\omega_S = \frac{1}{a^2}\sqrt{\frac{D}{\rho h}}\sqrt{\frac{\psi^2 + \frac{NL^2}{D}(1+\phi^2\psi)}{1+\phi^2\psi}} \qquad [4.18]$$

According to equation [4.18], the size or non-local effect has a significant influence on the vibration response of a uniaxially in-plane prestressed graphene sheet. The frequency values obtained with the non-local plate model are always smaller than that obtained with the local plate model. Also, the effect of the non-local parameter on the frequency of the second mode of vibration is higher compared to the fundamental mode of vibration. This implies that the mode shape of the second highly of vibration will be highly influenced by the non-local parameter.

A higher in-plane tensile preload N yields higher natural frequencies, while a higher in-plane compressive preload yields lower natural frequencies. It is also found that with an increase in the compressive preload, the decreasing rates of the eigen-values or the natural frequencies go faster, and the plate is more likely to reach the flexural buckling state $\omega_S = 0$. Furthermore, as the size of plate structures decreases and the non-local effects increase, the allowed values for the initial compressive stresses (critical buckling stress) are reduced. When the initial compressive stresses reach their limit values, the vibrational mode is buckled and the graphene structures cannot serve as vibration components any more. Therefore, the magnitudes of the axial compressive stresses are more critical for small-sized graphene structures. The values of the compressive preload for the resonance mode of vibration are a very important parameter in the design of a nanoplate-based device, because most of them worked on the first resonance modes.

4.5. In-plane vibration

Here, we extend the study to in-plane vibration [MUR 09b]. Axial or in-plane vibrations may be important. Axial or in-plane vibration experiments can be used for the determination of Young's modulus of carbon nanotubes (CNTs) and graphene sheets. Furthermore, graphene sheets can be used for microelectromechanical and nanoelectromechanical devices and be subjected to in-plane vibration. Generally, the flexural modes occur at low frequencies. However, vibrating nanoplates (SLGSs) may also have in-plane modes at relatively high frequencies and can be of very practical significance in high-operating frequencies. Graphene sheets when used as electromechanical resonators can be externally excited and exhibit in-plane vibrations.

Based on the non-local elasticity, the in-plane vibration of graphene can be written as:

$$C_{11}\frac{\partial^2 u}{\partial x^2} + C_{12}\frac{\partial^2 v}{\partial x \partial y} + C_{66}\left(\frac{\partial^2 v}{\partial x \partial y} + \frac{\partial^2 u}{\partial y^2}\right) = \rho h[1 - (e_0 l_c)^2 \nabla^2]\frac{\partial^2 u}{\partial t^2} \qquad [4.19]$$

$$C_{66}\left(\frac{\partial^2 v}{\partial x^2} + \frac{\partial^2 u}{\partial x \partial y}\right) + C_{12}\frac{\partial^2 u}{\partial x \partial y} + C_{22}\frac{\partial^2 v}{\partial y^2} = \rho h[1 - (e_0 l_c)^2 \nabla^2]\frac{\partial^2 v}{\partial t^2} \qquad [4.20]$$

where:

$$C_{11} = C_{22} = \frac{Eh}{(1-\nu^2)}, \quad C_{12} = C_{21} = \frac{\nu Eh}{(1-\nu^2)}, \quad C_{66} = Gh \qquad [4.21]$$

The separation of variable solution for equations [4.19]–[4.20] can be written as:

$$u(x,y) = Ae^{\mu x}e^{\lambda y}e^{i\omega t}; \;\; v(x,y) = Be^{\mu x}e^{\lambda y}e^{i\omega t} \qquad [4.22]$$

Here, μ and λ are the constants, ω represents the circular frequency and i is the conventional imaginary number $\sqrt{-1}$.

The expression of free in-plane vibration (FIV) frequency of graphene sheets can be written as [MUR 09b]:

$$\beta = \frac{\omega a}{c} = \sqrt{\frac{n^2\pi^2R^2 + m^2\pi^2}{1+\phi^2(n^2\pi^2R^2 + m^2\pi^2)}} \qquad [4.23]$$

Equation [4.23] gives the dimensionless frequency parameter taking into account the non-local effects. It is clear from relation [4.23] that when $\phi = 0$, we arrive at FIV of classical rectangular plate.

In terms of frequency ratio, equation [4.23] can be expressed as:

$$\text{Frequency Ratio} = \frac{\beta_{NL}}{\beta_L} = \sqrt{\frac{1}{1+\phi^2(n^2\pi^2R^2+m^2\pi^2)}} \qquad [4.24]$$

The influence of small-length scale on the FIV of graphene sheets is discussed here. The non-local solution for FIV is smaller than the classical (local) solutions. This is attributed to the effect of small-length scale. In addition, increasing the non-local parameter decreases the frequency ratio (natural frequency) of FIV. This implies that increasing the non-local parameter leads to a decrease in the stiffness of structure for in-plane vibration. Furthermore, as the length of the graphene increases the frequency ratio increases. This is obvious because with an increase in length, the influence of non-local effect reduces. According to Murmu and Pradhan [MUR 09b], at approximately $L \geq 50$ nm all results converge to the local frequency (e_0l_c=0). This implies that the non-local effect diminishes with an increase in the graphene length and vanishes after a certain length. This may be interpreted as the wavelength becoming larger with an increase in side length which decreases the effect of the small-length scale. For wave modes (m=1, n=1), the highest difference in frequency ratio values for different non-local parameters is observed for L between 5 and 10 nm.

The non-local effects are highly prominent in FIV of higher wave modes. In addition, the highest differences in frequency ratio values for different non-local parameters are found at larger lengths. Furthermore, at higher wave modes all results converge to the local frequency (e_0l_c=0) at higher lengths. It is seen that the effect of small-length scale is higher for higher wave modes. This phenomenon is due to the effect of a small wavelength at higher wave numbers. At smaller wavelengths, interactions between atoms increase which leads to an increase in the non-local effects.

4.6. Buckling of graphene sheets

Consider a rectangular SLGS of dimension $(a \times b)$ in the x and y directions and subjected to an external in-plane load. The load can be uniaxial as well as biaxial. Using the concept of non-local elasticity, governing equations for determining the buckling load can be derived. Some of the relations are presented below.

4.6.1. *Uniaxial buckling*

Based on the non-local elasticity theory, the governing equations of buckling of rectangular SLGS at room temperature are expressed as [PRA 09b]:

$$D\nabla^2.\nabla^2 w + P[1-(e_0 l_c)^2\nabla^2]\frac{\partial^2 w}{\partial^2 x} = 0 \quad [4.25]$$

For multilayer graphene sheets, if we ignore the Vander wall forces and consider it to be a thick plate, equation [4.25] would have to be modified. The non-local plate theory had to be developed based on the Mindlin plate theory.

Consider that the boundary conditions of SLGS are simply supported as in the case of free vibration. The boundary condition is thus expressed by$w(0,y) = w(a,y) = w(x,0) = w(x,b) = 0$ and non-local moment $M(0,y) = M(a,y) = M(x,0) = M(x,b) = 0$.

We assume the solution of equation [4.25] to be:

$$w = W_{mn} sin\frac{m\pi x}{a} sin\frac{n\pi y}{b} \quad [4.26]$$

For simply-supported SLGS with aspect ratio, $R = a/b$, the buckling load can be expressed as:

$$P = \frac{D}{L^2}\frac{\psi^2}{(m\pi)^2[1+\phi^2\psi]} \quad [4.27]$$

If a macroscopic (say thin graphite) plate is used, the expression applied to get the buckling load is:

$$P = \frac{D}{L^2}\frac{\psi^2}{(m\pi)^2} \quad [4.28]$$

The expression is the cause of the effect that $\phi = 0$ in equation [4.28]. In fact, ϕ is not exactly zero but $\phi \longrightarrow 0$, $(l_R \longrightarrow 0)$ and is very small and is neglected in classical elasticity theory.

For buckling analysis, we consider the SLGSs as an idealized non-local nanoplate and we assume the following properties. The Young's modulus of the SLGS is considered as $E = 1.06$ TPa, the Poisson ratio $\upsilon = 0.25$, and the thickness of the SLGS is taken as $h = 0.34$ nm. We see that if the buckling loads are obtained based on the present non-local elasticity theory, the buckling loads are influenced by size effects. As in the vibration phenomenon, according to equation [4.28] the non-local buckling loads of graphene sheets are highly sensitive on the single non-local parameter $e_o l_c$. Consider the non-local parameter $e_o l_c$ varied from 0 to 2 nm. Comparing the two relations, the difference of local (equation [4.28]) and non-local buckling loads results (equation [4.27]) is amplified depending on the value of the non-local parameter $e_o l_c$. Another important observation that can be made is that the buckling loads of the SLGS based on non-local elasticity theory are always smaller compared to solutions based on local elasticity theory. This can be attributed to the fact that for the same buckling loads with different the non-local parameters $e_o l_c$, the bending stiffness of the graphene sheets considering the non-local model is lower than its local counterpart. Such a phenomenon arises due to small size effects and will be discussed in Chapters 6 and 7.

4.6.2. *Graphene sheets embedded in an elastic medium*

In the previous section, the single-layer sheet was free and there was no effect of any surrounding. Suppose that a representative graphene sheet is embedded into an elastic medium [PRA 09b]. And we need to find the buckling loads of the graphene sheet based on non-local elasticity theory. We consider that the elastic medium is represented by the Winkler foundation model and the Pasternak foundation model.

Winkler foundation model

The equation of buckling load of the graphene sheet using the Winkler model:

$$D\nabla^2.\nabla^2 w + P[1-(e_0 l_c)^2\nabla^2]\frac{\partial^2 w}{\partial^2 x} + [1-(e_0 l_c)^2\nabla^2]K_w w = 0 \qquad [4.29]$$

Pasternak foundation model

The equation of buckling load of the graphene sheet using the Pasternak model:

$$D\nabla^2.\nabla^2 w + P[1-(e_0 l_c)^2\nabla^2]\frac{\partial^2 w}{\partial^2 x} + [1-(e_0 l_c)^2\nabla^2]\left(K_w w - K_S\frac{\partial^2 w}{\partial^2 x}\right) = 0 \quad [4.30]$$

The terms K_W and K_s are the constants to each elastic medium, known as the Winkler modulus and the shear foundation model, respectively.

Similar to section 4.6.1, for a simply-supported graphene plate in an elastic medium, the buckling loads can be expressed as follows.

Winkler foundation model

$$P = \frac{D}{L^2}\frac{\psi^2+\frac{K_w a^4}{D}(1+\phi^2\psi)}{(m\pi)^2[1+\phi^2\psi]} \quad [4.31]$$

Pasternak foundation model

$$P = \frac{D}{L^2}\frac{\psi^2+\frac{K_w a^4}{D}(1+\phi^2\psi)+\frac{K_S a^2}{D}\psi(1+\phi^2\psi)}{(m\pi)^2[1+\phi^2\psi]} \quad [4.32]$$

According to equation [4.32], non-local buckling loads are sensitive to the elastic medium (surrounding matrix). The surrounding matrix is modeled as the Winkler elastic medium. If the SLGS is embedded into a highly stiff elastic medium, the buckling loads will be higher value. As the stiffness of the elastic medium (surrounding matrix) increases, the buckling loads of the SLGS will also increase. This increasing trend of buckling loads with a surrounding matrix is noted to be influenced significantly by the non-local parameter ($e_o l_c$). For lower values of $e_o l_c$, the frequencies are higher while they are lower for high $e_o l_c$. In other words, this interprets that if the graphene sheets are embedded into a soft elastic medium, buckling loads will be quite low for very small-size graphene sheets compared to a hard elastic medium.

Consider that the surrounding matrix is modeled as a Pasternak elastic medium. If the value of shear modulus increases, the buckling loads of

SLGS increase with the Winkler stiffness constant. The buckling loads obtained from the Pasternak foundation model are relatively larger than those obtained from the Winkler foundation model. Similar non-local effects are seen here. For a higher $e_o l_c$ value, the buckling loads with the Pasternak foundation model are lower. Unlike the Winkler foundation model, the increases in buckling loads are generally linear.

4.7. Summary

Using the advanced continuum theory of non-local plate theory, applying as the governing equations for a graphene (idealized plate) structure, a lot of studies have been conducted by researchers. The studies include bending, vibration and buckling analysis of graphene, which has been briefly discussed in this chapter. It is important to note that the local elasticity is able to impart close prediction of graphene behavior when compared to MD simulation results. However, the local results are not accurate as we ignore the important size effects at the nanoscale. From non-local plate theories, the studies on bending, vibration and buckling analysis of graphene have predicted larger deflection, smaller frequencies and buckling loads in comparison to classical plate theories. There are, however, some broad important key issues of non-local elasticity plate theories that still require attention:

1) What is the exact constant (universal) value of non-local parameter for graphene? Or is the value of non-local parameter not constant for graphene?

2) Will the single non-local parameter able to describe all the nanoscale effects in the graphene sheets?

3) How far are the structural predictions from reality? Do they really behave this way or the other way (lesser deflection, larger frequencies and buckling loads) if modified couple stress theory is used?

The third question can be justified as follows. Sections 4.3, 4.4 and 4.6 on the mechanism of non-local effect in graphene (section 4.6) is the manifestation of the effect of nanoscale. However, the source was not discussed. The reason for decreased frequencies and buckling load is the consideration of *size-independent constant Young's modulus*. According to non-local elasticity theory, the stress at a point in a body not only depends on the strain at that point but also on all strains in the body. The differential form of non-local constitutive relation is $[1-(e_0 a)^2\nabla^2]\sigma_{\mathrm{NL}} = E\varepsilon$. Here a=lc from the relation, we obtain the expression of Young's modulus as $E_{NL} = \frac{[1-(e_0 a)^2 \nabla^2]\sigma_{\mathrm{NL}}}{\varepsilon}$. For a nanostructure, if the stress at a point (atom) behaves as local elastic stress, then we have $\sigma_{\mathrm{NL}} = \sigma_{\mathrm{L}} = E\varepsilon$, and the size-independent Young's modulus is obtained from $E_{\mathrm{L}} = \frac{\sigma_{\mathrm{L}}}{\varepsilon}$. Considering the term containing non-local effects is positive, we have E_{L} (size-independent) $>E_{\mathrm{NL}}$ (size-dependent), it follows that the bending rigidities (stiffness) as D_{L} (size-independent) $>D_{\mathrm{NL}}$ (size-dependent).

This implies that we are using a constant *size-independent Young's modulus* in all the governing equations of vibration and buckling presented in this chapter.

5

One-Dimensional Double-Nanostructure-Systems

Non-local analysis of one-dimensional nanostructure systems such as coupled double-carbon nanotubes are introduced in this chapter. The chapter addresses three important and fundamental aspects related to double one-dimensional nanostructures: (a) longitudinal vibration of coupled double-nanorods, (b) transverse vibration of coupled double-nanobeams and (c) axial instability of double-nanobeams. The theoretical development of the first two vibration phenomena, i.e. (a) and (b), are addressed in Chapters 2 and 3, respectively. This chapter begins by investigating the longitudinal vibration of a double-nanorod-system (DNRS). The DNRS are important in nanooptomechanical systems (NOMS). For the development of the governing equations, Eringen's non-local elasticity is utilized. Numerical results are presented for coupled double-carbon-nanotube-rod system. This study highlights that the non-local effect considerably influences the axial vibration of DNRS. Likewise the analysis on longituidinal vibration, non-local analysis dealing with flexural vibration and buckling double-nanobeam-system (DNBS) are also discussed in the subsequent sections.

5.1. Background

Nanomaterials are observed to have promising mechanical, chemical, electrical, optical and electronic properties. Such nanomaterials include carbon nanotubes, graphene sheets, zinc oxide (ZnO) nanowires, boron nitride (BN) nanotubes and nanosheets. The excellent properties of these nanomaterials have led to their multiple usages in the field of nanoelectronics, nanodevices, nanosensors, nanooscillators, nanoactuators, nanobearings and micromechanical resonators, as transporters of drugs, hydrogen storage, electrical batteries, nanocomposites and nanooptomechanical systems (NOMS). However, for further effective potential applications of these nanostructures, proper physical, chemical and mechanical understanding of the nanomaterials is essential.

Nanomaterials are the basis material of many nanoscale objects. Nanoscale objects are referred to as nanostructures. Recently, various one-dimensional nanostructures have been realized. They include nanodots, nanorods, nanowires, nanobelts, nanotubes, nanobridges and nanonails, nanowalls, nanohelixes and seamless nanorings. Among all of these the one-dimensional nanostructures, nanotubes, nanorods and nanowires are the most widely studied. This is because of the easy material formation and device applications.

Nanorods are one-dimensional nanoobjects ranging from 1 to 3,000 nm. These miniscule dimension nanorods or nanowires can be grown using various methods. A typical nanorod is shown in Figure 5.1. The popular methods include (1) vapor phase synthesis [WAG 70], (2) metal-organic chemical vapor deposition [PAR 02] and (3) hydrothermal synthesis [CHO 04]. Nanorods have found applications in a variety of nanodevices. These include ultraviolet photodetectors, nanosensors, transistors, diodes and LED arrays.

Figure 5.1. *ZnO nanorods growing on central nanowire cores. The length of these side nanorods ranges from 100 to 3,000 nm, while the width of the nanowires' core is 10–20 nm [WEN 03]*

The theory of non-local elasticity [ERI 83] is being increasingly used for efficient analysis of nanostructures, i.e. nanorods, nanobeams, nanorings, carbon nanotubes, nanoplates and graphenes, etc. Deotare *et al.* [DEO 09] introduced the non-local elasticity theory to nanotechnology. He developed a non-local Benoulli/Euler beam model and studied a case for a cantilever beam. Lu [LU 07b] utilized the Euler–Bernoulli non-local beam model and studied the dynamic analysis of axially prestressed micro/nanobeam structures based on non-local beam theory. Wang *et al.* [WAN 06] studied the elastic buckling phenomenon of microrods and nanorods/tubes based on the non-local Timoshenko beam theory. A detailed formulation for bending, vibration and buckling of carbon nanotubes was proposed using the non-local Euler–Bernoulli and Timoshenko theory by Reddy and Pang [RED 08]. Wang and Duan [WAN 08] investigated the free vibration problem of nanorings/arches using non-local elasticity theory. Axial frequencies using non-local continuum rod theories were presented by Aydogdu [AYD 09]. He further proposed a generalized non-local beam theory to study bending, buckling and free vibration of nanobeams [AYD 09]. Heireche *et al.* [HEI 08] analyzed the sound wave propagation in single-walled carbon nanotubes (SWCNTs) using non-local elasticity. Vibration and instability of tubular microbeams and nanobeams conveying fluid were studied using the theory of non-local elasticity [WAN 09]. Yan *et al.* carried out non-local elastic analysis of axially compressed buckling of triple-walled carbon nanotubes under the effect of temperature field. Torsional buckling of carbon nanotubes using non-local elasticity shell models was presented by Khademolhosseini *et al.* [KHA 10]. The non-local elasticity was also used in the nonlinear vibration analysis of carbon nanotubes [YAN 10]. The majority of the literature on non-local elasticity exists pertaining to the free transverse vibration of micro/nanobeams (CNTs), nanoplates (graphene sheets) and nanoshells (CNTS). This is because the transverse modes are more prone to becoming excited by common types of external excitations. On the contrary, only limited works on non-local elasticity have been devoted to the longitudinal vibration of nanobeams and nanoplates. Aydogdu [AYD 09] developed a non-local elastic rod model and applied it to investigate the small-scale effect on the axial vibration of clamped-clamped and clamped-free nanorods. Recently, Filiz and Aydogdu [FIL 10] applied the axial vibration of non-local rod theory to carbon nanotube heterojunction systems.

Like the transverse vibration, axial vibration experiments can also be used for the determination of Young's modulus of CNTs. Generally, the flexural modes occur at low frequencies. However, vibrating nanobeams (CNTs) may also have longitudinal modes at relatively high frequencies and can be of very practical significance in high-operating frequencies. Nanorods, when used as electromechanical resonators, can be externally excited and exhibit axial vibrations. Furthermore, for a moving nanoparticle inside a SWCNT, the SWCNT generally vibrates both in the transverse and longitudinal directions. The longitudinal vibration is generated because of the existing friction between the outer surface of the moving nanoparticle and the inner surface of the SWCNT. It is also reported [SAP 06] that transport measurements on suspended SWCNTs show signatures of phonon-assisted tunneling, influenced by longitudinal vibration (stretching) modes.

Therefore, there is a strong scientific need to gain an understanding of the longitudinal vibration of complex nanorods system and the mathematical modeling of such phenomena. In this section, we present, from a mathematical point of view, the longitudinal vibration of a double nanorod system (DNRS). Small-scale effects on the DNRS are investigated by employing non-local continuum mechanics. Governing equations for DNRS are developed using Eringen's non-local elasticity theory. An analytical method is proposed for solving the non-local frequencies of axially vibrating DNRS. Clamped-clamped and clamped-free boundary conditions are employed in and their explicit relationships have been derived. It has been shown that the non-local effect is quite significant in the longitudinal vibration studies and needs to be included in the continuum model of DNRS. The present results could bring helpful insights while investigating multiple-nanorod-system-models for nanosensor applications and in NOMS.

5.2. Revisiting non-local rod theory

The nanorod is assumed to be slender. In the one-dimensional form, the constitutive relation is [AYD 09]:

$$\sigma(x) - (e_0 a)^2 \sigma''(x) = E\varepsilon(x) \quad [5.1]$$

where the symbol $''$ in equation [5.1] denotes the derivation operator $\frac{\partial^2}{\partial x^2}$. The term E is the conventional Young's modulus of the nanostructure component (nanorod and nanobeam).

We represent the following stress resultant N as:

$$N = \int_A \sigma(x) dA \qquad [5.2]$$

where A is the cross-sectional area of the nanorod. Coupling equations [5.2] and [5.1] yields the following non-local stress resultant expression:

$$N - (e_0 a)^2 N''(x) = EAu'(x, t) \qquad [5.3]$$

Consider a slender nanorod of length L. The Euler–Lagrange equation (the equation of motion) for an axially vibrating nanorod in the domain $0 < x < L$ can be expressed as:

$$N' + f(x, t) = m\ddot{u}(x, t) \qquad [5.4]$$

where $\ddot{}$ implies the time derivative $\frac{\partial^2}{\partial t^2}$. Here, m and $u(x, t)$ are the mass and axial displacements of the nanorod and $f(x, t)$ is the axially distributed force. The expression for the axially vibrating nanorod can be derived as:

$$-EAu''(x, t) - f(x, t) + (e_0 a)^2 f''^{(x,t)} + m_1 \ddot{u}_1(x, t) - (e_0 a)^2 m_1 \ddot{u}_1''(x, t) = 0 \qquad [5.5]$$

Next, we utilize equation [5.5] for the development of coupled DNRS.

5.2.1. *Equations of motion of double-nanorod-system*

Consider a DNRS as shown in Figure 5.2. The nanorods (nanorod-1 and nanorod-2) are considered to be of length L. Axially distributed springs attach the two nanorods. The springs are used to substitute the effect of the elastic medium, Langevin forces or van der Waals forces (vdW) between the two nanorods. The springs are assumed to have a stiffness, k. In general, the two nanorods are different where the length, mass per unit length and axial rigidity of the ith rod are L_i, m_i and E_iA_i (i=1, 2), respectively. These parameters are assumed to be constant along each nanorod. The longitudinal displacements over the two nanorods are denoted by $u_1(x,t)$ and $u_2(x,t)$, respectively (Figure 5.2).

Using the non-local continuum rod theory (equation [5.5]), the governing equations for the longitudinal vibration of DNRS can be expressed as:

Nanorod-1

$$-E_1A_1u_1''(x,t) + k[u_1(x,t) - u_2(x,t)] - (e_0a)^2k[u_1''(x,t) - u_2''(x,t)] + m_1\ddot{u}_1(x,t) - (e_0a)^2m_1\ddot{u}_1''(x,t) = f(x,t) - (e_0a)^2f''(x,t) \quad [5.6]$$

Nanorod-2

$$-E_2A_2u_2''(x,t) - k[u_1(x,t) - u_2(x,t)] + (e_0a)^2k[u_1''(x,t) - u_2''(x,t)] + m_2\ddot{u}_2(x,t) - (e_0a)^2m_2\ddot{u}_2''(x,t) = 0 \quad [5.7]$$

where x is the axial position along the nanorods. Dots $(\cdot)$ and primes $('')$ denote partial derivatives with respect to time t and position coordinate x, respectively. With a simple manipulation of variables, the equations can be uncoupled and general analysis can be used to determine the solutions.

$$E_1A_1 = E_2A_2 = \bar{e} \equiv \text{constant} \quad [5.8]$$

$$m_1 = m_2 = m \equiv \text{constant} \quad [5.9]$$

$$f(x,t) = 0 \ \text{(free-vibration)} \quad [5.10]$$

where $\bar{e}$ and m denote the axial rigidity and mass per unit length, respectively. It should be noted that $\bar{e}$ and e_0 are different and should not be confused.

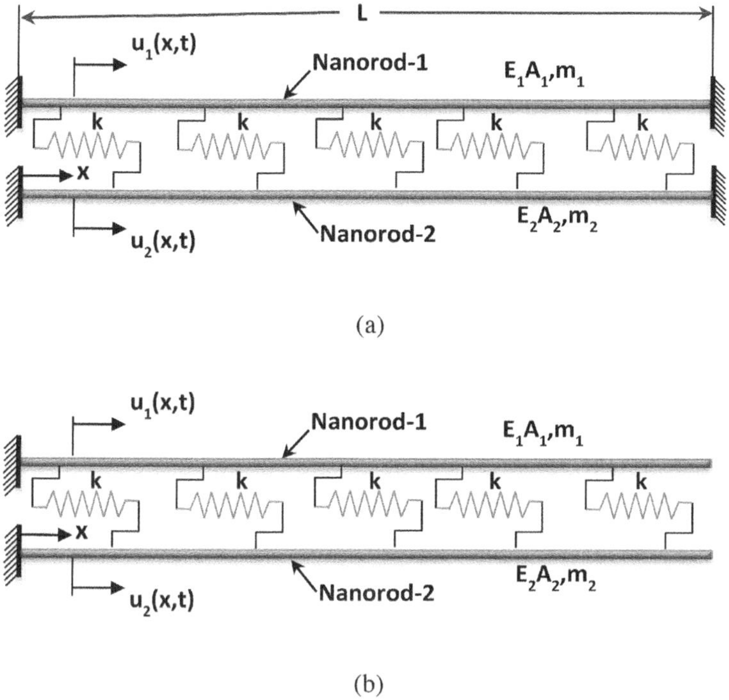

Figure 5.2. *Double nanorod configuration. a) Clamped-clamped boundary condition and b) clamped-free (cantilever) boundary condition*

Considering equations [5.6]–[5.7] and the assumptions from equations [5.8]–[5.10], we obtain:

Nanorod-1

$$-\bar{e}u_1''(x,t) + k[u_1(x,t) - u_2(x,t)] - (e_0a)^2k[u_1''(x,t) - u_2''(x,t)] + m\ddot{u}_1(x,t) - (e_0a)^2m\ddot{u}_1''(x,t) = 0 \quad [5.11]$$

Nanorod-2

$$-\bar{e}u_2''(x,t) - k[u_1(x,t) - u_2(x,t)] + (e_0a)^2k[u_1''(x,t) - u_2''(x,t)] + m\ddot{u}_2(x,t) - (e_0a)^2m\ddot{u}_2''(x,t) = 0 \quad [5.12]$$

For the DNRS, we denote:

$$u(x,t) = u_1(x,t) - u_2(x,t), \quad [5.13]$$

such that

$$u_1(x,t) = u(x,t) + u_2(x,t) \quad [5.14]$$

Here, $u(x,t)$ is the relative displacement of the primary nanorod with respect to the secondary nanorod. Using the definitions in equations [5.13] and [5.14] and substituting in equations [5.11] and [5.12], we obtain:

$$-\bar{e}u''(x,t) + 2ku(x,t) - 2(e_0a)^2ku''(x,t) + m\ddot{u}(x,t) - (e_0a)^2m\ddot{u}''(x,t) = 0 \quad [5.15]$$

$$-\bar{e}u_2''(x,t) + m\ddot{u}_2(x,t) - (e_0a)^2m\ddot{u}_2''(x,t) = ku(x,t) - (e_0a)^2ku''(x,t) \quad [5.16]$$

When the non-local effects are ignored ($e_0a = 0$) and a single rod is considered, the above equations revert to the equations of classical rod theory. For the present analysis of coupled DNRS, we will primarily be dealing with equation [5.15].

5.2.2. *Solution methodology*

Assuming that the relative motion $u(x,t)$ is one of its natural modes of vibration, the general solution of equation [5.15] can be written as:

$$u(x,t) = U(x)e^{i\omega t} \quad [5.17]$$

where ω is the natural frequency and $U(x)$ is the corresponding mode shape of the DNRS.

Substituting equation [5.17] into equation [5.15], we obtain:

$$U''(x) + \beta^2\, U(x) = 0 \quad [5.18]$$

where

$$\beta^2 = \frac{m\omega^2 - 2k}{\bar{e} + 2(e_0a)^2k - (e_0a)^2m\omega^2} \quad [5.19]$$

5.2.3. *Clamped-clamped boundary condition*

Now, we assume that all the ends of the DNRS are clamped (Figure 5.2(a)). The solution of equation [5.18] can be expressed in the form:

$$U(x) = A_1 \cos\beta x + A_2 \sin\beta x \qquad [5.20]$$

The boundary conditions of clamped-clamped case (for both nanorods) are given as:

Nanorod-1

$$u_1(0,t) = 0; \quad u_1(L,t) = 0 \qquad [5.21]$$

Nanorod-2

$$u_2(0,t) = 0; \quad u_2(L,t) = 0 \qquad [5.22]$$

It should be noted that in the present case (all ends clamped), the local and non-local boundary conditions are equivalent.

We define:

$$u(0,t) = u_1(0,t) - u_2(0,t), \qquad [5.23]$$

$$u(L,t) = u_1(L,t) - u_2(L,t), \qquad [5.24]$$

And using equation [5.20], we obtain:

$$U(0) = 0, \quad U(L) = 0 \qquad [5.25]$$

Substituting equation [5.25] into equation [5.20] yields the following expression:

$$A_1 = 0, \quad A_2 \sin\beta L = 0, \qquad [5.26]$$

Therefore, the frequency equation is:

$$\sin\beta L = 0, \qquad [5.27]$$

and their roots are:

$$\beta L = r\pi, \qquad r = 1,2 \ldots. \qquad [5.28]$$

The combination of equation [5.28] and equation [5.19] leads to:

$$\frac{m\omega^2-2k}{\bar{e}+2(e_0a)^2k-(e_0a)^2m\omega^2}=\left(\frac{r\pi}{L}\right)^2 \qquad [5.29]$$

From equation [5.29], the natural frequency can be obtained explicitly as:

$$(\omega)_r=\sqrt{\frac{\left(\frac{\pi r}{L}\right)^2(\bar{e}+2k(e_0a)^2)+2k}{m\left(1+(e_0a)^2\left(\frac{\pi r}{L}\right)^2\right)}}\,, \qquad r=1,2\ldots \qquad [5.30]$$

The ratio of the local and non-local longitudinal natural frequencies of the DNRS can be given as:

$$\text{Ratio}=\left(\frac{\omega_{NL}}{\omega_L}\right)_r=\sqrt{\frac{1}{1+\left(\frac{e_0a}{L}\right)^2(\pi r)^2}\left(1+\frac{2\left(\frac{e_0a}{L}\right)^2(\pi r)^2}{\frac{\bar{e}}{kL^2}(\pi r)^2+2}\right)}\,, \quad r=1,2\ldots \qquad [5.31]$$

5.2.4. *Clamped-free (cantilever) boundary condition*

The boundary conditions of the clamped-free case (for both nanorods) are given as:

Nanorod-1

$$u_1(0,t)=0;\ N_1(L,t)=\bar{e}u'_1(L,t)+m(e_0a)^2\ddot{u}'_1+(e_0a)^2k[u'_1(L,t)-u'_2(L,t)]=0 \qquad [5.32]$$

Nanorod-2

$$u_2(0,t)=0;\ N_2(L,t)=\bar{e}u'_2(L,t)+m(e_0a)^2\ddot{u}'_2-(e_0a)^2k[u'_1(L,t)-u'_2(L,t)]=0 \qquad [5.33]$$

Using equation [5.14], we write:

$$u(0,t)=u_1(0,t)-u_2(0,t)=0, \qquad [5.34]$$

$$N_1(L,t) - N_2(L,t) = \bar{e}u'_1(L,t) + m(e_0a)^2\ddot{u}'_1 + (e_0a)^2k[u'_1(L,t) - u'_2(L,t)] - \bar{e}u'_2(L,t) - m(e_0a)^2\ddot{u}'_2 + (e_0a)^2k[u'_1(L,t) - u'_2(L,t)] = 0 \quad [5.35]$$

Simplifying the above equations, we obtain:

$$U(0) = 0, \qquad U'(L) = 0 \quad [5.36]$$

It should be noted that for the present case (C-F), the local and non-local boundary conditions reduce to similar expressions.

Substituting equation [5.36] into equation [5.20], we obtain:

$$A_1 = 0, \qquad A_2\cos\beta L = 0, \quad [5.37]$$

Therefore, the frequency equation is:

$$\cos\beta L = 0, \quad [5.38]$$

The roots of this transcendental equation are:

$$\beta L = (2r-1)\frac{\pi}{2}, \qquad r = 1,2 \ldots. \quad [5.39]$$

From equation [5.19], we obtain:

$$\frac{m\omega^2 - 2k}{\bar{e} + 2(e_0a)^2k - (e_0a)^2m\omega^2} = \left[(2r-1)\frac{\pi}{2L}\right]^2 \quad [5.40]$$

Thus, the natural frequency of the clamped-free DNRS is given by:

$$(\omega)_r = \sqrt{\frac{\left[\frac{(2r-1)\pi}{2L}\right]^2(\bar{e} + 2k(e_0a)^2) + 2k}{m\left(1 + (e_0a)^2\left[\frac{(2r-1)\pi}{2L}\right]^2\right)}}, \; r = 1,2 \ldots. \quad [5.41]$$

The ratio of the local and non-local longitudinal natural frequencies of the system is given by:

$$\text{Ratio} = \left(\frac{\omega_{NL}}{\omega_L}\right)_r = \sqrt{\frac{1}{1+\left(\frac{e_0a}{L}\right)^2\left[\frac{(2r-1)\pi}{2}\right]^2}\left(1+\frac{2\left(\frac{e_0a}{L}\right)^2\left[\frac{(2r-1)\pi}{2}\right]^2}{\frac{\bar{e}}{kL^2}\left[\frac{(2r-1)\pi}{2}\right]^2+2}\right)},$$

$$r = 1,23, \ldots. \qquad [5.42]$$

5.2.5. *Longitudinal vibration of auxiliary (secondary) nanorod*

Using the vibration analysis as above, the longitudinal vibration of the secondary nanorod (nanorod-2) can be obtained. For the longitudinal vibration of the secondary nanorod, the relative axial displacements between the two nanorods are absent, that is $(u_1 - u_2 = 0)$. We apply the same procedure for solving equation [5.16].

Clamped-clamped boundary conditions:

The eigenvalues for the all-ends-clamped boundary conditions are obtained as:

$$\beta_{2r} = \frac{m\omega^2}{\bar{e}-(e_0a)^2m\omega^2} = \left(\frac{r\pi}{L}\right)^2 \qquad [5.43]$$

From this, the natural frequencies can be obtained as:

$$\omega_{2r} = \sqrt{\frac{\left[\frac{r\pi}{L}\right]^2\bar{e}}{m\left(1+(e_0a)^2\left[\frac{r\pi}{L}\right]^2\right)}} \qquad [5.44]$$

The ratio of the non-local and local frequencies is given by:

$$\text{Ratio} = \left(\frac{\omega_{\text{NL}}}{\omega_{\text{L}}}\right)_{2r} = \sqrt{\frac{1}{1+\left(\frac{e_0a}{L}\right)^2\left[\frac{r\pi}{L}\right]^2}} \qquad [5.45]$$

Clamped-free boundary conditions:

The eigenvalues for the clamped-free boundary conditions for both nanorods are given as:

$$\beta_{2r} = \frac{m\omega^2}{\bar{e}-(e_0 a)^2 m\omega^2} = \left[\frac{(2r-1)\pi}{2L}\right]^2 \quad [5.46]$$

The natural frequencies for the clamped-free boundary condition are given as:

$$\omega_{2r} = \sqrt{\frac{\left[\frac{(2r-1)\,\pi}{2L}\right]^2 \bar{e}}{m\left(1+(e_0 a)^2\left[\frac{(2r-1)\,\pi}{2L}\right]^2\right)}} \quad [5.47]$$

The ratio of the non-local and local frequencies is obtained as:

$$\text{Ratio} = \left(\frac{\omega_{\text{NL}}}{\omega_{\text{L}}}\right)_{2r} = \sqrt{\frac{1}{1+\left(\frac{e_0 a}{L}\right)^2\left[\frac{(2r-1)\pi}{2}\right]^2}} \quad [5.48]$$

The frequency ratios in equations [5.45] and [5.48] turn out to be the same as that of the axial vibration of a single nanorod [AYD 09] for the corresponding boundary conditions. Here, it should be interpreted that the DNRS vibrates in-phase motion, and that the whole DNRS can be considered as a single nanorod. For in-phase-type vibration, the DNRS would become independent of the effect of coupling medium (springs, attractive van der Waal forces and Langevin forces).

5.3. Axial vibration of double-rod system

The non-local theory for DNRS developed here is a generalized theory and can be applied for the longitudinal vibration analysis of coupled carbon nanotubes, double Zno nanorods systems and double-nanobeam-systems for NOMS application.

The results for longitudinal vibration of DNRSs obtained using the non-local elasticity theory are compared with the previous classical solutions [ERO 04]. In [ERO 04], vibration of scale-free double-rod-system is studied. For the comparative study, we consider the following arbitrary values, $L =$ 1 nm, $m = 10^{-9}$kg/m, $\bar{e} = 1$ nN and $e_0 a = 0 - 2$ nm. Both low and high stiffness values of the coupling springs are considered, i.e. $K = 8$ N/nm and $K = 80$ N/nm, respectively. Here, it should be noted that similar numerical magnitudes were taken in [ERO 04], but the length was in a larger scale (in meters).

For a further realistic approach, we consider the properties of the nanorod as that of a SWCNT [LU 97]. An armchair SWCNT with chirality (5, 5) is considered. The radius of each individual nanorod (SWCNT) is assumed as 0.34 nm. Young's modulus, *E*, is considered as 0.971 TPa [LU 97]. Here, the term "DNRS" would denote the case of a double-carbon-nanotube-rod system.

To illustrate the influence of the small-length scale (non-local effects) on the longitudinal vibration of DNRS, the frequency ratio is analyzed for different (1) lengths, (2) scale coefficients or non-local parameter, (3) boundary conditions and (4) modes of vibration. Both clamped-clamped and clamped-free boundary conditions (Figure 5.2) are employed in the study. Different values of length of the nanorod are considered for the computation. Lengths are taken in the range of L=2–30 nm. Both the nanotubes (nanorod-1 and nanorod-2) are assumed to have the same geometrical and material properties.

5.3.1. *Effect of the non-local parameter in the clamped-type DNRS*

Figure 5.3a depicts the frequency ratio percent (FRP) versus the length of the DNRS for different non-local parameters (e_0a). FRP is defined as:

$$\text{FRP} = (1 - \text{Ratio}) \times 100 \qquad [5.49]$$

where Ratio is defined as the ratio of non-local frequency to local frequency (ω_{NL}/ω_L). All the four-boundary conditions of the DNRS in this figure are considered to be clamped (C) (Figure 5.2a). For the choice of the non-local parameter, we take e_0a = 0.0 nm, 0.5 nm, 1.0 nm, 1.5 nm and 2.0 nm. We can see that when ($e_0a = 0$), the non-local rod model reverts to classical rod theory. The axially distributed springs have been assumed to be a stiffness value of k=2 N/nm. The stiffness of the axially distributed springs will depend on the stiffness of the chosen elastic medium enclosed between the nanorods or atomistic forces. If there is no intermediate elastic medium, the twin nanorods will be subjected to van der Waals forces or Langevin forces. The stiffness of the springs will then depend on the perpendicular distance between the two nanorods.

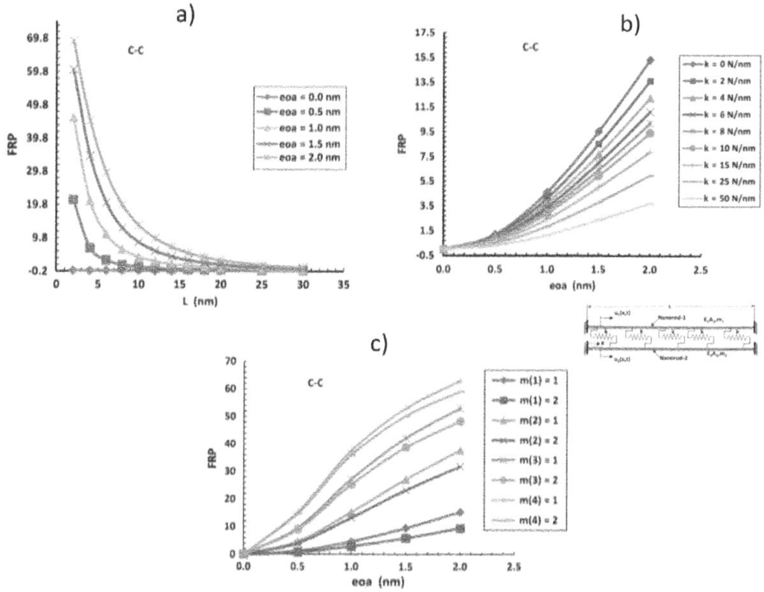

Figure 5.3. *a) Variation in frequency ratio percent with length of nan-orods in DNRS for different non-local parameters and clamped–clamped boundary condition, b) Variation in frequency ratio percent with non-local parameter for different stiffness's of springs and clamped–clamped boundary condition in DNRS, c) Variation in frequency ratio percent with non-local parameter for first eight modes of vibration and clamped–clamped. For a color version of the figure, see www.iste.co.uk/murmu/non-local.zip*

From Figure 5.3(a), it is observed that the non-local frequency solution of DNRS is smaller than the classical (local) frequency solution. This is attributed to the effect of small-length scale. In addition, increasing the non-local parameter decreases the natural frequency (increases the FRP). This can be accounted for by the fact that the non-local parameter leads to a decrease in the overall stiffness of DNRS. The small-scale effects make the DNRS "softer". As the length of the nanorods in DNRS increases, the FRP decreases. This is because with the increase in length, the influence of non-local effect reduces. With the further increase in length, the curves drop and become flat in nature. At larger lengths, there is no difference between the frequencies by the local and non-local rod models. At approximately $L \geq 30$ nm, all results converge to the local frequency (e_0a=0). This implies that the non-local effect diminishes with the increase in the DNRS length and practically disappears after a certain length. Similar findings are reported by other authors [AYD 09]. This may be interpreted by the fact that the wavelength gets larger with the increase in the length, which in turn decreases the effect of the small-length scale.

5.3.2. *Coupling spring stiffness in DNRS*

To show the effect of the stiffness of the springs on the longitudinal vibration of DNRS, FRP versus non-local parameter e_0a for various stiffness values has been plotted and shown in Figure 5.3(b). The stiffness of the springs is assumed to be in the range k=0 N/nm–50 N/nm. Figure 5.3(b) shows that as the non-local parameter increases, the FRP increases for all stiffness of the springs considered. This implies that the natural frequencies of the DNRS decrease with the increasing non-local effects (same as Figure 5.3(a)). It is noted that with the increasing value of the spring stiffness, the FRP decreases (i.e. frequency increases). It is interesting to note that non-local effects are more pronounced in nanorod without the presence of the springs or DNRS in in-phase-type vibration compared to strongly coupled nanorods (in out-of-phase vibration). The stiffness of the coupling springs has a subduing effect on the small-scale effects for longitudinal vibrating DNRS.

5.3.3. *Higher modes of vibration in DNRS*

To show the effect of higher modes in the longitudinal vibration of DNRS, FRP versus the non-local parameter for the first eight natural frequencies is plotted and shown in Figure 5.3(c). The term $m(1)$ represents the first modes with two natural frequencies. The first lowest natural frequency $m(1)$ =1 is the frequency of the secondary nanorod (or double nanorod acting as single nanorod,$(u_1 - u_2 = 0)$). This happens when the DNRS is vibrating in-phase motion. The second natural frequency $m(1) =$ 2 is from the DNRS when $(u_1 - u_2 \neq 0)$. This happens for the case when DNRS vibrates in out-of-phase motion.

It is observed from Figure 5.3(c) that, with the increase in the non-local parameter, the value of FRP increases. This effect is more often seen for the higher modes of longitudinal vibration. Thus, the non-local effects for the longitudinal vibration of DNRS are more pronounced in higher modes. It should be noted that the first frequency of each mode is independent of the stiffness of the coupling springs (due to axial in-phase vibration).

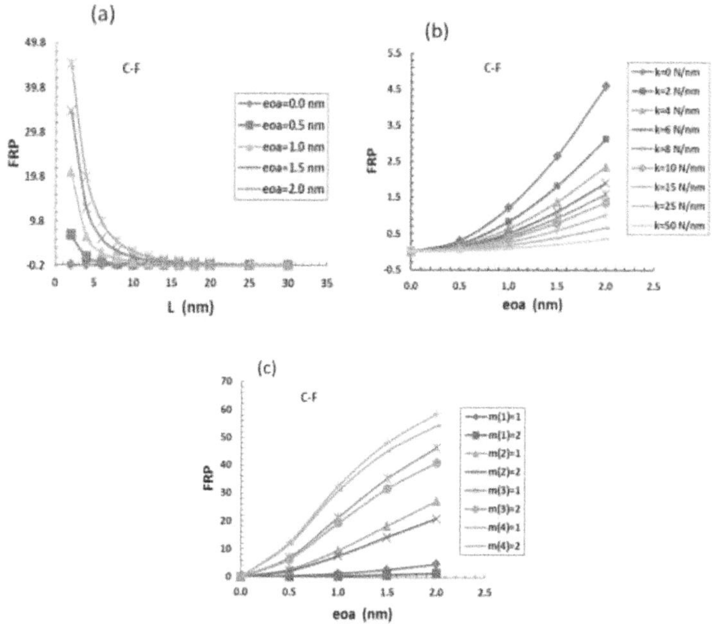

Figure 5.4. *a) Variation in frequency ratio percent with length of nanorods in DNRS for different non-local parameters and clamped-free boundary condition, b) variation in frequency ratio percent with nonlocal parameter for different stiffnesses of springs and clamped-free boundary condition in DNRS, c) variation in frequency ratio percent with non-local parameter for first eight modes of vibration and clamped-free boundary condition. For a color version of the figure, see www.iste.co.uk/murmu/non-local.zip*

5.3.4. *Effect of non-local parameter, spring stiffness and higher modes in cantilever-type-DNRS*

Figures 5.4 a-c show the variation of FRP for the cantilever-type-DNRS, analogous to what is described in Figures 5.3 a-c for the clamped-type DNRS. Figures 5.4 a-c show the plot for clamped-free boundary conditions for both nanorod-1 and nanorod-2. It can be observed from Figure 5.4a that with the increase in the length of DNRS, the FRP decreases. Approximately at $L \geq 20$ nm, all results converge to the local frequency ($e_0a = 0$). It is thus concluded that the convergence of local and non-local models is faster (with respect to length) in the cantilever boundary condition than in the clamped-clamped boundary condition. In addition, it is also noted that with the increasing non-local parameter, the natural longitudinal frequency of the DNRS decreases. These non-local frequency results are opposite to that of the transverse vibration of nanobeams with cantilever boundary conditions [LU 06]. In nanobeams, fundamental frequencies increase with increasing non-local parameters, while in nanorods fundamental frequencies decrease with increasing non-local parameters.

In Figure 5.4b, the FRP is plotted against the non-local parameter for the clamped-free boundary conditions. The stiffness of the springs has a subduing effect on the small-scale effects. From Figure 5.4c, it is seen that the non-local effect in the clamped-free boundary condition is more prominent for the higher modes of longitudinal vibration than in the all-clamped DNRS.

5.4. Summary

A non-local rod model for the longitudinal vibration of DNRS is developed based on Eringen's non-local continuum mechanics. In this model, the small-scale effects are explicitly taken into account. An exact analytical method is proposed for solving the non-local frequencies of axially vibrating DNRS. Clamped-clamped and clamped-free boundary conditions are employed and their explicit relationships have been derived. It is found that the longitudinal vibration frequencies of DNRS are overestimated by the classical (local) nanorod model as it disregards the effect of the small-length scale. It is found that the stiffness of the coupling springs in DNRS has a subduing effect on the small-scale effects. For cantilever boundary conditions, unlike the transverse vibration of nanobeams, the fundamental natural frequency of axially vibrating nanorods has a decreasing nature with the increasing non-local parameter. Similar to the transverse vibration of nanobeams, the non-local effects in longitudinal vibration of DNRS are more pronounced in the higher modes of vibration. The present study may bring helpful insights while investigating multiple-nanorod-system-models for future NOMS applications.

5.5. Transverse vibration of double-nanobeam-systems

Vibration analysis of double-nanobeam-systems is considered here. Double-nanobeam-systems are important in NOMS and sensor applications. Expressions for free bending-vibration of double-nanobeam-system are established within the framework of Eringen's non-local elasticity theory. An analytical method is developed for determining the natural frequencies of the non-local double-nanobeam-system. Explicit closed-form expressions for natural frequencies are derived for the case when all four ends are simply supported. The study highlights that the small-scale effects considerably influence the transverse vibration of double-nanobeam-systems. The non-local natural frequencies of double-nanobeam-system are smaller when compared to the corresponding local frequency values. The small-scale effects in the vibrating system are higher with increasing values of non-local parameters for the case of in-phase modes of vibration than in the out-of-phase modes of vibration. The increase in the stiffness of the coupling springs in double-nanobeam-system reduces the non-local effects during the out-of-phase modes of vibration.

5.5.1. *Background*

Recently, nanomaterials have intensely stimulated the interest of the scientific researcher's communities in physics, chemistry and engineering. These nanomaterials have special properties resulting from their nanoscale dimensions. Common examples of materials that exhibit interesting properties on the nanoscale are nanoparticles, nanowires and nanotubes (i.e. carbon nanotubes and ZnO nanotubes), etc. These nanomaterials have promising mechanical (tensile strength), chemical, electrical, optical and electronic properties. Because of many desirable properties, these nanomaterials are perceived to be the components for various nanoelectromechanical systems (NEMS) and nanocomposites:

Non-local elasticity accounts for the small-scale effects arising at the nanoscale level. Recent literature [SUD 03, WAN 06, WAN 07, LU 07, RED 07, HEI 08, RED 08, AYD 09, LEE 10, MUR 09a, MUR 09b, PRA 10, MUR 01, EIS 87] shows that the theory of non-local elasticity is being increasingly used for reliable and quick analysis of nanostructures, i.e. nanobeams, nanoplates, nanorings, carbon nanotubes, graphenes, nanoswitches and microtubules.

Extensive studies of non-local beams using non-local elasticity theory can be found in the literature. Reddy [RED 07] proposed the elaborate expressions of non-local beam theories for bending, vibration and buckling phenomena. Wang *et al.* [WAN 08] presented a beam bending solution based on the non-local Timoshenko beam theory. A formulation of third-order non-local beam theories based on Eringen's non-local continuum theory and the Reddy and Leung higher order beam models for bending, buckling and vibration of nanobeams was reported by Niu *et al.* [NIU 09]. Lim and Wang [LIM 07] introduced the exact variational non-local stress modeling with asymptotic higher order strain gradients for nanobeams. These non-local beam theories are generally utilized for the analysis of carbon nanotubes.

In the non-local elasticity theory, the small-scale effects are captured by assuming that the stress at a point is a function of the strains at all points in the domain. The majority of the existing works on non-local elasticity pertain to the free transverse vibration of single nanobeams. Though the mechanical studies of nanobeams may include vibration of multiple-walled nanotubes, the study of discrete multiple-nanobeam-system is particularly limited in the literature. However, for the scale-free (classical) double-beam-systems, some numbers of work on the vibration analysis have been reported [ABU 06, ZHA 08, LI 07, ONI 03].

The study of double-nanobeam-system is extremely important in the area of NOMS. The employment of double-nanobeam-systems has been reported by various researchers. Frank *et al.* [FRA 10] presented a dynamically reconfigurable photonic crystal nanobeam cavity. Their work involved two closely situated parallel vibrating clamped double-nanobeam-systems (Figure 5.5).

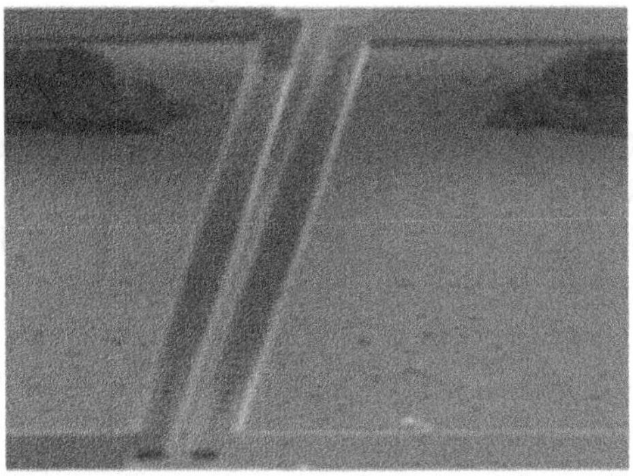

Figure 5.5. *Scanning electron microscope image of a double-nanobeam-system [FRA 10]*

Eichenfield *et. al.* [EIC 09] described the design, fabrication and measurement of a cavity NOMS. The NOMS consists of two closely separated coupled nanobeams. The researchers [EIC 09] fabricated the low-dimension double-nanobeam-system by depositing stoichiometric silicon nitride using low-pressure-chemical-vapor-deposition on a silicon wafer. Deotare *et al.* [DEO 09] studied the coupled photonic crystal nanobeam cavities consisting of two parallel suspended nanobeams separated by a small gap. The use of vibration properties in double-nanobeam-system has also been reported by Lin *et al.* [LIN 10]. The authors studied the coherent mixing of mechanical excitations in nanooptomechanical structures. Chan *et al.* [CHA 09] used the double-nanobeam-system to perform optical and mechanical design of a zipper photonic crystal optomechanical cavity. Most of the works reported here are experimental and computational works.

Therefore, based on the above discussion there is a strong encouragement to gain an understanding of the entire subject of vibration of complex-nanobeam-system and the mathematical modeling of such phenomena. In this chapter, an investigation is carried out to understand the small-scale effects in the free bending-vibration of a non-local double-nanobeam-system (NDNBS). Equations for free bending-vibration of a double-nanobeam-system (NDNBS) are formulated within the framework of Eringen's non-local elasticity. The two nanobeams are assumed to be attached by distributed vertical transverse springs. These springs may represent the stiffness of an enclosed elastic medium or Van der Waals forces or forces due to optomechanical coupling between the two nanobeams. An exact analytical method is shown for solving the non-local frequencies of transversely vibrating NDNBS. The simplification in the computation is achieved based on the change of variables to decouple the set of two second-order partial differential equations. It is assumed that the two nanobeams in the NDNBS are identical, and the boundary conditions on the same side of the system are the same. Simply-supported boundary conditions are employed in this study. Explicit expressions for the natural frequencies of NDNBS are derived. Results are obtained for various vibration phases of the NDNBS. The vibration phases include in-phase and out-of-phase modes of vibration. The effects of (1) non-local parameter, (2) stiffness of the springs and (3) the higher modes on the resonance frequency of the NDNBS are discussed.

5.5.2. *Non-local double-nanobeam-system*

Consider the non-local double nanobeam system (NDNBS) as shown in Figure 5.6. The two nanobeams of the NDNBS are referred to as nanobeam-1 and nanobeam-2. The nanobeams are considered to be of length L. Vertically distributed springs couple the two nanobeams. The springs may be used to substitute the electrostatic force [FRA 10], elastic medium [MUR 09], Van der Waals forces or forces due to optomechanical coupling between the two nanobeams. The springs have a stiffness k. In [FRA 10], it is reported that by applying a potential difference directly across the nanobeams, an attractive electrostatic force can be induced between the two nanobeams. Thereby, the equivalent spring stiffness can be varied between the nanobeams.

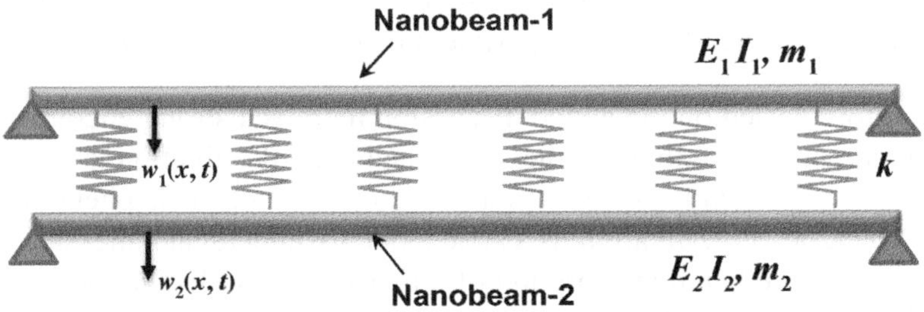

Figure 5.6. *The configuration of a simply-supported double-nanobeam-system (NDNBS)*

Generally, the two nanobeams are different where the length, mass per unit length and bending rigidity of the ith beam are L_i, m_i and E_iI_i (i=1, 2), respectively. These parameters are assumed to be constant along each nanobeam. The bending displacements over the two nanobeams are denoted by $w_1(x,t)$ and $w_2(x,t)$, respectively (Figure 5.6).

The governing non-local equations for NDNBS can be written as:

Nanobeam-1

$$E_1I_1w_1''''(x,t) + k[w_1(x,t) - w_2(x,t)] - (e_0a)^2k[w_1''(x,t) - w_2''(x,t)] + m_1\ddot{w}_1(x,t) - (e_0a)^2m_1\ddot{w}_1''(x,t) = 0 \qquad [5.50]$$

Nanobeam-2

$$E_2I_2w_2''''(x,t) - k[w_1(x,t) - w_2(x,t)] + (e_0a)^2k[w_1''(x,t) - w_2''(x,t)] + m_2\ddot{w}_2(x,t) - (e_0a)^2m_2\ddot{w}_2''(x,t) = 0 \qquad [5.51]$$

where dots (·) and primes (‘) denote partial derivatives with respect to time t and position coordinate x, respectively.

We assume that:

$$E_1 I_1 = (E_2 I_2) = EI \equiv \text{constant} \quad [5.52]$$

$$m_1 = m_2 = m \equiv \text{constant} \quad [5.53]$$

Considering equations [5.50] and [5.51] and using the assumptions from equations [5.52] and [5.53], we obtain:

Nanobeam-1

$$EIw_1''''(x,t) + k[w_1(x,t) - w_2(x,t)] - (e_0 a)^2 k[w_1''(x,t) - w_2''(x,t)] + m\ddot{w}_1(x,t) - (e_0 a)^2 m\ddot{w}_1''(x,t) = 0 \quad [5.54]$$

Nanobeam-2

$$EIw_2''''(x,t) - k[w_1(x,t) - w_2(x,t)] + (e_0 a)^2 k[w_1''(x,t) - w_2''(x,t)] + m\ddot{w}_2(x,t) - (e_0 a)^2 m\ddot{w}_2''(x,t) = 0 \quad [5.55]$$

For the NDNBS, we employ a change of variables by considering $w(x,t)$ as the relative displacement of the nanobeam-1 with respect to the nanobeam-2:

$$w(x,t) = w_1(x,t) - w_2(x,t), \quad [5.56]$$

such that:

$$w_1(x,t) = w(x,t) + w_2(x,t) \quad [5.57]$$

Subtracting equation [5.55] from equation [5.54] gives:

$$EI[w_1''''(x,t) - w_2''''(x,t)] + 2k[w_1(x,t) - w_2(x,t)] - 2(e_0 a)^2 k[w_1''(x,t) - w_2''(x,t)]$$

$$+m[\ddot{w}_1(x,t) - \ddot{w}_2(x,t)] - (e_0 a)^2 m[\ddot{w}_1''(x,t) - \ddot{w}_2''(x,t)] = 0 \quad [5.58]$$

Finally we obtain:

$$EIw''''(x,t) + 2kw(x,t) - 2(e_0a)^2kw''(x,t) + m\ddot{w}(x,t) - (e_0a)^2m\ddot{w}''(x,t) = 0 \quad [5.59a]$$

and

$$EIw_2''''(x,t) \quad + m\ddot{w}_2(x,t) - (e_0a)^2m\ddot{w}_2''(x,t) = kw(x,t) - (e_0a)^2kw''(x,t) \quad [5.59b]$$

It should be noted that when the non-local effects are ignored ($e_0a = 0$) and a single nanobeam is considered, the above equations revert to the equations of classical Euler–Bernoulli beam theory. For the present analysis of coupled NDNBS, we see the simplicity in using equation [5.59a].

5.6. Vibration of non-local double-nanobeam-system

Assuming that the relative motion $w(x,t)$ is one of its natural modes of vibration, the general solution of equation [5.59a] is written as:

$$w(x,t) = W(x)e^{i\omega t} \quad [5.60]$$

Here, ω is the frequency, $W(x)$ is the corresponding deformation shape of the NDNBS and i is the conventional imaginary number $\sqrt{-1}$. Substituting equation [5.60] into equation [5.59a] yields:

$$A_1W''''(x) + A_2W''(x) - A_3W(x) = 0 \quad [5.61]$$

where the coefficients are:

$$A_1 = EI; A_2 = m\omega^2(e_0a)^2 - 2k(e_0a)^2; \quad A_3 = m\omega^2 - 2k \quad [5.62]$$

The general solution of equation [5.61] can be given as:

$$W(x) = C_1\sin\alpha x + C_2\cos\alpha x + C_3\sinh\beta x + C_4\cosh\beta x \quad [5.63]$$

where

$$\alpha^2 = \frac{1}{2A_1}\left(A_2 + \sqrt{{A_2}^2 + 4A_1A_3}\right) \quad [5.64]$$

$$\beta^2 = \frac{1}{2A_1}\left(A_2 - \sqrt{{A_2}^2 + 4A_1A_3}\right) \qquad [5.65]$$

The terms C_1, C_2, C_3 and C_4 are to be determined from the boundary conditions.

5.7. Boundary conditions in non-local double-nanobeam-system

Now, we present the mathematical expressions of the boundary conditions in NDNBS. The boundary conditions of simply-supported conditions are described here. At each end of the nanobeams in NDNBS, the displacement and the non-local moments are considered to be zero. They can be mathematically expressed as:

Nanobeam-1

$$w_1(0,t) = 0; \qquad [5.66a]$$

$$M_1(0,t) = -EI{w_1}''(0,t) + (e_0a)^2 m\ddot{w}_1(0,t) + (e_0a)^2 k[w_1(0,t) - w_2(0,t)] = 0 \qquad [5.66b]$$

and

$$w_1(L,t) = 0; \qquad [5.67a]$$

$$M_1(L,t) = -EI{w_1}''(L,t) + (e_0a)^2 m\ddot{w}_1(L,t) + (e_0a)^2 k[w_1(L,t) - w_2(L,t)] = 0 \qquad [5.67b]$$

Nanobeam-2

$$w_2(0,t) = 0; \qquad [5.68a]$$

$$M_2(0,t) = -EI{w_2}''(0,t) + (e_0a)^2 m\ddot{w}_1 - (e_0a)^2 k[w_1(0,t) - w_2(0,t)] = 0 \qquad [5.68b]$$

and

$$w_2(L,t) = 0; \qquad [5.69a]$$

$$M_2(L,t) = -EIw_2''(L,t) + (e_0a)^2 m\ddot{w}_1 - (e_0a)^2 k[w_1(L,t) - w_2(L,t)] = 0 \quad [5.69b]$$

Using equation [5.56], the boundary condition simplifies to:

$$w(0,t) = w_1(0,t) - w_2(0,t) = 0; \quad [5.70]$$

$$M_1(0,t) - M_2(0,t) = -EIw_1''(0,t) + (e_0a)^2[m\ddot{w}_1(0,t) + k[w_1(0,t) - w_2(0,t)]] + EIw_2''(0,t) - (e_0a)^2[m\ddot{w}_1 - k[w_1(0,t) - w_2(0,t)]] = 0; \quad [5.71]$$

$$w(L,t) = w_1(L,t) - w_2(L,t) = 0; \quad [5.72a]$$

$$M_1(L,t) - M_2(L,t) == -EIw_1''(L,t) + (e_0a)^2[m\ddot{w}_1(L,t) + k[w_1(L,t) - w_2(L,t)]] + EIw_2''(L,t) - (e_0a)^2[m\ddot{w}_1 - k[w_1(L,t) - w_2(L,t)]] = 0 \quad [5.72b]$$

Simplifying the boundary conditions effectively reduces to:

$$W(0) = 0 \text{ and } W''(0) = 0, \quad W(L) = 0 \text{ and } W''(L) = 0, \quad [5.73]$$

The above boundary conditions, therefore, simplify to that of the classical single nanobeam. Here, it should be noted that the local and non-local boundary conditions are effectively the same.

Similarly, the boundary conditions of all ends clamped can be presented as:

For nanobeam-1

$$w_1(0,t) = 0; \quad w_1'(0,t) = 0 \quad [5.74]$$

$$w_1(L,t) = 0; \quad w_1'(L,t) = 0 \quad [5.75]$$

For nanobeam-2

$$w_2(0,t) = 0; \quad w_2'(0,t) = 0 \quad [5.76]$$

$$w_2(L,t) = 0; \quad w_2'(L,t) = 0 \quad [5.77]$$

Using equation [5.14], the boundary condition of the clamped case becomes:

$$w(0,t) = w_1(0,t) - w_2(0,t) = 0, \quad [5.78]$$

$$w'(0,t) = w_1'(0,t) - w_2'(0,t) = 0, \quad [5.79]$$

$$w(L,t) = w_1(L,t) - w_2(L,t) = 0, \quad [5.80]$$

$$w'(L,t) = w_1'(L,t) - w_2'(L,t) = 0, \quad [5.81]$$

Simplifying the effective boundary conditions reduces to:

$$W(0) = 0 \text{ and } W'(0) = 0, \quad W(L) = 0 \text{ and } W'(L) = 0, \quad [5.82]$$

Similar procedures can be applied for other mixed boundary conditions such as the clamped-free case.

5.8. Exact solutions of the frequency equations

A. Both nanobeams of NDNBS are vibrating out-of-phase; $(w_1 - w_2 \neq 0)$.

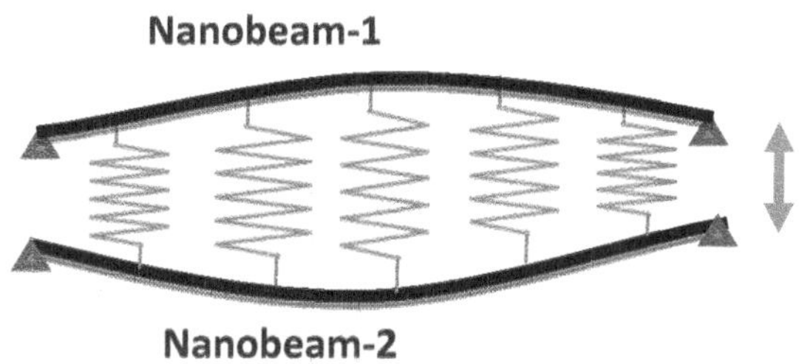

Figure 5.7. *Out-of-phase vibration of the double-nanobeam-system* $(w_1 - w_2 \neq 0)$

Consider the case of the NDBNS when both nanobeams are vibrating with in-phase sequence and out-of-phase sequence. The configuration of the NDNBS with out-of-phase sequence of vibration $(w_1 - w_2 \neq 0)$ is shown in Figure 5.7. In this section, we solve the frequency for the out-of-phase vibration. Here, we consider the case when all the ends have simply-supported boundary conditions (Figure 5.6). For simply-supported cases, the use of the boundary conditions equation yields:

$$C_2 = 0, C_4 = 0 \quad [5.83]$$

and

$$\begin{bmatrix} \sin\alpha L & \sinh\beta L \\ EI\alpha^2 \sin\alpha L & EI\beta^2 \sinh\beta L \end{bmatrix} \begin{Bmatrix} C_1 \\ C_3 \end{Bmatrix} = \begin{Bmatrix} 0 \\ 0 \end{Bmatrix} \quad [5.84]$$

For the non-trivial solution of equation [5.84], the determinant is zero and that yields:

$$\sin\alpha L(EI\beta^2 sinh\beta L \; - EI\alpha^2 \, sinh\beta L) = 0 \quad [5.85]$$

Therefore, the frequency equation is:

$$\sin \alpha L = 0 \quad [5.86]$$

which implies:

$$\alpha L = n\pi, \qquad n = 1,2 \ldots.. \quad [5.87]$$

From equation [5.61]:

$$2A_1\alpha^2 = A_2 + \sqrt{A_2{}^2 + 4A_1A_3} \quad [5.88]$$

which yields:

$$A_1\alpha^4 - A_2\alpha^2 - A_3 = 0 \quad [5.89]$$

Here, we define the frequency, stiffness and non-local parameter as:

$$\Omega = \omega L^2 \sqrt{\frac{m}{EI}}, \quad K = \frac{kL^4}{EI}, \quad \mu = \frac{e_0 a}{L} \quad [5.90]$$

Using equations [5.64] and [5.90], the natural frequency of NDNBS is:

$$\Omega_n = \sqrt{\frac{(n\pi)^4 + 2K + 2K(\mu)^2(n\pi)^2}{1 + (\mu)^2(n\pi)^2}}, \quad n = 1,2 \ldots \quad [5.91]$$

Similarly, for the case of an all ends clamped boundary condition, the roots of the following transcendental equation along with equation [5.64] give the natural frequencies of the NDNBS:

$$-2 + 2\cos\alpha L\,\cos\beta L + \left(\frac{\alpha}{\beta} - \frac{\beta}{\alpha}\right)\sin\alpha L\,\sin\beta L = 0 \qquad [5.92]$$

B. Both nanobeams of NDNBS are vibrating in phase; $(w_1 - w_2 = 0)$.

Next, the in-phase sequence of vibration will be considered (Figure 5.8). For the present NDNBS, the relative displacements between the two nanobeams are absent $(w_1 - w_2 = 0)$. Here, we solve equation [5.59b] for the vibration of NDNBS. The vibration of nanobeam-2 would represent the vibration of the coupled vibrating system. We apply the same procedure for solving equation [5.59a]. Similarly, we can obtain the natural frequencies. The natural frequencies for the NDNBS in this case can be expressed as:

$$\Omega_n = \sqrt{\frac{(n\pi)^4}{1+(\mu)^2(n\pi)^2}}, \quad n = 1,2\,... \qquad [5.93]$$

For this case, the vibration of NDNBS is independent of the stiffness of the connecting springs and therefore the NDNBS can be effectively treated as a single nanobeam.

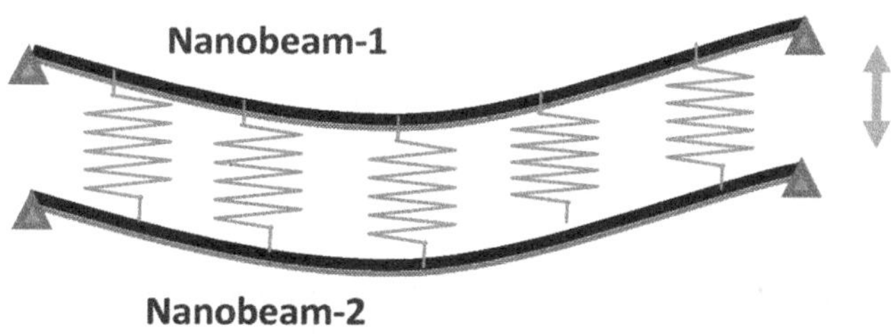

Figure 5.8. *In-phase vibration of the double-nanobeam system* $(w_1 - w_2 = 0)$

C. One nanobeam of NDNBS is stationary; $(w_2 = 0)$

Consider the case of NDNBS when one of the two nanobeams (i.e. nanobeam-2) is stationary $(w_2 = 0)$. The schematic diagram is shown in Figure 5.9.

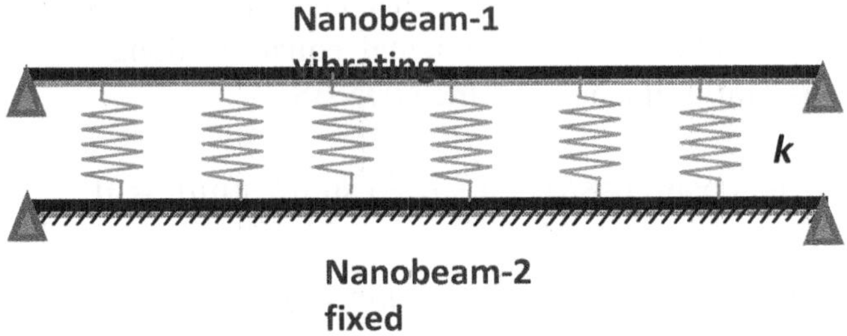

Figure 5.9. *Vibration of NDNBS with one nanobeam stationary*

Using the equations from non-local elasticity the governing equation for the NDNBS in this case reduces to:

$$EIw''''(x,t) + kw - (e_0a)^2kw''(x,t) + m\ddot{w}(x,t) - (e_0a)^2m\ddot{w}''(x,t) = 0 \quad [5.94]$$

and the equivalent non-local boundary conditions are:

$$w(0) = 0 \text{ and } w''(0) = 0, \quad w(L) = 0 \text{ and } w''(L) = 0, \quad [5.95]$$

In this case, the NDNBS behaves as if the nanobeam is embedded into an elastic medium. The elastic medium can be modeled as Winkler elastic foundation [WAN 06]. The stiffness of the elastic medium is denoted by k. By following the same procedure as the solution of equation [5.19a], the non-local frequency of NDNBS can be explicitly expressed as:

$$\Omega_r = \sqrt{\frac{(r\pi)^4 + K + K(\mu)^2(r\pi)^2}{1+(\mu)^2(r\pi)^2}}, r = 1,2 \ldots \quad [5.96]$$

where K is the stiffness parameter of the coupling springs and μ is the non-local parameter as defined in equation [5.90]. In fact, when one of the nanobeams (i.e. nanobeam-2) in NDNBS is fixed ($w_2 = 0$), the NDNBS behaves as a nanobeam on an elastic medium.

5.9. Discussions

1) The non-local theory for NDNBS illustrated here is a generalized theory and can be applied for the bending-vibration analysis of coupled carbon nanotubes, double ZnO nanobeam systems and double-nanobeam-systems for NOMS application [FRA 10, EIC 09, 48-50]. The reliability of

non-local elasticity theory in the analysis of nanostructures (nanotubes and graphene sheet) can be observed in various earlier works [SUD 03, WAN 06, WAN 07, LU 07, RED 07, HEI 08, RED 08, AYD 09, LEE 10, MUR 09a, MUR 09b, PRA 10, MUR 01, EIS 87, 37-43].

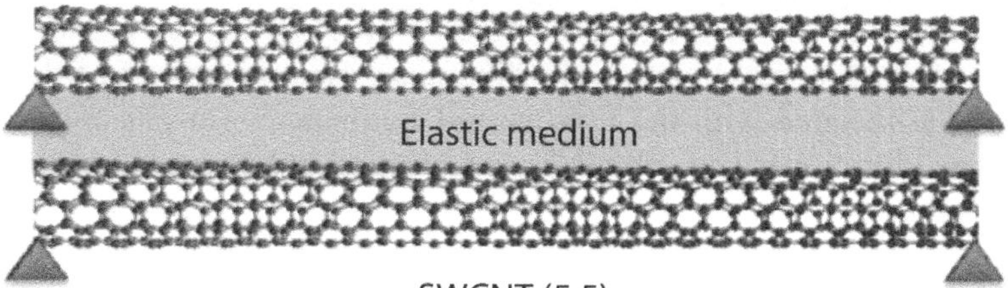

Figure 5.10. *Coupled carbon nanotube system*

For the present study, the properties of the nanobeams are considered to be that of a SWCNT [LU 97]. An armchair SWCNT with chirality (5, 5) is considered (Figure 5.10). The radius of each individual SWCNT is assumed as 0.34 nm. Young's modulus, E, is taken as 0.971 TPa [LU 97]. The length is taken as 100 nm. The frequency results of the NDNBS are presented in terms of the frequency parameters (equation [5.51]). The non-local parameter and the stiffness of the springs are computed as given in equation [5.50]. Different values of spring parameters, K, are considered. Spring stiffness represents the stiffness of the enclosing elastic medium. Both high and large stiffness of springs are assumed. Values of K range from 5 to 500. Both the nanotubes (nanotube-1 and nanotube-2) are assumed to have the same geometrical and material properties.

In general, the non-local parameters may be taken as $e_0 = 0.3917$ and a =0.142 nm (distance between carbon-carbon atoms). For carbon nanotubes and graphenes, the range of $e_0 a = 0 - 2.0$ nm has been widely used [WAN 07, LU 07, RED 07] and [MUR 09a]. In the present study, we take the scale coefficient μ or non-local parameter in the similar range as $\mu = 0 - 1$].

5.9.1. *Effect of small scale on vibrating NDNBS*

Figure 5.11 shows the variation of the frequency parameter with the scale coefficient for different cases of NDNBS. Three different cases of NDNBS are considered. Case 1, case 2 and case 3 depict the conditions (1) when both the nanotubes vibrate in the out-of-phase sequence $(w_1 - w_2 \neq 0)$ (2) when one of the nanotubes in NDNBS is stationary $(w_2 = 0)$ (3) when both the nanotubes vibrate with in-phase sequence $(w_1 - w_2 = 0)$, respectively. The results for the frequency parameter Ω are in the dimensionless form as in equation [5.50]. The stiffness parameter of the springs is assumed to be constant (K=10). Unless stated, the frequency parameter would denote the parameter associated with the first natural frequencies (in-phase and out-of-phase-type vibration). The SWCNTs are referred to here as nanobeams.

Figure 5.11a shows that as the values of the non-local parameter μ increase, the frequency parameter Ω decreases. The decreasing trend of the frequency parameter against increasing non-local parameter happens for all the three cases considered (i.e. case 1, case 2 and case 3). This decrease in value of the frequency parameters is due to the assimilation of small-scale effects in the NDNBS. Therefore, by the non-local elastic model the size effects are reflected in the NDNBS.

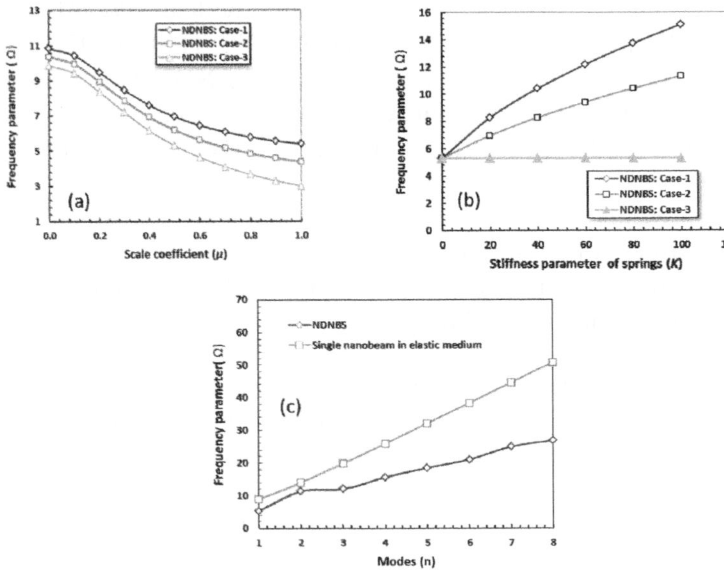

Figure 5.11. *The variation of the frequency parameter as a function of the spring parameter and the scale-coefficient (μ), a) for the 1st in-phase and out of phase buckling of NDNBS, b) for the 2nd in-phase and out of phase buckling of NDNBS, c) for the 3rd in phase and out of phase buckling of NDNBS. For a color version of the figure, see www.iste.co.uk/murmu/non-local.zip*

On comparison of the three cases of NDNBS (i.e. case 1, case 2 and case 3), the frequency parameter for case 1 (NDNBS vibrating with out-of-phase sequence) is larger than the frequency parameter for case 2 and case 3. The relative higher frequency in case 1 (equation [5.51]) is due to the coupling effect of the spring and the two nanobeams (nanotubes). The presence of springs for case 1 makes the NDBNS stiffer and increases the stiffness of the system. For case 2, the stiffness effect due to the auxiliary nanobeam (nanobeam 2) is absent. Thereby, there is an effective lower stiffness parameter in case 2.

In addition, it is seen that the values of the frequency parameter for case 2 are larger than the values of the frequency parameter for case 3. In case 3, the frequency parameter is relatively low because the NDBNS becomes independent of the effect of the spring stiffness. The NDNBS vibrates in the in-phase sequence. For case 3, the NDNBS becomes similar to the frequency of the single nanobeam without the effect of an elastic medium. In other words, the whole NDNBS can be treated as a vibrating single nanobeam and the coupling internal structure is effectless (case 3). Furthermore, it is important to note that the above-mentioned frequency behavior for NDBNS (Figure 5.11a) is amplified for larger values of the scale coefficient values.

In general, it is worth noticing that the small-scale effects in NDNBS are higher with an increasing non-local parameter in the in-phase vibration than in the out-of-phase vibration. This is because the stiffness of the springs in out-of-phase vibration reduces the non-local effects. For a different set of boundary conditions, this behavior may be different. As small-scale effects reduce the natural frequencies of the vibrating nanosystem, nanoresonators can be designed so as to vibrate in the out-of-phase modes. Thus, the nanoresonators would have higher resonant frequencies. This implies the significance of the use of small-scale effects in NDNBS. If some damping properties are present within the NDNBS, then the damping behavior could be effective in out-of-phase mode vibration (similar to stiffening NDNBS). Thus, one of the nanobeams would act as a vibration absorber. However, the damping behavior would be ineffective in the in-phase mode vibration. This vibration behavior will be important for study in NOMS.

5.9.2. *Effect of the stiffness of the coupling springs on NDNBS*

To illustrate the effect of the smaller and higher values of the stiffness of the coupling springs, curves have been plotted for the frequency parameter against the stiffness parameter of springs. Figure 5.11b depicts the variation of the frequency parameter (Ω) with the stiffness parameter of the springs (K) for the three cases of NDBNS. The non-local parameter μ is considered as 0.5. As the stiffness parameter of the coupling springs increases, the frequency parameter increases. The difference of frequencies between case 1 and case 2 is amplified for higher values of stiffness parameter of springs. Case 3 of NDNBS is unchangeable with increasing stiffness of springs due to the in-phase behavior of vibration.

5.9.3. *Analysis of higher modes of NDNBS*

To see the effect of higher natural frequencies of NDNBS, curves have been plotted for frequency parameters against the first eight natural frequencies (Figure 5.11c). Curves have been illustrated for general NDNBS (both nanobeams vibrating) and a single nanobeam embedded into an elastic medium. The non-local parameter μ is considered as 0.5. The first, second, third and fourth pair of natural frequencies consist of in-phase and out-of-phase modes of NDNBS, respectively. In in-phase mode, the two nanobeams vibrate in the same direction while in out-of-phase mode the beam vibrates in opposite direction. In in-phase modes of vibration, the natural frequencies are independent of stiffness of springs. The corresponding pairs of in-phase and out-of-phase modes are also sometimes referred to as submodes. Here, the non-local parameter is assumed to be μ =0.5. From Figure 5.11c, it is observed that the frequency parameter increases with an increase in number of modes (wave numbers). The higher natural frequencies of NDNBS are less than the NDNBS with one nanobeam fixed. This is due to inclusion of both in-phase and antiphase modes in the system of NDNBS.

For the comprehension of the vibration of NDNBS with small-scale effects, the natural frequencies of the double nanobeam system for different non-local parameters and stiffness parameters are plotted in a three-dimensional graph. Figures 5.12 a-c show the plots for the first, second and third in-phase and out-of-phase vibrations, respectively. The stiffness parameter K of the coupling springs is considered in the range of 0–500.

The non-local parameter μ is varied from 0 to 1. From Figure 5.12a, it is observed that the increase in the non-local parameter has a reducing effect on the natural frequencies of the NDNBS. However, this reduction cannot be experienced at higher stiffness parameter values. The stiffness of the springs has a reducing effect on the small-scale effects of the NDNBS. It should be noted that for in-phase vibration, the natural frequencies are independent of coupling springs. Figure 5.12b shows the plot for second in-phase and out-of-phase vibrations, respectively.

Compared to first natural frequencies (Figure 5.12a), the frequencies for second natural frequencies (Figure 5.12b) are more affected by scale effects. This is in line with the transverse vibration of uncoupled single nanobeams. In addition, it is observed that the stiffness effect is less in both second in-phase and out-of-phase modes of vibration. The stiffness effect is further reduced between third in-phase and out-of-phase natural frequencies (Figure 5.12c). Thus, it can be concluded that although the small-scale effects are more in higher frequencies (modes), the effect of stiffness of coupling springs reduces the non-local effects.

The present work provides an analytical solution that serves as a benchmark for further investigation of more complex n-nanobeam systems such as triple-nanobeam systems (TNBS). Different boundary conditions at the ends will result in different vibration behaviors. This would find an application in the design of nanoresonators. The present work could also be useful in the study of double-nanoplate system for future NOMS studies.

5.10. Summary

In this section, theoretical non-local elasticity is developed for the free bending-vibration of a double-nanobeam-system (NDNBS). An exact analytical method is developed for determining the non-local frequencies of transversely vibrating NDNBS. The in-phase and out-of-phase vibrations are examined in detail. The study shows that non-local effects are important in the transverse vibration of NDNBS. Non-local effects reduce the frequencies of the NDNBS. Increasing the stiffness of the springs in NDNBS reduces the non-local effects. The small-scale effects in NDNBS are more prominent with the increasing non-local parameter in the in-phase vibration than in the out-of-phase vibration. The frequencies in the in-phase vibration are independent of the stiffness of the springs. The NDNBS can then be treated as a single nanobeam. However, the frequencies of NDNBS in the out-of-phase vibration increase with the increasing stiffness parameter of the coupling springs. Furthermore, this work provides an analytical solution which could serve as the starting point for further investigation of more complex n-nanobeam systems.

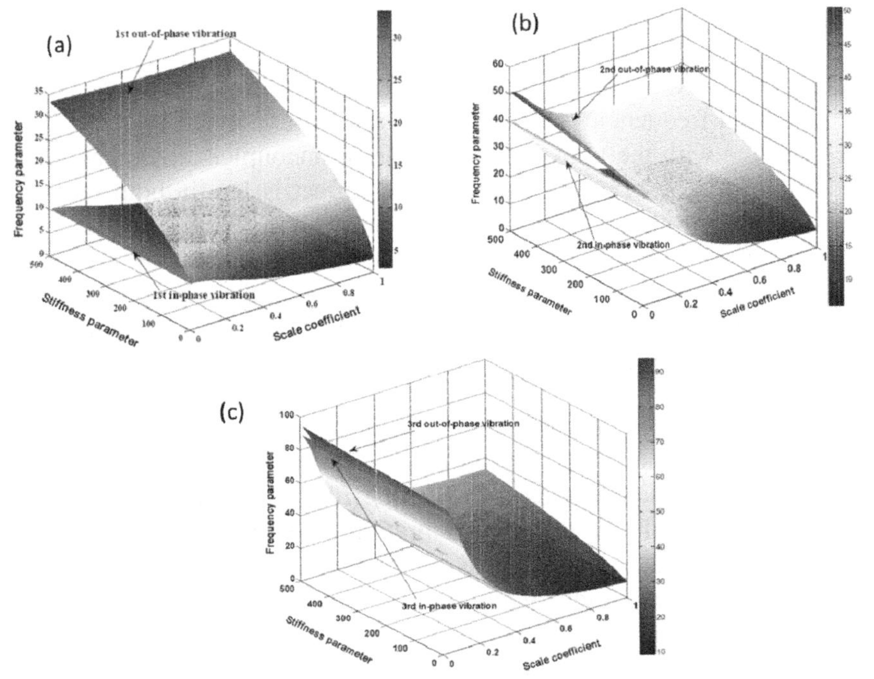

Figure 5.12. *The variation of the buckling load as a function of the spring stiffness (K) and the scale-coefficient (μ), a) for the 1st in-phase and out-of-phase vibration of NDNBS , b) for the 2nd in-phase and out-of-phase vibration of NDNBS, c) for the 3rd in-phase and out-of-phase vibration of NDNBS. For a color version of the figure, see www.iste.co.uk/murmu/non-local.zip*

5.11. Axial instability of double-nanobeam-systems

This section considers the axial instability of double-nanobeam-systems. Eringen's non-local elasticity is utilized for modeling the double-nanobeam-systems. The non-local theory accounts for the small-scale effects arising at the nanoscale. The small-scale effects substantially influence the instability of double-nanobeam-systems. Results reveal that the small-scale effects are higher with increasing values of non-local parameter for the case of in-phase (synchronous) buckling modes than the out-of-phase (asynchronous) buckling modes. The increase in the stiffness of the coupling elastic medium in a double-nanobeam-system reduces the small-scale effects during the out-of-phase (asynchronous) buckling modes. Analysis of the scale effects in higher buckling loads of a double-nanobeam-system with synchronous and asynchronous modes is also discussed in this section. The theoretical development presented herein may serve as a reference for non-local theories as applied to the instability analysis of complex-nanobeam-system such as complex carbon nanotube system.

5.11.1. *Background*

Based on the above discussion, in this section an investigation is carried out to illustrate the small-scale effects in the axial instability of non-local double-nanobeam-system (NDNBS) subjected to longitudinal compression. The understanding of axial instability or buckling of NDNBS is important from structural integrity of the nanosystem. Furthermore, this chapter provides a unique yet simple method of obtaining the exact solution for the buckling of a double-nanobeam system. Equations for the buckling load of a a double-nanobeam-system (NDNBS) are formulated within the framework of Eringen's non-local elasticity [ERI 83]. The two nanobeams are assumed to be attached by distributed vertical transverse springs. These springs may represent the stiffness of an enclosed elastic medium, forces due to nanooptomechanical effects or Van der Waals forces. An exact analytical method is proposed for solving the compressive load of NDNBS at which buckling occurs. Simply-supported boundary conditions are employed in this study. The results are obtained for various buckling modes of the NDNBS. The buckling phenomenon includes in-phase (synchronous) and out-of-phase (asynchronous) modes of buckling. The effects of (1) non-local parameter or scale coefficient, (2) stiffness of the springs and (3) the higher modes on the buckling behavior of the NDNBS are discussed. This work is aimed at providing an analytical scale-based non-local approach which could serve as the starting point for further investigation of more scale-dependent complex n-nanobeam systems.

5.11.2. *Buckling equations of non-local double-nanobeam-systems*

Consider a non-local double-nanobeam-system (NDNBS) under compression as shown in Figure 5.13. The two nanobeams are denoted as nanobeam-1 and nanobeam-2. The nanobeams are assumed to be slender and satisfy the Euler–Bernoulli beam theory. Vertically distributed springs attach the two nanobeams. The stiffness of the springs is equivalent to the Winkler constant in a Winkler foundation model. The springs can be used to represent an elastic medium, the forces due to nanooptomechanical effects [FRA] or Van der Waals forces between the two nanobeams. These forces arise when the dimension of the system approaches the nanoscale. The springs are assumed to have stiffness, k. The two nanobeams are different where the length, mass per unit length and bending rigidity of the ith beam are L_i ,m_i and $E_iI_i(i$=1, 2), respectively. These parameters are assumed to be constant along each nanobeam. The bending displacements over the two nanobeams are denoted by $w_1(x,t)$ and $w_2(x,t)$, respectively (Figure 5.13).

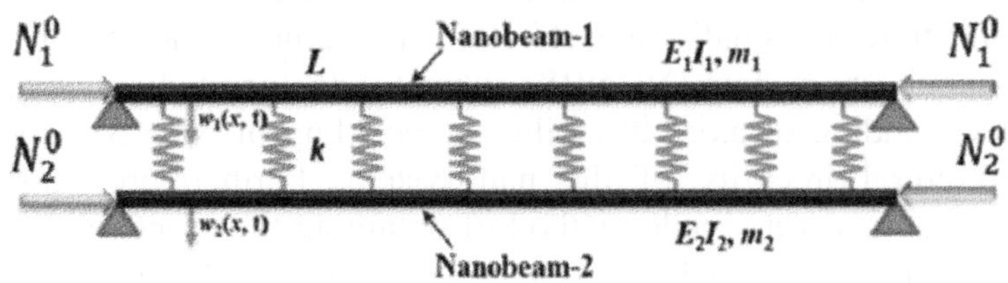

Figure 5.13. *Schematic diagram of elastically connected double-nanobeam-system subjected to compressive axial load*

Let the nanobeams subjected to axial forces N_1^0 and N_2^0 be:

$$N_1^0 = A_1\sigma_x^0 ; \qquad N_2^0 = A_2\sigma_x^0 \qquad [5.97]$$

Employing the non-local elastic theory, the equations of motion of two compressed nanobeams can be given by:

Nanobeam-1

$$E_1I_1w_1''''(x,t) + k[w_1(x,t) - w_2(x,t)] - (e_0a)^2k[w_1''(x,t) - w_2''(x,t)] + N_1^0w_1''(x,t) - (e_0a)^2N_1^0w_1''''(x,t)_1 = 0 \qquad [5.98]$$

Nanobeam-2

$$E_2I_2w_2''''(x,t) - k[w_1(x,t) - w_2(x,t)] + (e_0a)^2k[w_1''(x,t) - w_2''(x,t)] + N_2^0w_2''(x,t) - (e_0a)^2N_2^0w_2''''(x,t) = 0 \quad [5.99]$$

Primes (') denote partial derivatives with respect to position coordinate x, respectively. For the complete derivation of the equation of motion of a single non-local Euler–Bernoulli beam, we can see [RED 07].

We assume that both the nanobeams and the axial forces within them are identical, that is:

$$E_1I_1 = (E_2I_2) = EI \equiv \text{constant} \quad [5.100]$$

$$N_1^0 = N_2^0 = \widehat{N} \equiv \text{constant (uniformly prestressed)} \quad [5.101]$$

Considering equations [5.98]–[5.99] and assumptions from equations [5.100]–[5.101], we obtain the individual equations:

Nanobeam-1

$$EIw_1''''(x,t) + k[w_1(x,t) - w_2(x,t)] - (e_0a)^2k[w_1''(x,t) - w_2''(x,t)] + \widehat{N}w_1''(x,t) - (e_0a)^2\widehat{N}w_1''''(x,t) = 0 \quad [5.102]$$

Nanobeam-2

$$EIw_2''''(x,t) - k[w_1(x,t) - w_2(x,t)] + (e_0a)^2k[w_1''(x,t) - w_2''(x,t)] + \widehat{N}w_2''(x,t) - (e_0a)^2\widehat{N}w_2''''(x,t) = 0 \quad [5.103]$$

For the NDNBS, we propose a change of variables:

$$w(x,t) = w_1(x,t) - w_2(x,t), \quad [5.104]$$

such that:

$$w_1(x,t) = w(x,t) + w_2(x,t) \quad [5.105]$$

Here, $w(x,t)$ is the relative displacement of the main nanobeam with respect to the auxiliary beam.

Subtracting equation [5.103] from equation [5.103] gives:

$$EI[w_1''''(x,t) - w_2''''(x,t)] + 2k[w_1(x,t) - w_2(x,t)] - 2(e_0a)^2k[w_1''(x,t) - w_2''(x,t)] + \widehat{N}(w_1''(x,t) - w_2''(x,t)) - (e_0a)^2\widehat{N}(w_1''''(x,t) - w_2''''(x,t)) = 0 \quad [5.106]$$

Simplifying we obtain two equations:

$$EIw''''(x,t) + 2kw(x,t) - 2(e_0a)^2kw''(x,t) + \widehat{N}w''(x,t) - (e_0a)^2\widehat{N}w''''(x,t) = 0 \quad [5.107]$$

$$EIw_2''''(x,t) \quad + \widehat{N}w_2''(x,t) - (e_0a)^2\widehat{N}w_2''''(x,t) = kw(x,t) - (e_0a)^2kw''(x,t) \quad [5.108]$$

When the non-local effects are ignored ($e_0a = 0$) and a single nanobeam is considered, the above equations revert to the equations of the classical Euler–Bernoulli beam theory. For the present analysis of coupled NDNBS, we see the simplicity in using equation [5.107]. Here, we will be dealing with equation [5.107] for coupled NDNBS.

5.12. Non-local boundary conditions of NDNBS

Now, we present the mathematical expressions of the boundary conditions in NDNBS. The boundary conditions for the simply-supported case are described here. At each end of the nanobeams in NDNBS, the displacement and the non-local moments are considered to be zero [RED 07]. They can be mathematically expressed as:

Nanobeam-1: at $x = 0$

$$w_1(0,t) = 0; \quad [5.109]$$

$$M_1(0,t) = -EIw_1''(0,t) + (e_0a)^2k[w_1(0,t) - w_2(0,t)] + (e_0a)^2\widehat{N}w''_1(0,t) = 0$$

Nanobeam-1: at $x = L$

$$w_1(L,t) = 0; \quad [5.110]$$

$$M_1(L,t) = -EIw_1''(L,t) + (e_0 a)^2 k[w_1(L,t) - w_2(L,t)] + (e_0 a)^2 \widehat{N} w''_1(L,t) = 0 \quad [5.111]$$

Nanobeam-2: at $x = 0$

$$w_2(0,t) = 0; \quad [5.112]$$

$$M_2(0,t) = -EIw_2''(0,t) - (e_0 a)^2 k[w_1(0,t) - w_2(0,t)] + (e_0 a)^2 \widehat{N} w''_2(0,t) = 0 \quad [5.113]$$

Nanobeam-2: at $x = L$

$$w_2(L,t) = 0; \quad [5.114]$$

$$M_2(L,t) = -EIw_2''(L,t) - (e_0 a)^2 k[w_1(L,t) - w_2(L,t)] + (e_0 a)^2 \widehat{N} w''_2(0,t) = 0 \quad [5.115]$$

NDNBS: at $x = 0$

$$w(0,t) = w_1(0,t) - w_2(0,t) = 0; \quad [5.116]$$

$$M_1(0,t) - M_2(0,t) = -EIw_1''(0,t) + (e_0 a)^2[k[w_1(0,t) - w_2(0,t)]] + EIw_2''(0,t) - (e_0 a)^2[-k[w_1(0,t) - w_2(0,t)]] + (e_0 a)^2 \widehat{N}[w''_1(0,t) - w''_2(0,t)] = 0; \quad [5.117]$$

NDNBS: at $x = L$

$$w(L,t) = w_1(L,t) - w_2(L,t) = 0; \quad [5.118]$$

$$M_1(L,t) - M_2(L,t) = -EIw_1''(L,t) + (e_0 a)^2[k[w_1(L,t) - w_2(L,t)]] + EIw_2''(L,t) - (e_0 a)^2[-k[w_1(L,t) - w_2(L,t)]] + (e_0 a)^2 \widehat{N}[w''_1(L,t) - w''_2(L,t)] = 0; \quad [5.119]$$

Simplifying the boundary conditions effectively reduces to:

$$w(0) = 0 \text{ and } w''(0) = 0, \quad w(L) = 0 \text{ and } w''(L) = 0, \quad [5.120]$$

Here, it can be seen that the boundary conditions due to local elasticity and non-local elasticity are equivalent.

5.13. Buckling states of double-nanobeam-system

5.13.1. *Out-of-phase buckling load: (w_1-w_2 ≠ 0)*

We obtain the expression of the buckling load when the double nanobeam system is buckled. We consider both the cases of the NDBNS when the nanobeams buckle with in-phase (synchronous) and out-of-phase (asynchronous) sequence.

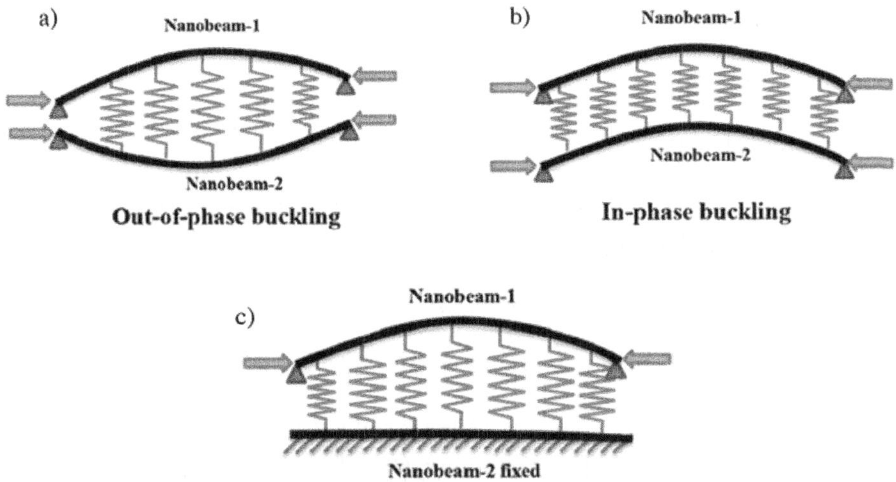

Figure 5.14. *a) Out-of-phase buckling of double-nanobeam-system, b) in-phase buckling of double-nanobeam-system, c) buckling of double-nanobeam-system when one nanobeam is fixed*

The configuration of NDNBS with the out-of-phase sequence of buckling ($w_1 - w_2 \neq 0$) is shown in Figure 5.14a. In this section, we evaluate the critical buckling load for the out-of-phase-type buckling. Using the boundary condition (equation [5.32]), let us assume that the buckling mode is:

$$w = W\sin\left(\frac{m\pi}{L}\right)x \qquad [5.121]$$

Substituting equation [5.121] into equation [5.107] yields:

$$EI\left(\frac{m\pi}{L}\right)^4 w + 2kw + 2(e_0a)^2k\left(\frac{m\pi}{L}\right)^2 w - \widehat{N}\left(\frac{m\pi}{L}\right)^2 w - (e_0a)^2\widehat{N}\left(\frac{m\pi}{L}\right)^4 w = 0 \qquad [5.122]$$

We introduce the following parameters for the sake of simplicity and generality:

$$K = \frac{kL^4}{EI}, \quad \mu = \frac{e_0a}{L}, \quad F = \frac{\widehat{N}L^2}{EI} \qquad [5.123]$$

Solving [5.122] and using [5.123] we obtain the expression of buckling load in out-of-phase sequence.

$$F = \frac{(m\pi)^4+2K+2K(\mu)^2(m\pi)^2}{(m\pi)^2[1+(\mu)^2(m\pi)^2]}, \quad m = 1,2,3 \ldots \qquad [5.124]$$

5.13.2. *In-phase buckling state: ($w_1 - w_2 = 0$)*

Here, the in-phase (synchronous) sequence of buckling state will be considered. The schematic illustration is shown in Figure 5.14b. For the present NDNBS, the relative displacements between the two nanobeams are absent, i.e. ($w_1 - w_2 = 0$).

Here, we solve equation [5.108] for the in-phase sequence of buckling. The NDNBS can be considered to be one of the nanobeams (i.e. nanobeam-2). We apply the same procedure for solving equation [5.108]. The buckling load of the NDNBS is evaluated as:

$$F = \frac{(m\pi)^2}{[1+(\mu)^2(m\pi)^2]}, \quad m = 1,2,3 \ldots \qquad [5.125]$$

For this case, the buckling phenomenon in the NDNBS is independent of the stiffness of the connecting springs and therefore the NDNBS can be effectively treated as a single nanobeam.

5.13.3. *One nanobeam is fixed:* $(w_2 = 0)$

Consider the case of NDNBS when one of the two nanobeams (i.e. nanobeam-2) is stationary ($w_2 = 0$). The schematic configuration is shown in Figure 5.14(c).

Using the equations from non-local elasticity, the governing equation for the NDNBS in this case reduces to:

$$EIw''''(x,t) + kw - (e_0a)^2kw''(x,t) + \widehat{N}w'' - (e_0a)^2\widehat{N}w\ 0 \qquad [5.126]$$

and the boundary conditions are expressed as:

$$w(0) = 0 \text{ and } w''(0) = 0, \quad w(L) = 0 \text{ and } w''(L) = 0, \qquad [5.127]$$

Here, it is worth noting that, in this case, the NDNBS behaves as if the nanobeam is embedded in an elastic medium (Figure 5.14(c)). The elastic medium can be modeled as the Winkler elastic foundation. The stiffness of the elastic medium is denoted by *k*. By following the same procedure as the solution of equation [5.107], the explicit non-local buckling load of NDNBS can easily be obtained. The buckling load is evaluated as:

$$F = \frac{(r\pi)^4 + K + K(\mu)^2(r\pi)^2}{(r\pi)^2[1+(\mu)^2(r\pi)^2]}, \quad r = 1,2,3\ldots \qquad [5.128]$$

In fact, when one of the nanobeams (i.e. nanobeam-2) in NDNBS is fixed ($w_2 = 0$), the NDNBS behaves as nanobeam on an elastic medium.

5.14. Coupled carbon nanotube systems

The non-local theory for NDNBS illustrated here is a generalized theory and can be applied for the instability analysis of coupled carbon nanotubes, double ZnO nanobeam systems and in nanocomposites. The applicability of non-local elasticity theory in the analysis of single nanostructures (nanotubes and graphene sheet) has been well established in various previous works [PED 03, SUD 03, WAN 06, WAN 07, LU 07, RED 07, HEI 08, RED 08, AYD 09, LEE 10, MUR 09a, MUR 09b, PRA 10, MUR 01]. Here, we present the non-local behavior of double-nanostructure-system.

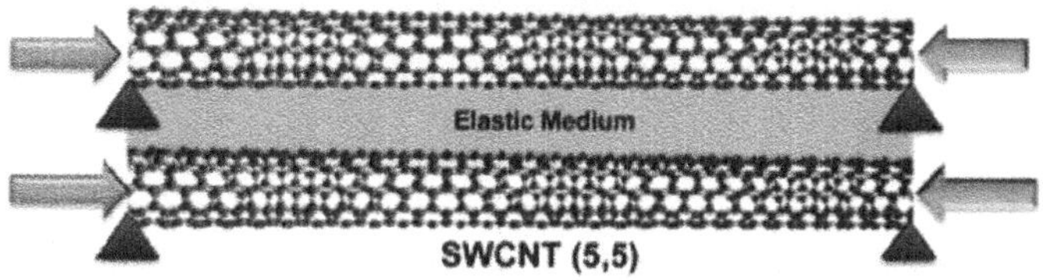

Figure 5.15. *Schematic diagram of coupled-carbon-nanotubes-system under compression load*

For the present study, we assume two carbon nanotubes being elastically attached by an elastic medium (Figure 5.15). An armchair SWCNT with chirality (5, 5) is considered. The radius of each individual SWCNT is assumed as 0.34 nm. Young's modulus, E, is taken as 0.971 TPa. The length is taken as 20 nm. Density ρ is taken as 2,300 kg/m^3. The buckling load of the NDNBS is presented in terms of the load parameters. The non-local parameter and the stiffness of the springs are computed as given in equation [5.123]. Spring stiffness represents the stiffness of the enclosing elastic medium. Different values of spring stiffness parameters, K, are considered. This is because the elastic medium can be of low as well as of high stiffness. The values of K range from 1 to 100. Both the nanotubes (nanotube-1 and nanotube-2) are assumed to have the same geometrical and material properties. It should be noted that the coupled carbon nanotube system is different from the conventional double-walled carbon nanotubes.

The non-local parameters are generally taken as $e_0 = 0.39$ [ERI 83] and a =0.142 nm (distance between carbon-carbon atoms). For carbon nanotubes and graphene sheets, the range of $e_0 a = 0 - 2.0$ nm has been widely used. For generality, in the present study, we take the scale coefficient μ or non-local parameter in the range as $\mu = 0 - 1$.

5.15. Results and discussions on the scale-dependent buckling phenomenon

To show the influence of small scale on the critical buckling load of the coupled-carbon-nanotube-systems subjected to compressive forces, curves have been plotted for the buckling load against the scale coefficient (non-local parameter, μ). To quantify the small-scale effect, we introduce a parameter buckling load reduction percent (BLRP). BLRP is defined as:

$$\mathrm{BLRP} = \left(\frac{F_{\text{Local Theory}} - F_{\text{Non-local Theory}}}{F_{\text{Local Theory}}}\right) \times 100 \qquad [5.129]$$

Figure 5.16 shows the variation of BLRP against the scale coefficient (μ). From Figure 5.16, it can be observed that as the scale coefficient μ increases the BLRP increases. This implies that for an increasing scale coefficient, the value of the buckling load decreases. The reduction in buckling load is due to the incorporation of non-local effects in the material properties of the carbon nanotubes. The non-local effect reduces the stiffness of the material and hence the comparative lower buckling loads.

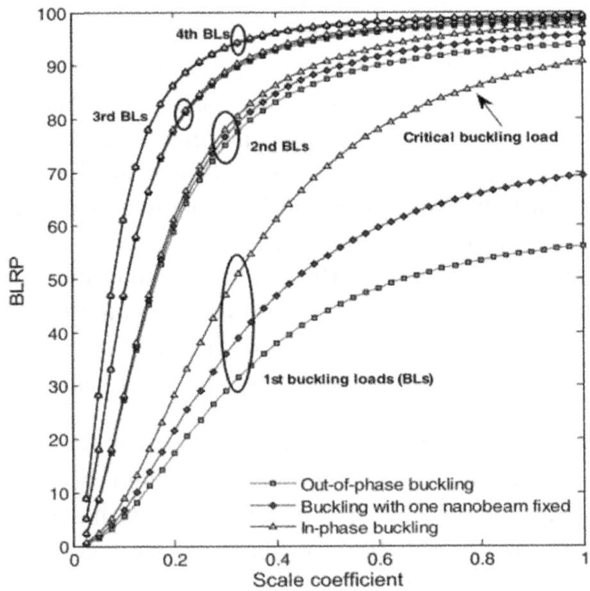

Figure 5.16. *Effect of scale-coefficient (μ = e0a/L) on Buckling Load Reduction Percent (BLRP) for critical and higher buckling loads in coupled-SWCNT-systems*

Three cases of axial instability are considered here; *case 1*: out-of-phase buckling (asynchronous); *case 2*: buckling with one SWCNT fixed; *case 3*: in-phase (synchronous) buckling. The stiffness parameter of the coupling springs between SWCNTs is assumed to be K=30. Comparing the three cases of coupled-carbon nanobeam system, we observe that the BLRP for case 3 (in-phase buckling) is larger than the BLRP for case 1 (out-of-phase buckling) and case 2 (one-SWCNT fixed). In other words, the scale coefficient significantly reduces the in-phase buckling load (thus higher BLRP) compared to other cases considered. The relatively higher BLRP in case 3 is due to the absence of coupling effect of the spring and the two nanobeams (nanotubes). In addition, it can be seen that the values of the BLRP for case 2 (one-SWCNT fixed) are larger than the values of the BLRP for case 1 (out-of-phase buckling). For case 2, the coupled carbon nanotube system becomes similar to the buckling phenomenon of the single SWCNT with the effect of elastic medium.

To see the influence of small-scale effects on the higher buckling loads of the coupled system, curves have also been plotted for BLRP against the scale coefficient with higher buckling loads. From Figure 5.16, we observe that with the increase in higher buckling loads, the BLRP for all cases of buckling phenomenon increases. This implies that the higher buckling loads of the coupled system are significantly reduced due to the non-local effects. These results are in line with the earlier results on non-local elasticity [RED 07]. Furthermore, it can also be noted that the difference between the in-phase-type buckling state, out-of-phase-type buckling and buckling with one SWCNT fixed becomes lower for higher modes of buckling loads. Thus, it can be concluded that although the small-scale effects are more in higher modes of buckling (modes), the effect of stiffness of coupling springs reduces the non-local effects. The lowest buckling mode (critical buckling load) is the buckling mode in first in-phase-type buckling (Figure 5.16). The in-phase and out-of-phase buckling modes are sometimes referred to as submodes.

To illustrate the influence of stiffness of the springs (elastic medium) on the critical buckling load of the coupled-carbon-nanotube-systems, curves have been plotted for the buckling load against the scale coefficient. Spring stiffness represents the stiffness of the enclosing elastic medium. Different values of the stiffness parameter of the coupling springs are considered. Figure 5.17(a–f) depicts the stiffness of the springs on the critical buckling loads of coupled systems. The stiffness parameters of the coupling springs are taken as K=2, 10, 20, 30 50, 100. As the stiffness parameter of the coupling springs increases, the BLRP decreases. Considering all values of the stiffness parameter, and comparing the three cases of coupled-carbon nanobeam system, it is noted that the BLRP for case 3 (in-phase buckling) is larger than the BLRP for case 1 (out-of-phase buckling) and case 3 (one-

SWCNT fixed). These different changes of BLRP with the increasing scale coefficient for the three different cases are more amplified as the stiffness parameter of the springs increases. For case 1 (out-of-phase buckling) and case 2 (one SWCNT fixed), the BLRP reduces with increasing values of stiffness parameter. This observation implies that case 1 (out-of-phase buckling) and case 2 (one SWCNT fixed) are less affected by scale effects. Comparing case 1 and case 2, it can be seen that the BLRP is lower for out-of-phase buckling than for buckling in case 2. Thus, the out-of-phase buckling phenomenon is less affected by the small scale or non-local effects. This out-of-phase buckling phenomenon can be attributed to the fact that the coupling springs in the vibrating system dampen the non-local effects. In-phase buckling of a coupled-system is unchangeable with increasing stiffness of springs. This is accounted for due to the in-phase buckling mode of behavior. For the in-phase type of buckling, the coupled system behaves as if a single SWCNT without the effect of internal elastic medium. In other words, the whole coupled system can be treated as a single nanoelement and the coupling internal structure is effectless.

Finally, for the comprehension of the buckling phenomenon of NDNBS with small-scale effects, the buckling load parameter F of the double nanobeam system for different non-local parameters (scale coefficient) and spring stiffness parameter (K) is plotted in a three-dimensional graph. The stiffness parameter K of the coupling springs is considered in the range of 0–100. Figures 5.18 a-c show the plots for the first, second and third in-phase and out-of-phase buckling phenomenon, respectively. The non-local parameter μ is varied from 0 to 1. From Figure 5.18(a), it is observed that the increase in the non-local parameter has a reducing effect on the non-local buckling load of the NDNBS. However, this reduction cannot be experienced at higher stiffness parameter values. The stiffness of the springs has a reducing effect on the small-scale effects of the NDNBS. It should be noted that for in-phase buckling, the loads are independent of coupling springs. Figure 5.18(b) shows the plot for the second in-phase and out-of-phase buckling, respectively. Compared to the first buckling loads (Figure 5.18(a)), the buckling loads for second modes (Figure 5.18(b)) are more affected by the scale effects. This is in line with the buckling of uncoupled single nanobeam. In addition, it is observed that the stiffness effect is less in both second in-phase and out-of-phase modes of buckling.

The stiffness effect is further reduced between the third in-phase and out-of-phase buckling loads (Figure 5.18(c)). Thus, it can be concluded that although the small-scale effects are more in higher modes, the effect of stiffness of coupling springs reduces the non-local effects.

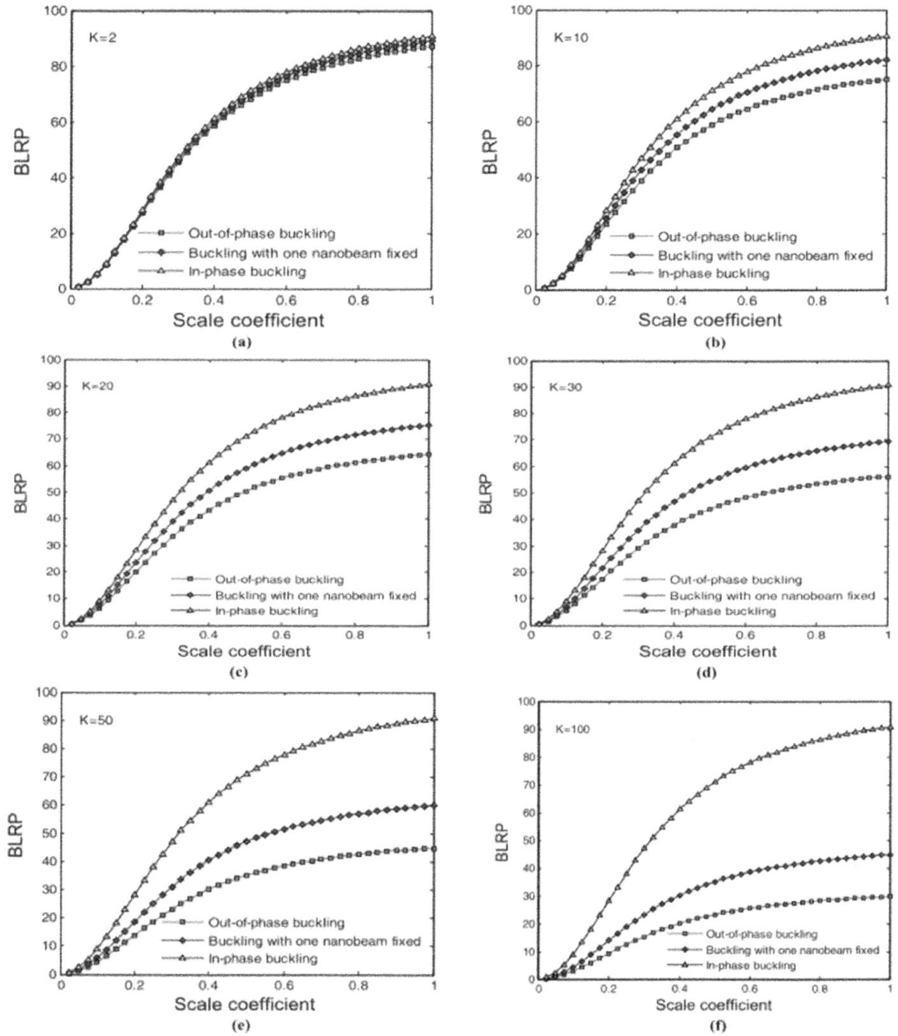

Figure 5.17. *Effect of scale-coefficient ($\mu = e_0a/L$) on Buckling Load Reduction Percent (BLRP) for higher different values of stiffness (K) of springs in coupled-SWCNT-systems; (a) K = 2; (b) K = 10; (c) K = 20; (d) K = 30; (e) K = 50; (f) K = 100*

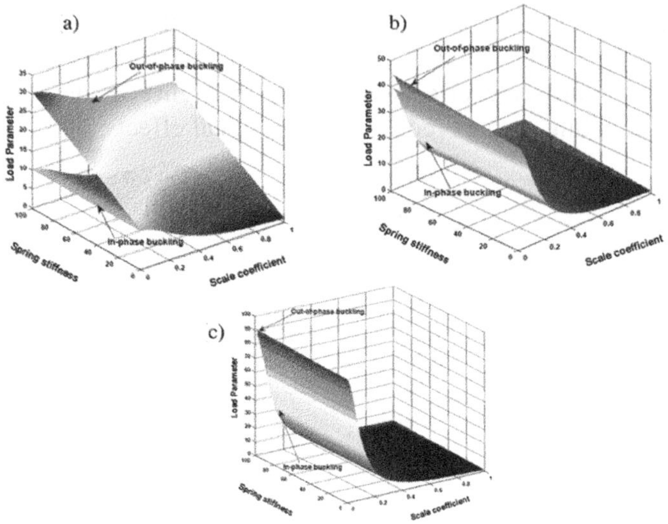

Figure 5.18. *The variation of the buckling load as a function of the spring stiffness (K) and the scalecoefficient (μ), a) for the 1st in-phase and out-of-phase buckling of NDNBS , b) for the 2nd in-phase and out-of-phase buckling of NDNBS, c) for the 3rd in-phase and out-of-phase buckling of NDNBS*

5.15. Summary

Axial elastic instability study of non-local double-nanobeam-system is presented. Double SWCNT system coupled by elastic medium is considered. Scale effects in the in-phase (synchronous) and the out-of-phase (asynchronous) buckling phenomenon are examined in details. The study shows that non-local or scale effects are important in the instability of double-nanobeam-system. Non-local effects reduce the buckling loads in the system. Increasing the stiffness of the springs in coupled nanosystems reduces the non-local effects. The small-scale effects in coupled nanosystems are more prominent with the increasing non-local parameter in the in-phase buckling than in the out-of-phase condition. The buckling loads are independent of the stiffness of the springs in the in-phase-type buckling. However, the buckling load of coupled nanosystems in the out-of-phase buckling increases with the increasing stiffness parameter of the coupling springs. The non-local theory for non-local double-nanobeam-system illustrated here is a generalized theory and can be applied for the instability analysis of short coupled carbon nanotubes, double ZnO nanobeam systems and complex nanosystems.

6

Double-Nanoplate-Systems

Mechanical investigation of the study of elastically connected double-plate systems is important for both theoretical and practical reasons. Many important structures can be modeled as composite structures. Similar to macroplates, an important technological extension of the concept of the single-nanoplate-system would be that of the complex-nanoplate-systems. Complex-nanoplate-systems may find applications in nano-optomechanical systems (NOMS). Vibration of double-nanobeam systems in NOMS is also an example. Vibration analysis of double-nanoplate systems with small aspect ratio is very relevant to NOMS. Elastically connected double-nanoplate systems can also be used for acoustic and vibration isolation. Furthermore, complex-nanoplate-systems can have important applications in nanosensors and nanocomposites. Vibration characteristics as well as buckling of multiple nanoplates (graphene sheets (GSs)) dispersed in nanocomposites can be important from the structural performance of nanocomposites. Vibration and buckling analysis of double-nanoplate-systems coupled with elastic mediums is worth understanding in this respect. In this chapter, we cover the non-local elasticity analysis of double-nanoplate-systems. The next chapter covers the analyisis on multiple-nanostructures. Theoretical aspects related to both vibration phenomenon as well as buckling of double-nanoplate-systems are introduced here.

6.1. Double-nanoplate-system

A large amount of research has been conducted on one-dimensional nanostructures such as carbon nanotubes (CNTs) using non-local elasticity. Recently, structural analysis of two-dimensional nanoplates via non-local elasticity has been gaining widespread popularity. Nanoplates are important nanostructures. Nanoplates such as graphene are one of the prominent new materials for next-generation nanoelectronic devices. Studies on GSs include vibration and buckling studies on single-walled and multi-walled GSs. A schematic model of two-dimensional nanostructures (graphene) reinforced nanocomposites is shown in Figure 6.1.

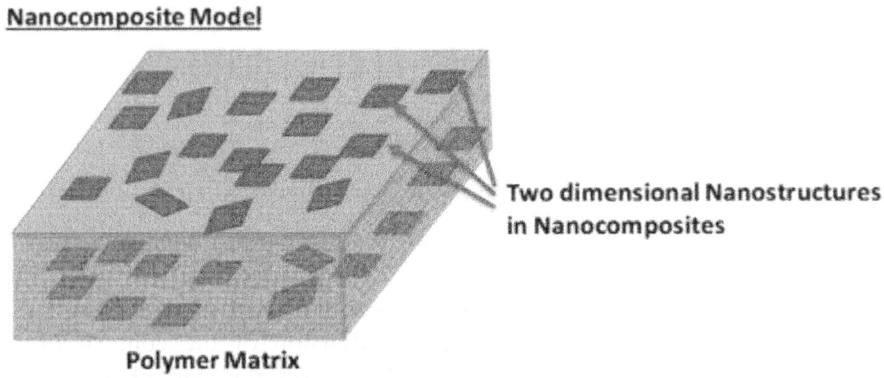

Figure 6.1. *Nanoplates dispersed in polymer matrix in nanocomposites*

Double-nanoplate-systems have important practical applications. In this chapter, theoretical models of vibration and buckling of nanoplates of double-nanoplates-systems are presented. Double-nanoplate-systems could exist embedded in matrix of composite as shown in Figure 6.1. The two nanoplates are elastically bonded by an enclosing elastic medium. It should be noted that similar models can be developed for a different problem of double-layered and multi-layered nanoplates with constant Van der Waals forces between nanoplates. Here, we have an enclosing elastic medium to act as the Van der Waals forces. The elastic medium is modeled by vertical springs. First, the vibration of double-nanoplate-systems is presented in sections 6.2–6.5. Next, the buckling of double-nanoplate-systems is presented in the remaining sections. Expressions for free bending-vibration of double-nanoplate-systems are established within the framework of Eringen's non-local elasticity by updating Kirchhoff's plate theory in section 6.3. An analytical method is introduced for determining the natural frequencies of non-local double-nanoplate-system (NDNPS). Explicit closed-form expressions for natural frequencies are derived for the case when all four ends are simply supported. Furthermore, section 6.5 presents a unique yet simple method for obtaining the exact solution of the free vibration of double-nanoplate systems. The simplification in the computation is achieved based on the change of variables to decouple the set of two fourth-order partial differential equations. Two single-layered GSs enclosed by polymer matrix are considered. Isotropic assumption of GSs is considered in the study. The results highlights that the small scale or non-local effects considerably influence the transverse vibration of NDNPS. Furthermore, the effect of aspect ratio and higher modes on the natural frequencies of NDNPS are discussed. This theory, developed through sections 6.2–6.5, may provide an analytical scale-based non-local approach which could serve as the starting point for further investigation of more complex *n*-nanoplates systems.

6.2. Vibration of double-nanoplate-system

Consider a double-nanoplate-system which is transversely vibrating. The two nanoplates are assumed to be bonded by an enclosing elastic medium as discussed earlier. Here the double nano-plate-system is not considered as a single equivalent system but as a coupled plate system. A situation of this type could arise in multiple GSs dispersed in nanocomposites. An analytical method is shown for determining the natural frequencies of NDNPS. Explicit closed-form expressions for natural frequencies are derived in detail for the case when all four ends are simply supported. Two single-layered GSs coupled within a bonding polymer matrix are considered. It is seen that the small-scale effects considerably influence the transverse vibration of NDNPS. We can see that the small-scale effects in NDNPS are higher with the increasing values of non-local parameter for the case of synchronous modes of vibration than in the asynchronous modes. The increase in the stiffness of the coupling springs in NDNPS reduces the small-scale effects during the asynchronous modes of vibration. Furthermore, the effect of aspect ratio and higher modes on the natural frequencies of NDNPS are presented.

6.3. Equations of motion for non-local double-nanoplate-system

Consider the NDNPS as shown in Figure 6.2(a). The two nanoplates of the NDNPS are referred to as nanoplate-1 and nanoplate-2. The two nanoplates are coupled by an elastic medium (polymer matrix). Here, we will develop equations based on the change of variables. For mathematical modeling, it is assumed that vertically distributed identical Winkler springs couple the two nanoplates (Figure 6.2(b)). In general, the springs may be used to substitute the elastic medium, due to the nano-optomechanical effect or Van der Waals forces between the two nanoplates. The springs are assumed to have stiffness k. Different values of k for different polymer matrices can be used for the study. The nanoplates are considered to be of length L and width W. Generally, the two nanoplates are different where the length, width, mass per unit length and bending rigidity of the ith plate are L_i, W_i, m_i and D_i (i=1, 2), respectively. These parameters are assumed to be constant along each nanoplate.

The bending displacements over the two nanoplates are denoted by $w_1(x, y, t)$ and $w_2(x, y, t)$, respectively (Figure 6.2).

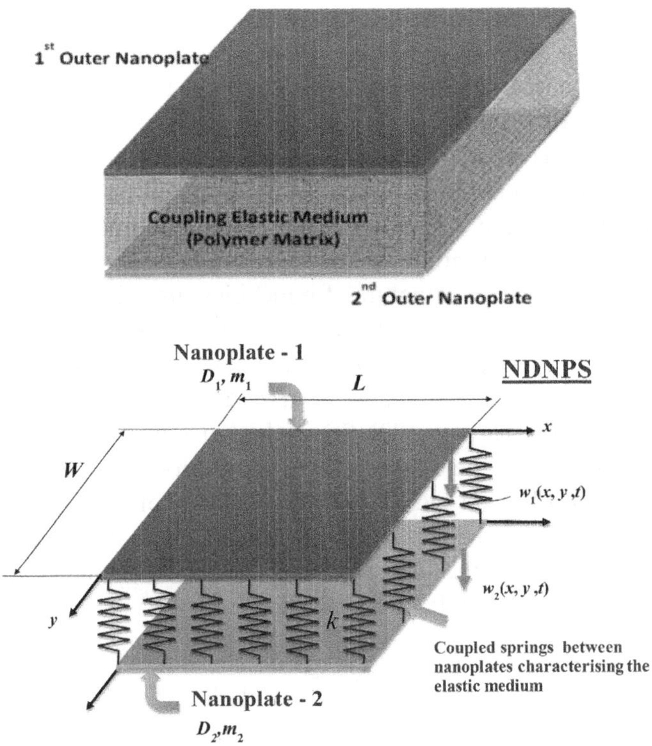

Figure 6.2. *a) Double-nanoplate-system coupled by an elastic medium, b) mathematical idealization of nanoplate-system characterized by the coupling springs*

Using non-local plate equations, the individual governing equations for NDNPS can be written as:

Nanoplate-1

$$D_1\nabla^4 w_1(x,y,t) + \rho_1 h_1 \ddot{w}_1(x,y,t) + k[w_1(x,y,t) - w_2(x,y,t)] - k(e_0 a)^2\nabla^2[w_1(x,y,t) - w_2(x,y,t)] - (e_0 a)^2\rho_1 h_1 \nabla^2 \ddot{w}_1(x,y,t) = f(x,y,t) - (e_0 a)^2\nabla^2 f(x,y,t) \quad [6.1]$$

where the bending rigidity of nanoplate-1 can be expressed as:

$$D_1 = E_1 h_1^3/12(1 - \upsilon_1^2). \quad [6.2]$$

The terms E, h and υ are the Young's modulus, thickness and Poisson's ratio of the naonoplate. The term $f(x,y,t)$ is the forcing function. Subscript 1 denotes the properties of the nanoplate-1.

Nanoplate -2

$$D_2\nabla^4 w_2(x,y,t) + \rho_2 h_2 \ddot{w}_2(x,y,t) - k[w_1(x,y,t) - w_2(x,y,t)] + k(e_0 a)^2\nabla^2[w_1(x,y,t) - w_2(x,y,t)] - (e_0 a)^2\rho_2 h_2 \nabla^2 \ddot{w}_2(x,y,t) = 0 \quad [6.3]$$

where the bending rigidity of nanoplate-2 is expressed as:

$$D_2 = E_2 h_2^3/12(1 - \upsilon_2^2) \quad [6.4]$$

For the present analysis, we assume that:

$$D_1 = D_2 = D \equiv \text{constant}$$

$$\rho_1 h_1 = \rho_2 h_2 = \rho h \equiv \text{constant}$$

$$f(x,y,t) = 0 \qquad \text{(free vibration)} \quad [6.5]$$

Substituting the assumptions (equation [6.5]), we obtain:

Nanoplate -1

$$D\nabla^4 w_1(x,y,t) + \rho h \ddot{w}_1(x,y,t) + k[w_1(x,y,t) - w_2(x,y,t)] - k(e_0 a)^2\nabla^2[w_1(x,y,t) - w_2(x,y,t)] - (e_0 a)^2\rho h \nabla^2 \ddot{w}_2(x,y,t) = 0 \quad [6.6]$$

Nanoplate -2

$$D\nabla^4 w_2(x,y,t) + \rho h\ddot{w}_2(x,y,t) - k[w_1(x,y,t) - w_2(x,y,t)] + k(e_0a)^2\nabla^2[w_1(x,y,t) - w_2(x,y,t)] - (e_0a)^2\rho h\nabla^2\ddot{w}_2(x,y,t) = 0 \quad [6.7]$$

Next, for the NDNPS we employ a change of variables by considering $w(x,y,t)$ as the relative displacement of the nanoplate-1 with respect to the nanoplate-2:

$$w(x,y,t) = w_1(x,y,t) - w_2(x,y,t), \quad [6.8]$$

such that, for nanoplate-1, the displacement is expressed as:

$$w_1(x,y,t) = w(x,y,t) + w_2(x,y,t) \quad [6.9]$$

Subtracting:

$$D\nabla^4[w_1(x,y,t) - w_2(x,y,t)] + 2k[w_1(x,y,t) - w_2(x,y,t)] + \rho h[\ddot{w}_1(x,y,t) - \ddot{w}_2(x,y,t)] - 2k(e_0a)^2\nabla^2[w_1(x,y,t) - w_2(x,y,t)] - \rho h(e_0a)^2\nabla^2[\ddot{w}_1(x,y,t) - \ddot{w}_2(x,y,t)] = 0 \quad [6.10]$$

Thus, we obtain:

$$D\nabla^4 w(x,y,t) + 2kw(x,y,t) + \rho h\ddot{w}(x,y,t) - 2k(e_0a)^2\nabla^2 w(x,y,t) - \rho h(e_0a)^2\nabla^2\ddot{w}(x,y,t) = 0 \quad [6.11]$$

and

$$D\nabla^4 w_2(x,y,t) + \rho h\ddot{w}_2(x,y,t) - (e_0a)^2\rho h\nabla^2\ddot{w}_2(x,y,t) = k(e_0a)^2 w(x,y,t) - k(e_0a)^2\nabla^2 w(x,y,t) \quad [6.12]$$

For the present analysis of coupled NDNPS, we observe the simplicity in using equations. [6.11] and [6.12]. It should be noted that when the non-local effects are ignored ($e_0a = 0$) and a single nanoplate is considered, the above equations revert to the equations of classical scale-free Kirchhoff's plate theory.

6.4. Boundary conditions in non-local double-nanoplate-system

Now, we present the explicit mathematical expressions of the boundary conditions of the double-nanoplate-system. Different boundary conditions can be studied. Here, we assume that all the edges in the nanoplate system are simply supported. At each end of the nanoplates in NDNPS, the displacement and the non-local moments are considered to be zero. They can be mathematically expressed as:

Nanoplate-1

Displacement condition:

$$w_1(0,y,t)=0;\ w_1(L,y,t)=0;\ w_1(x,0,t)=0;\ w_1(x,W,t)=0; \quad [6.13]$$

Non-local moment condition:

$$M_1(0,y,t)=0;\ M_1(L,y,t)=0;\ M_1(x,0,t)=0;\ M_1(x,W,t)=0; \quad [6.14]$$

Nanoplate-2

Displacement condition:

$$w_2(0,y,t)=0;\ w_2(L,y,t)=0;\ w_2(x,0,t)=0;\ w_2(x,W,t)=0; \quad [6.15]$$

Non-local moment condition:

$$M_2(0,y,t)=0;\ M_2(L,y,t)=0;\ M_2(x,0,t)=0;\ M_2(x,W,t)=0; \quad [6.16]$$

Now, we use equation [6.8] in the boundary conditions to obtain:

$$w(0,y,t)=w_1(0,y,t)-w_2(0,y,t)=0 \quad [6.17a]$$

$$w(L,y,t)=w_1(L,y,t)-w_2(L,y,t)=0 \quad [6.17b]$$

$$w(x,0,t)=w_1(x,0,t)-w_2(x,0,t)=0 \quad [6.17c]$$

$$w(x,W,t)=w_1(x,W,t)-w_2(x,W,t)=0 \quad [6.17d]$$

$$M(0,y,t)=M_1(0,y,t)-M_2(0,y,t)=0 \quad [6.18a]$$

$$M(L,y,t)=M_1(L,y,t)-M_2(L,y,t)=0 \quad [6.18b]$$

$$M(x,0,t)=M_1(x,0,t)-M_2(x,0,t)=0 \quad [6.18c]$$

$$M(x,W,t)=M_1(x,W,t)-M_2(x,W,t)=0 \quad [6.18d]$$

Similarly, we can obtain the non-local boundary conditions for all edges clamped (CCCC) of NDNPS:

$$w(0,y,t) = w_1(0,y,t) - w_2(0,y,t) = 0 \quad [6.19a]$$

$$w(L,y,t) = w_1(L,y,t) - w_2(L,y,t) = 0 \quad [6.19b]$$

$$w(x,0,t) = w_1(x,0,t) - w_2(x,0,t) = 0 \quad [6.19c]$$

$$w(x,W,t) = w_1(x,W,t) - w_2(x,W,t) = 0 \quad [6.19d]$$

$$w'(0,y,t) = w_1'(0,y,t) - w_2'(0,y,t) = 0 \quad [6.20a]$$

$$w'(L,y,t) = w_1'(L,y,t) - w_2'(L,y,t) = 0 \quad [6.20b]$$

$$w'(x,0,t) = w_1'(x,0,t) - w_2'(x,0,t) = 0 \quad [6.20c]$$

$$w'(x,W,t) = w_1'(x,W,t) - w_2'(x,W,t) = 0 \quad [6.20d]$$

Equations [6.11] and [6.12] along with boundary conditions will yield the frequencies of NDNPS.

6.5. Exact solutions of the frequency equations

6.5.1. *Both nanoplates of NDNPS are vibrating out-of-phase: ($w_1 - w_2 \neq 0$)*

Consider the case of the NDNPS when both the nanoplates are vibrating with in-phase sequence (synchronous) and out-of-phase (asynchronous) sequence. The configuration of the NDNPS with the out-of-phase sequence of vibration ($w_1 - w_2 \neq 0$) is shown in Figure 6.3. In this section, we solve the frequency for the out-of-phase (asynchronous) vibration. Here, we consider the case when all the ends have simply-supported boundary conditions. For the present case, we deal with equation [6.11].

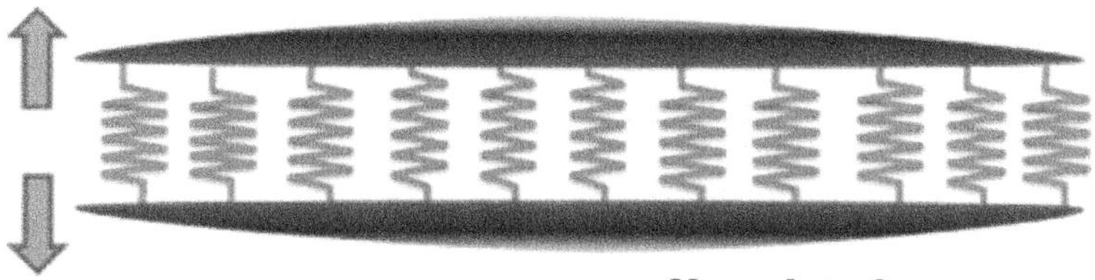

Figure 6.3. *Out-of-phase type or synchronous vibration of the double-nanoplate-system*

This equation system equation [6.11] with non-local boundary conditions can be solved by the Navier method assuming the solutions in the form:

$$w = \sum_{m=1}^{\infty} \sum_{n=1}^{\infty} W_{mn} \sin(a_m x)\, sin(b_n y) e^{i\omega_{mn} t} \quad [6.21]$$

where:

$$a_m = m\pi/L\,, \quad b_n = n\pi/W \quad [6.22]$$

and m and n are the half wave numbers.

Substituting equation [6.21] into equation [6.11] yields:

$$-D[(m\pi/L)^2 + (n\pi/W)^2]^2 W_{mn} - 2k(1 + (e_0 a)^2[(m\pi/L)^2 + (n\pi/W)^2])W_{mn} = -\rho h(1 + (e_0 a)^2[(m\pi/L)^2 + (n\pi/W)^2])\omega_{mn}^2 W_{mn} \quad [6.23]$$

We define the following parameters for the sake of convenience and generality:

$$\Omega_{\rm mn} = \omega_{mn} L^2 \sqrt{\frac{\rho h}{D}};\ R = L/W; \quad K = \frac{kL^2}{D}, \quad \mu = \frac{e_0 a}{L} \quad [6.24]$$

Using the parameters in equation [6.24] and substituting in equation [6.23], we obtain the natural frequencies of the NDNPS for asynchronous vibration as:

$$\Omega_{\mathrm{mn}} = \sqrt{\frac{[(m\pi)^2+R^2(n\pi)^2]^2+2K+2K\mu^2[(m\pi)^2+R^2(n\pi)^2]}{1+\mu^2[(m\pi)^2+R^2(n\pi)^2]}}, \quad m,n = 1,2 \ldots \qquad [6.25]$$

6.5.2. *Both nanoplates of NDNPS are vibrating in-phase: ($w_1 - w_2 = 0$)*

Next, the in-phase sequence (synchronous) of vibration is considered (Figure 6.4). For the present NDNPS, the relative displacements between the two nanoplates are absent($w_1 - w_2 = 0$). Here, we solve equation [6.12] for the vibration of NDNPS. The vibration of nanoplate-2 here would represent the vibration of the coupled vibrating system. We apply the same procedure for solving equation [6.12]. Using equation [6.21], we can obtain the natural frequencies. The natural frequencies for the NDNPS in this case can be expressed as:

$$\Omega_{\mathrm{mn}} = \sqrt{\frac{[(m\pi)^2+R^2(n\pi)^2]^2}{1+\mu^2[(m\pi)^2+R^2(n\pi)^2]}} \qquad m,n = 1,2,3 \ldots \qquad [6.26]$$

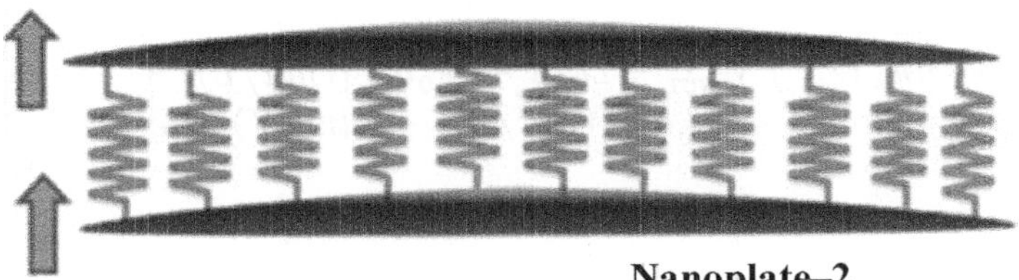

Figure 6.4. *In-phase type or synchronous vibration of the double-nanoplate-system*

For this case of vibration, we see that NDNBS is independent of the stiffness of the connecting springs and therefore the NDNPS can be effectively treated as a single nanoplate.

6.5.3. *One nanoplate of NDNPS is stationary: ($w_2(x,y,t) = 0$)*

Consider a special case of NDNPS when one of the two nanoplates (i.e. nanoplate-2) is stationary ($w_2 = 0$). The schematic diagram is shown in Figure 6.5.

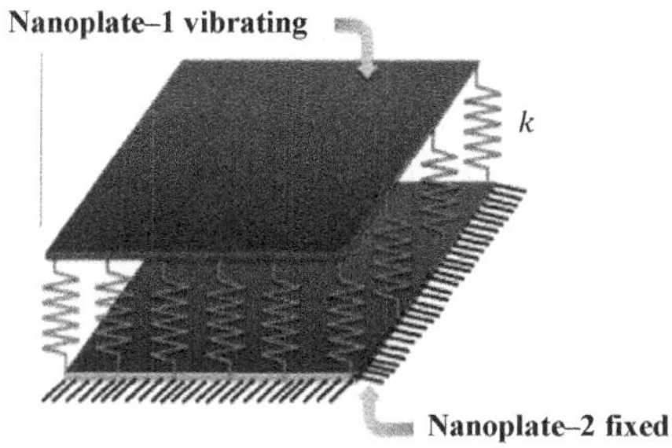

Figure 6.5. *Vibration of NDNPS with one nanoplate stationary*

For this case, using the equations from non-local elasticity the governing equation for the NDNPS is:

$$D\nabla^4 w(x,y,t) + kw(x,y,t) + m\ddot{w}(x,y,t) - k(e_0a)^2\nabla^2 w(x,y,t) - m(e_0a)^2\nabla^2\ddot{w}(x,y,t) = 0 \quad [6.27]$$

The boundary conditions are expressed as:

$$w(0,y,t) = 0;\ w(L,y,t) = 0;\ w(x,0,t) = 0;\ w(x,W,t) = 0; \quad [6.28]$$

and the non-local moment conditions are given by:

$$M(0,y,t) = 0;\ M(L,y,t) = 0;\ M(x,0,t) = 0;\ M(x,W,t) = 0; \quad [6.29]$$

In this case, the NDNPS behaves as if the nanoplate is embedded into an elastic medium. The elastic medium can be modeled as Winkler elastic foundation. The stiffness of the elastic medium is denoted by k. By following the same procedure as solution of equation [6.11], the non-local frequency of NDNPS can be explicitly expressed as:

$$\Omega_{mn} = \sqrt{\frac{[(m\pi)^2+R^2(n\pi)^2]^2 + K + K\mu^2[(m\pi)^2+R^2(n\pi)^2]}{1+\mu^2[(m\pi)^2+R^2(n\pi)^2]}} \quad [6.30]$$

where K is the stiffness parameter of the coupling springs and μ is the non-dimensional non-local parameter as defined in equation [6.24]. In fact, when one of the nanoplates (i.e. nanoplate-2) in NDNPS is fixed ($w_2 = 0$), the NDNPS behaves as a single nanoplate on an elastic medium.

6.5.4. Discussions

The non-local plate theory for NDNPS illustrated here is a generalized theory and can be applied for the bending-vibration analysis of coupled GSs, gold nanoplates, etc. The applicability of non-local elasticity theory in the analysis of nanostructures (nanotubes and GSs) can be observed in various earlier works [CHA 11, CHA 11, MUR 11, SIM 11, ERI 83, LU 06, YAN 10, MUR 09, MUR 09, MUR 10, WAN 08, RED 08, PED 03, WAN 06, LU 07, AYD 09, HEI 08, AYD 09, WAN 09, YAN 10, KHA 10, RED 07, WAN 08, NIU 09, SIM 10, SIM 11, HEI 08a, HEI 08b, TOU 09a, TOU 09b, LIM 07, PRA 09, AYD 09, FIL 10, AKS 11, WAN 10, NAR 11a, NAR 11b, JON 06].

As an illustration, the properties of the nanoplates are considered that of a single-layered GSs. The two single-layered GSs are coupled by embedded polymer as shown in Figure 6.6. Young's modulus of the GS is considered as $E = 1.06$ TPa, the Poisson ratio $\upsilon = 0.25$ and the mass density as $\rho = 2250 \frac{\text{kg}}{\text{m}^3}$. The thickness of the GS is taken as $h = 0.34$ nm.

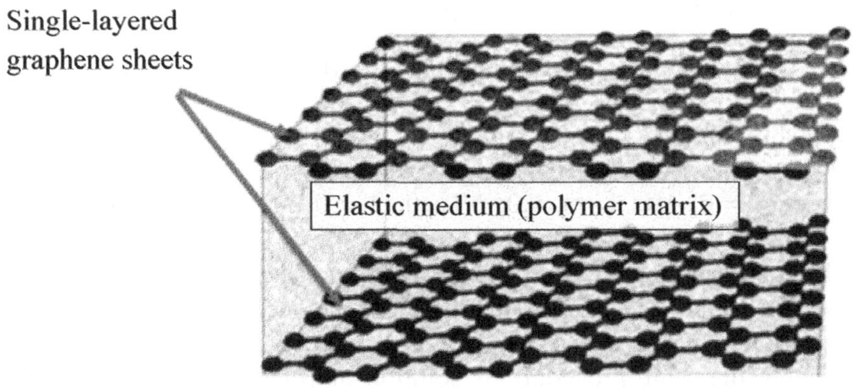

Figure 6.6. *Coupled graphene sheets as double-nanoplate-system in polymer matrix environment*

The frequency results of the NDNPS are presented in terms of the frequency parameters (equation [6.24]). The non-local parameter and the stiffness of the springs are computed as given in equation [6.24]. Different values of spring parameters, *K*, are considered. Spring stiffness represents the stiffness of the enclosing elastic medium. Both high and large stiffness of springs are assumed. Values of *K* range from 10 to 100. Both the GSs (GS-1 and GS-2) are assumed to have the same geometrical and material properties.

The non-local parameters are taken as $e_0 = 0.39$ [MUR 11] and $a = 0.142$ nm (distance between carbon-carbon atoms). For CNTs and graphenes, the range of $e_0 a = 0 - 2.0$ nm has been widely used. In the present study, we take the scale coefficient μ or non-local parameter in the similar range as $\mu = 0 - 1$.

To see the influence of small scale on the natural frequency of the coupled-GS-systems, curves have been plotted for frequency parameter and scale coefficient (non-local parameter, μ). The GSs are coupled by a polymer matrix of stiffness $K = 100$. To signify the small-scale effect, we introduce a parameter frequency reduction percent (FRP). FRP is defined as:

$$\text{FRP} = 100 \times \left(\frac{\Omega_{\text{Local}} - \Omega_{\text{Nonocal}}}{\Omega_{\text{Local}}}\right) \qquad [6.31]$$

Figure 6.7 shows the variation of the frequency parameter (within the terms of FRP) with the scale coefficient for different cases of NDNPS ($m = 1$, $n = 1$). Here, NDNPS denotes the coupled-GS-systems. The results for the frequency parameter Ω are in the dimensionless form as in equation [6.32]. The stiffness parameter of the springs is assumed to be constant ($K = 100$). Unless stated, the frequency parameter would denote the parameter associated with the first natural frequencies (in-phase and out-of-phase-type vibration). From Figure 6.7, it can be observed that as the scale coefficient μ increases the FRP increases. This implies that for increasing scale coefficient the value of frequency parameter decreases. The reduction in frequency parameter is due to the assimilation of small-scale effects in the NDNPS in the material properties of the GSs. The small-scale effect reduces the stiffness of the material and hence the comparative lower frequencies. Therefore, by the non-local elastic model the size effects are reflected in the NDNPS.

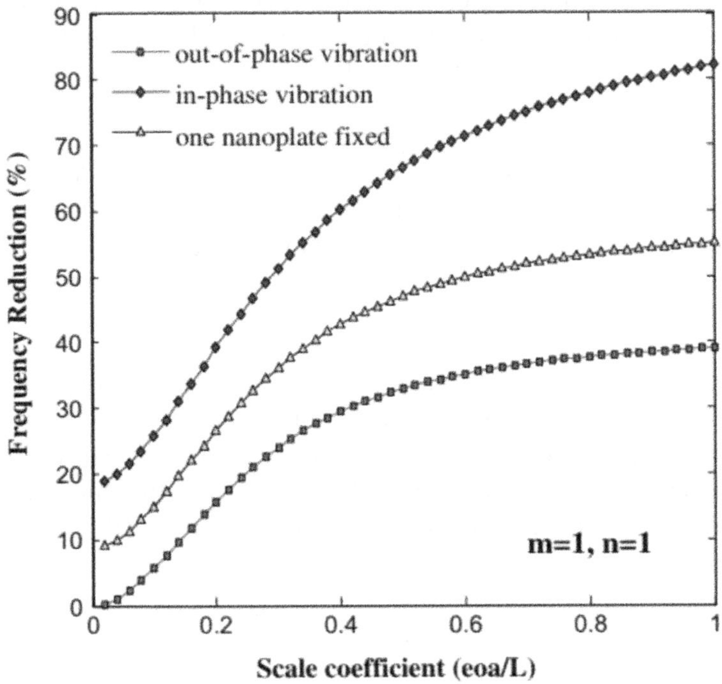

Figure 6.7. *Variation of frequency reduction percent with scale coefficient for out-of-phase, in-phase and one nanoplate fixed in NDNPS*

Three different cases of NDNPS are considered. Case 1, case 2 and case 3 depict the conditions (i) when both the GSs vibrate in the out-of-phase sequence($w_1 - w_2 \neq 0$), (ii) when one of the GSs in NDNPS is stationary ($w_2 = 0$) and (iii) when both the GSs vibrate with in-phase sequence ($w_1 - w_2 = 0$), respectively. Comparing the three cases of coupled-graphene-sheet-system, we observe that the FRP for case 3 (in-phase vibration) is larger than the FRP for case 1 (out-of-phase vibration) and case

2 (one-GS fixed). In other words, the scale coefficient significantly reduces the in-phase natural frequencies (thus higher FRP) compared to other cases considered. The relative higher FRP in in-phase vibration is due to the absence of coupling effect of the spring and the two nanoplates (GSs) and the whole NDNBS can be treated as a vibrating single GS. In general, it is worth noticing that the small-scale effects in NDNPS are higher with increasing non-local parameter in the in-phase vibration than in the out-of-phase vibration. This is because the stiffness of the springs in the out-of-phase (asynchronous) vibration reduces the non-local effects. In addition, it can be seen that the values of the FRP for case 2 (one-GS fixed) are larger than the values of the FRP for case 1 (out-of-phase vibration). For case 2, the coupled-graphene-sheet-system becomes similar to the vibration characteristic of the single GS with the effect of elastic medium.

To illustrate the influence of stiffness of the springs on the natural frequencies of the coupled-GS-systems, curves have been plotted for the FRP against the scale coefficient. Spring stiffness represents the stiffness of the enclosing elastic medium. Different values of stiffness parameter of the coupling springs are considered. Figure 6.8(a–f) depicts the influence of stiffness of the springs on the FRP of coupled systems. The stiffness parameters of the coupling springs are taken as $K = 0, 10, 20, 50, 80, 100$. As the stiffness parameter of the coupling springs increases, the FRP decreases. Aspect ratio is taken as unity. Considering all values of the stiffness parameter, and comparing the three cases of coupled-GS system, we note that the FRP for case 3 (in-phase vibration) is larger than the FRP for case 1 (out-of-phase vibration) and case 3 (one-GS fixed). These different changes of FRP with the increasing scale coefficient for the three different cases are more amplified as the stiffness parameter of the springs increases. For case 1 (out-of-phase vibration) and case 2 (one GS fixed), the FRP reduces with increasing values of stiffness parameter. This observation implies that case 1 (out-of-phase vibration) and case 2 (one GS fixed) are less affected by scale effects. Comparing the case 1 and case 2, it can be seen that the FRP is lesser for out-of-phase vibration than for vibration in case 2. Thus, the out-of-phase vibration is less affected by the small scale or non-local effects. This out-of-phase vibration can be attributed to the fact that the coupling springs in the vibrating system dampen the non-local effects.

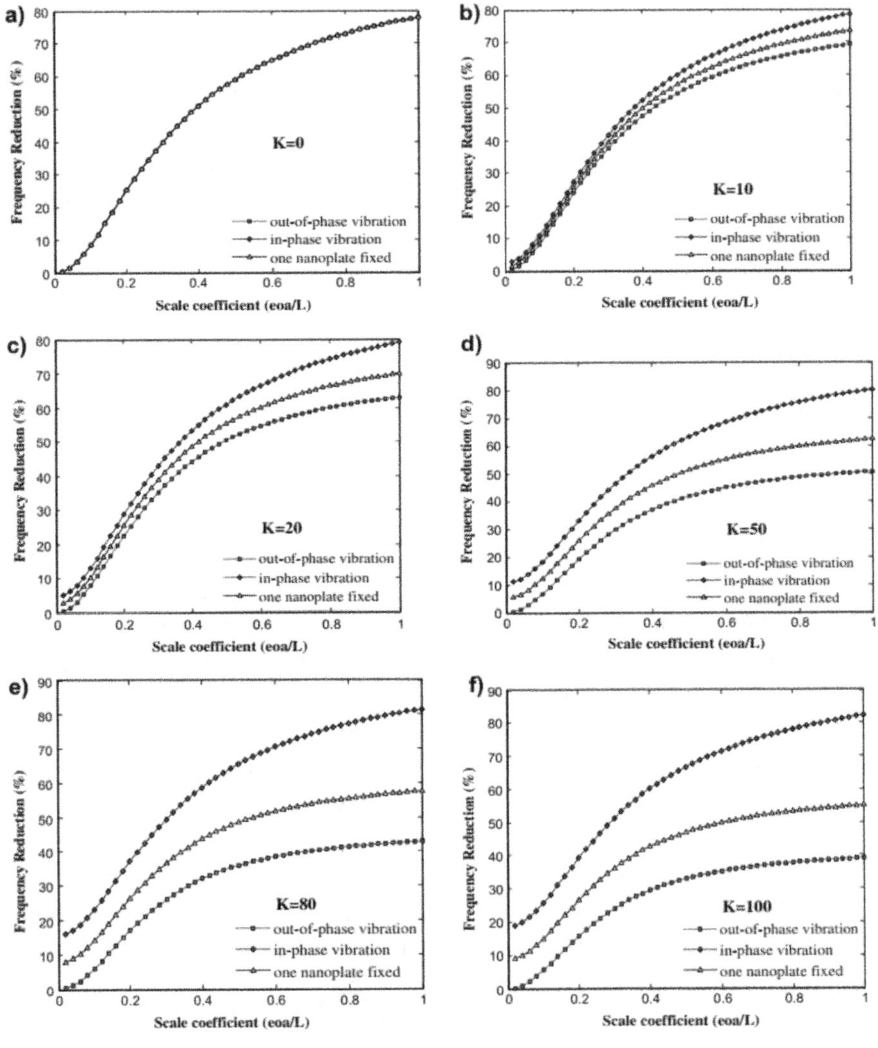

Figure 6.8. *Variation of frequency reduction percent with scale coefficient for different coupling springs in NDNPS: (a) K = 0, (b) K = 10, (c) K = 20, (d) K = 50, (e) K = 80, (f) K = 100. For a color version of this figure, see www.iste.co.uk/murmu/non-local.zip*

The in-phase vibration of a coupled-system is unchangeable with increasing stiffness of springs. This is accounted due to the in-phase vibration mode of behavior. For in-phase types of vibration, the coupled system behaves as if it had one single-walled graphene sheet (SWGS) without the effect of an internal elastic medium. In other words, the whole coupled system can be treated as a single nanoelement and the coupling an internal structure is effectless. In summary, it should be noted that the in-phase vibration of coupled-system is more affected by small-scale effects compared to out-of-phase vibration.

Next, we illustrate the influence of aspect ratios (L/W) of the nanoplates (GS) on the natural frequencies of the coupled-GS-systems. The GSs are assumed to be coupled by a polymer matrix of stiffness $K = 100$. Curves have been plotted for the FRP against the scale coefficient for different aspect ratios. Different values of aspect ratios of the NDNPS are considered. Figure 6.9(a–f) depicts the effect of aspect ratio on the FRP of coupled systems. The aspect ratios are taken as $L/W = 0.1, 0.5, 1, 2, 5, 10$.

From Figure 6.9, we see that with the increase in aspect ratios (L/W) of NDNPS, the FRP for all cases of vibration increases. We note that the difference between the in-phase-type vibration, out-of-phase-type vibration and vibration with one GS fixed becomes less for increasing aspect ratios (L/W) of NDNPS. Thus, it can be concluded that although the small-scale effects are more in higher aspect ratios (L/W) of NDNPS, the effects of stiffness of coupling springs are reduced in higher aspect ratios of NDNPS. And thus, less difference in curves between the in-phase-type vibration, out-of-phase-type vibration and vibration with one GS fixed becomes less for higher aspect ratios of NDNPS. The FRPs between the different cases of vibrations (in-phase, out-of-phase and with one GS stationary) are reduced in the following order:

$$(L/W = 0.1) > (L/W = 0.5) > (L/W = 1) > (L/W = 2) > (L/W = 5) > (L/W = 10).$$

To see the influence of small-scale effects on the higher natural frequencies of coupled system, curves have been plotted for FRP against scale coefficient with higher FRPS. Similarly to section 6.2, three cases of vibration characteristics are considered here; case 1: out-of-phase vibration; case 2: vibration with one GS fixed; case 3: in-phase vibration. The plots are shown in Figures 6.10(a–f). The stiffness parameter of the coupling springs between SWCNTs is assumed to be K=100. The higher natural frequencies are plotted for (m = 1, n = 2), (m = 2, n = 2), (m = 2, n = 3), (m = 3, n = 3), (m = 3, n = 4) and (m = 4, n = 4). Here, it should be noted that we are considering in-phase and out-of-phase vibration as submodes.

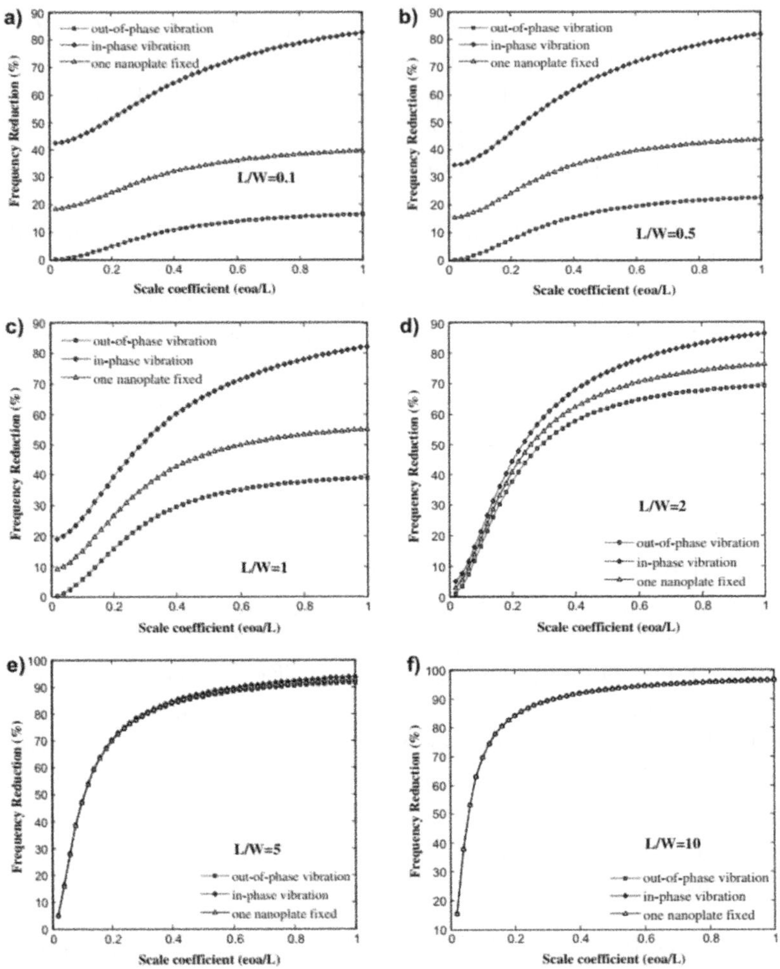

Figure 6.9. *Change of frequency reduction percent with scale coefficient for different aspect ratios of NDNPS: (a) L/W = 0.1, L/W = 0.5, L/W = 1, L/W = 2, L/W = 5, L/W = 10. For a color version of this figure, see www.iste.co.uk/murmu/non-local.zip*

From Figure 6.10, we see that with the increase in higher natural frequencies, the FRP for all cases of vibration increases. This implies that the higher natural frequencies of the coupled system are significantly reduced due to the non-local effects. These results are in-line with the earlier results on non-local elasticity [MUR 09]. Further, it can also be noticed that the difference between the in-phase-type vibration, out-of-phase-type vibration and vibration with one GS fixed becomes less for higher modes of vibration. Thus, it can be concluded that although the small-scale effects are more in higher modes of frequencies, the effects of stiffness of coupling springs are reduced in higher frequencies. And thus, less difference in curves between the in-phase-type vibration, out-of-phase-type vibration and vibration with one GS fixed becomes less for higher modes of vibration. The effect of stiffness of coupling springs in NDNPS is reduced in the following order of natural frequencies:

(m = 1, n = 2) > (m = 2, n = 2) > (m = 2, n = 3) > (m = 3, n = 3) > (m = 3, n = 4) > (m = 4, n = 4).

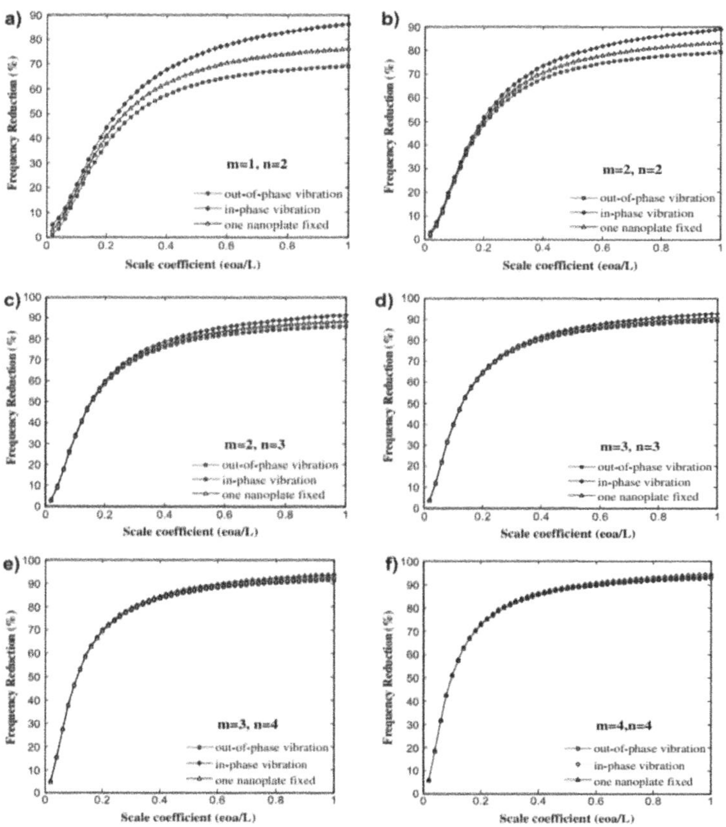

Figure 6.10. *Change of frequency reduction percent with scale coefficient for higher frequencies of NDNPS: (a) m = 1, n = 2; (b) m = 2, n = 2; (c) m = 2, n = 3; (d) m = 3, n = 3; (e) m = 3, n = 4; (f) m = 4, n = 4. For a color version of this figure, see www.iste.co.uk/murmu/non-local.zip*

6.5.5. *Non-local double-nanobeam-system versus non-local double-nanoplate-system*

Next, we present the vibration behavior of non-local double-nanoplate-systems (NDNPS) with respect to double-nanobeam system (NDNBS). Studies of double-nanobeam system by non-local elasticity can be seen in [6.1]. Similar to previous analysis, three cases of vibration are considered, out-of-phase vibration, in-phase vibration and one nano-entity fixed. Lower and higher stiffness of coupling springs are assumed here, i.e. $K = 10$ and $K = 100$. Figure 6.11 shows the change of FRP against scale coefficient for NDNBS and NDNPS with $K = 10$. From the figure, it is observed that the FRPs for double-nanoplate-systems are larger than the FRPs for double-nanobeam-systems. This is true for majority of scale coefficient or non-local parameter considered. However, it should be noted that if different cases of vibration (in-phase, out-of-phase and one nano-entity fixed) are considered, then double-beam-systems have prominent behavior compared to double-nanoplate-systems. The effect of spring stiffness is more prominent in double-nanobeam-systems which reduce the small-scale effect. However, for higher stiffness parameter, $K = 100$, FRP in double-nanoplate-systems would result in prominent difference in vibration with in-phase, out-of-phase and one nano-entity fixed. This is illustrated in Figure 6.12.

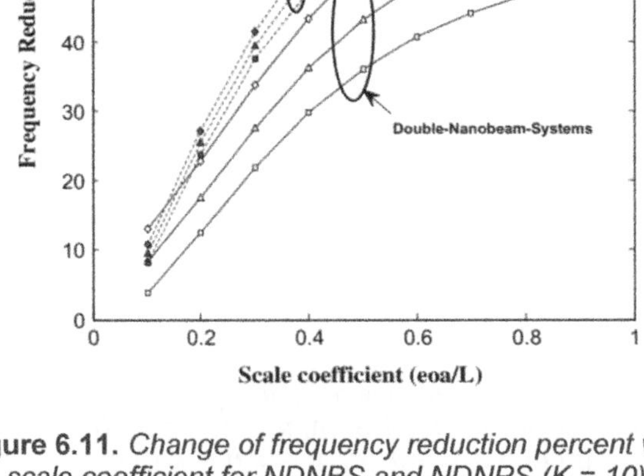

Figure 6.11. *Change of frequency reduction percent with scale coefficient for NDNBS and NDNPS (K = 10)*

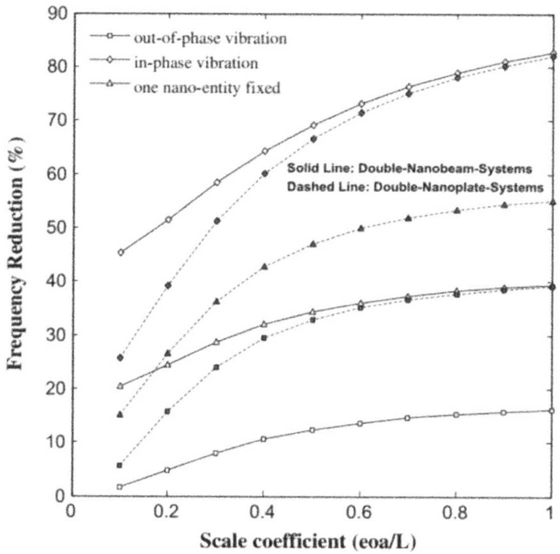

Figure 6.12. *Change of frequency reduction percent with scale coefficient for NDNBS and NDNPS (K = 100)*

In the end we say, further to nanocomposite applications, this double GS system can be used as nanoresonators. For a different set of boundary conditions, this behavior may be different. As small-scale effects reduce the natural frequencies of the vibrating nanosystem, nanoresonators can be designed so as to vibrate in the out-of-phase modes. Thus, nanoresonators would have higher resonant frequencies. This implies the significance of the use of small-scale effects in NDNPS. If some damping properties are present within the NDNPS, then the damping behavior could be effective in the out-of-phase mode vibration (similar to stiffening NDNPS). Thus, one of the nanoplates would act as a vibration absorber. However, the damping behavior would be ineffective in the in-phase mode vibration.

6.5.6. *Summary*

In this chapter, the expressions for free bending-vibration of bonded double-nanoplate-system are established utilizing non-local elasticity. A simple analytical method is introduced for determining the natural frequencies of bonded non-local double-nanoplate-system (NDNPS). Explicit closed-form expressions for natural frequencies are derived for the case when all four ends are simply supported. Two single-layered GSs coupled by polymer matrices are considered for the study. The double-nanoplate-system executes two kinds of vibrations: the synchronous vibrations with lower frequencies and the asynchronous vibrations with higher frequencies. The preceding section highlights that the small-scale effects considerably influence the transverse vibration of NDNPS. The small-scale effects in NDNPS are higher with increasing values of non-local parameter for the case of synchronous (in-phase) modes of vibration than in the asynchronous (out-of-phase) modes of vibration. The increase in the stiffness of the coupling springs in NDNPS reduces the small-scale effects during the asynchronous modes of vibration. The synchronous natural frequencies are not dependent on the stiffness parameter of the elastic medium. In this case, the double-nanoplate system oscillates as a single plate with the same natural frequencies. For rectangular NDNPS with increasing aspect ratios, FRP in the synchronous and asynchronous becomes similar.

6.6. Buckling behavior of double-nanoplate-systems

Buckling behavior in bonded uniaxially compressed double-nanoplate-systems is shown in this section. Both the synchronous and asynchronous-type buckling are considered in detail. The two nanoplates are assumed to be elastically bonded by polymer resin. The nanoscale effects of nanoplates are dealt with in the analysis by using non-local elasticity theory. The theory is utilized for deriving the expressions for buckling load of double-nanoplate-systems. A simple analytical method is introduced for determining the buckling load of non-local double-nanoplate-systems. Explicit closed-form expressions for buckling load are derived for the case when all four ends are simply supported. Single-layered graphene-sheets are considered as an example of nanoplates. The results highlight that the non-local effects considerably influence the buckling behavior of double-graphene-sheet-systems. Unlike buckling behavior of a single GS, the double-graphene-sheet-system undergoes both synchronous and asynchronous buckling. The non-local effects in double-graphene-sheet-system are higher with increasing values of non-local parameter for the case of synchronous buckling modes than in the asynchronous buckling modes. The increase in the stiffness of the coupling springs in double-graphene-sheet-system reduces the non-local effects during the asynchronous modes of buckling. Different aspect ratios of double-graphene-sheet-systems and higher buckling modes are also considered in this section.

6.6.1. *Background*

One important technological advance in the concept of the single nanobeams and nanoplates is that of the complex-nanobeam-systems and complex-nanoplate-systems. Examples of nanobeams include CNTs [IIJ 91], zinc oxide nanotubes [WAN 03], while examples of nanoplates include gold nanoplates [SHA 04] and GSs [GEI 07, GEI 09], etc. These structures are of nanodimension scale and possess advanced properties over traditional engineering materials. One simple example of complex-nanobeam-system is the double-nanobeam-system [MUR 10]. Double-nanobeam-systems are utilized in NOMS [FRA 10, EIC 09, DEO 09, LIN 10] application and nanocomposites.

Similarly to complex-nanobeam-system, complex-nanoplate-systems may find applications in NOMS and as acoustic and vibration isolators. One such example of complex-nanoplate-system would be the double-nanoplate-system [MUR 11]. It should be noted that the present complex systems are different from double-layered system, such as double-walled CNTs [CHA 11] and double-walled graphene systems [CHA 11]. As mentioned earlier these double-walled nanoentities are generally bonded by a constant Van der Waal force which is unlike the double-nanobeam-system or double-nanoplate-system. Double-nanoplate-system would have different bonding agents of varying stiffness modulus.

Double-nanoplate systems are important and can be found in nanocomposites structures such as multiple GSs dispersed in polymers. Though complex-nanoplate-systems such as double-graphene-sheet-systems are important in nanodevices and nanocomposites, only a few appear to relate to the elaborate study of its mechanical characteristics. Recently, Murmu and Adhikari [MUR 11] and Simsek [SIM 11] have addressed the mechanical studies of double-nanobeam-systems.

Understanding the buckling behavior and vibration characteristics of double-nanoplate-systems is important. Recently, the vibration behavior of such systems is considered [MUR 11]. In this section, we discuss the buckling behavioral aspect of double-nanoplate-systems considering non-local effects [ERI 83, LU 06, YAN 10, MUR 09, MUR 09, MUR 10, WAN 08, RED 08, PED 03, WAN 06, LU 07, AYD 09, HEI 08, AYD 09, WAN 09, YAN 10, KHA 10, RED 07, WAN 08, NIU 09, SIM 10, SIM 11, HEI 08a, HEI 08b, TOU 09a, TOU 09b, LIM 07, PRA 09, AYD 09, FIL 10,

AKS 11, WAN 10, NAR 11a, NAR 11b]. Non-local elasticity theory is considered. In non-local elasticity theory, the small-scale effects are captured by assuming that the stress at a point is a function of the strains at all points in the domain. Non-local theory considers long-range interatomic interactions and yields the results dependent on the size of a body. In the present problem, two single-layer GSs are considered. The present study can be used beyond GSs as nanoplates. The two single-layered GSs are elastically connected by an enclosing elastic medium, such as polymer resin. Expressions for the buckling load of a system are derived using non-local elasticity theory. An analytical method is introduced for determining the buckling load of non-local double-graphene-sheet-systems. Explicit closed-form expressions for buckling load are derived for the case when all four ends are simply supported. Furthermore, this section presents a unique yet simple method for obtaining the exact solution for the buckling load of double-graphene-sheet-systems. Attention is put on the non-local-scale effects in synchronous and asynchronous modes of buckling for various coupling springs, aspect ratio and higher buckling loads. The present study could be useful for nanocomposites.

6.6.2. *Uniaxially compressed double-nanoplate-system*

Consider a non-local double-nanoplate-system (NDNPS) [MUR 11] bonded by an elastic medium as shown in Figure 6.13. The two nanoplates are assumed to be coupled by polymer resin. The nanoplates are subjected to uniaxial compression. The two nanoplates of the NDNPS are referred to as nanoplate-1 and nanoplate-2.

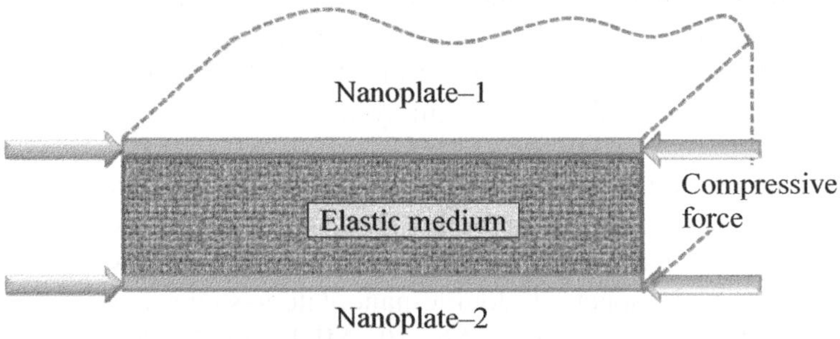

Figure 6.13. *Schematic configuration of uniaxially bonded double-nanoplate-system*

For mathematical modeling of NDNPS, the elastic medium is represented by vertically distributed identical springs as shown in Figure 6.14. In general, the springs may be used to substitute the elastic medium, forces due to nano-optomechanical effect [FRA 10, EIC 09, DEO 09, LIN 10], or Van der Waals forces between the two nanoplates. The springs are assumed to have a stiffness k. Different values of k for different stiffness of the polymer matrix can be used for the study. The nanoplates are considered to be of length L and width W. Generally, the two nanoplates are different where the length, width, mass per unit length and bending rigidity of the ith plate are L_i, W_i, m_i and D_i ($i = 1, 2$), respectively. These parameters are assumed to be constant along each nanoplate. The nanoplate-1 and nanoplate-2 are uniaxially compressed by force N_1 and N_2, respectively. The bending displacements over the two nanoplates are denoted by $w_1(x, y, t)$ and $w_2(x, y, t)$, respectively.

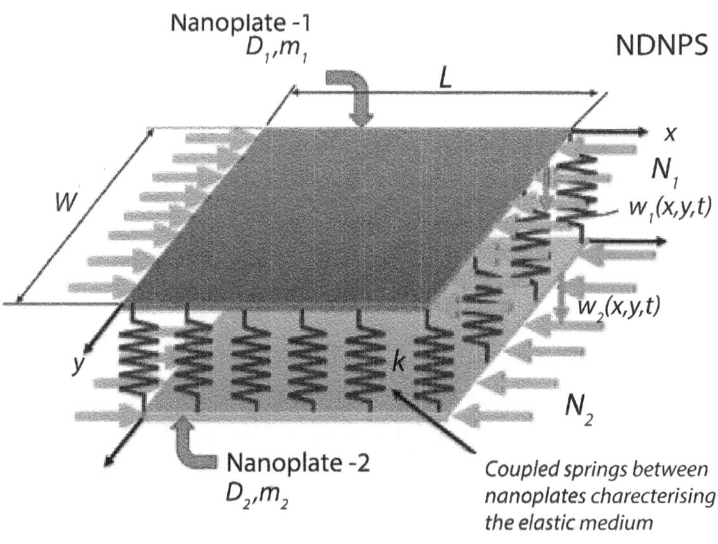

Figure 6.14. *Double-nanoplate-system characterized by coupling vertical springs of stiffness, k*

The individual governing equations of uniaxially compressed nanoplates for NDNPS based on the theory of non-local elasticity can be written as:

Nanoplate -1

$$D_1\nabla^4 w_1(x,y,t) + k[w_1(x,y,t) - w_2(x,y,t)] - k(e_0a)^2\nabla^2[w_1(x,y,t) - w_2(x,y,t)] - N_1 w''_1(x,y,t) + N_1(e_0a)^2\nabla^2 w''_1(x,y,t) = 0 \qquad [6.32a]$$

Nanoplate -2

$$D_2\nabla^4 w_2(x,y,t) - k[w_1(x,y,t) - w_2(x,y,t)] + k(e_0a)^2\nabla^2[w_1(x,y,t) - w_2(x,y,t)] - N_2 w''_2(x,y,t) + N_2(e_0a)^2\nabla^2 w''_2(x,y,t) = 0 \qquad [6.32b]$$

$$N_1 = N_2 = N \equiv \text{constant} \quad \text{(uniformly compressed)} \qquad [6.33]$$

Substituting the assumptions (equation [6.33] into equations. [6.32a, [6.32b] and [6.5], we obtain:

Nanoplate -1

$$D\nabla^4 w_1(x,y,t) + k[w_1(x,y,t) - w_2(x,y,t)] - k(e_0a)^2\nabla^2[w_1(x,y,t) - w_2(x,y,t)] - -Nw_1''(x,y,t) + N(e_0a)^2\nabla^2 w_1''(x,y,t) = 0 \qquad [6.34]$$

Nanoplate -2

$$D\nabla^4 w_2(x,y,t) - k(e_0a)^2[w_1(x,y,t) - w_2(x,y,t)] + k(e_0a)^2\nabla^2[w_1(x,y,t) - w_2(x,y,t)] - Nw''_2(x,y,t) + N(e_0a)^2\nabla^2 w''_2(x,y,t) = 0$$

[6.35]

Next, for the NDNPS, we employ a change of variables by considering $w(x,y,t)$ as the relative displacement of the nanoplate-1 with respect to the nanoplate-2 [MUR 11]

$$w(x,y,t) = w_1(x,y,t) - w_2(x,y,t), \qquad [6.36]$$

such that for nanoplate-1, the displacement is expressed as:

$$w_1(x,y,t) = w(x,y,t) + w_2(x,y,t) \qquad [6.37]$$

Subtracting equation [6.35] from [6.34] would lead to:

$$D\nabla^4[w_1(x,y,t) - w_2(x,y,t)] + 2k[w_1(x,y,t) - w_2(x,y,t)] - 2k(e_0a)^2\nabla^2[w_1(x,y,t) - w_2(x,y,t)] - N[w_1''(x,y,t) - w''_2(x,y,t)] + N(e_0a)^2\nabla^2[w_1''(x,y,t) - w''_2(x,y,t)] = 0 \quad [6.38]$$

The use of equation [6.36] in equation [6.38]:

$$D\nabla^4 w(x,y,t) + 2kw(x,y,t) - 2k(e_0a)^2\nabla^2 w(x,y,t) - Nw''(x,y,t) + N(e_0a)^2\nabla^2\, w''(x,y,t) = 0 \quad [6.39]$$

and equation [6.35] becomes

$$D\nabla^4 w_2(x,y,t) - Nw_2''(x,y,t) + N(e_0a)^2\nabla^2 w_2''(x,y,t) = k(e_0a)^2 w(x,y,t) - k(e_0a)^2\nabla^2 w(x,y,t) \quad [6.40]$$

For the present analysis of coupled NDNPS, we see the simplicity in using equations [6.39] and [6.40]. It should be noted that when the non-local effects are ignored ($e_0a = 0$) and a single nanoplate is considered, the above equations revert to the equations of Kirchhoff's classical plate theory.

6.6.3. *Buckling states of double-nanoplate-system*

Here, we explicitly consider the different cases of non-local buckling which would take place in double-nanoplate-systems [MUR 11] under uniaxial compressive force. The cases studied will be nanoplates buckling with out-of-phase (asynchronous) sequence, in-phase (synchronous) sequence; and when one of the nanoplates is considered to be fixed.

6.6.3.1. *Both nanoplates buckling asynchronously: (w_1-$w_2 \neq 0$)*

Figure 6.15 shows the configuration of double-nanoplate-system with the asynchronous (out-of-phase) sequence of buckling ($w_1 - w_2 \neq 0$). It should be noted that the figure represents first out-of-phase buckling phenomenon. In out-of-phase sequence of buckling, the nanoplates are buckled in opposite directions.

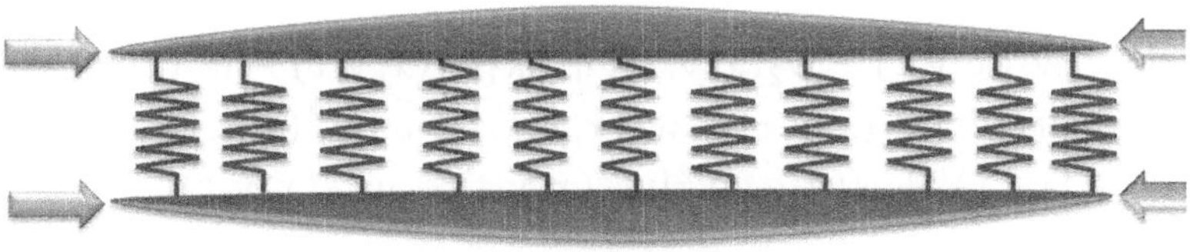

Figure 6.15. *Asynchronous-type buckling of the double-nanoplate-system*

In this section, we evaluate the buckling load for the out-of-phase (asynchronous) type buckling. We will use equation [6.39] for the buckling solution of double-nanoplate-system.

We assume that the buckling mode of the double-nanoplate-system as [JON 06]

$$w = W_{mn} sin\frac{m\pi x}{L} sin\frac{n\pi y}{W} \qquad [6.41]$$

Substituting equation [6.41] into equation [6.39] yields:

$$D\left[\left(\frac{m\pi}{L}\right)^2 + \left(\frac{n\pi}{W}\right)^2\right]^2 w + 2kw + 2k(e_0 a)^2\left[\left(\frac{m\pi}{L}\right)^2 + \left(\frac{n\pi}{W}\right)^2\right] w + N\left(\frac{m\pi}{L}\right)^2 w + N(e_0 a)^2\left(\frac{m\pi}{L}\right)^2\left[\left(\frac{m\pi}{L}\right)^2 + \left(\frac{n\pi}{W}\right)^2\right] w = 0 \qquad [6.42]$$

For the sake of simplicity, we introduce the following parameters:

$$K = \frac{kL^4}{D}; \quad \widehat{N} = -\frac{NL^2}{D}; \; \mu = \frac{e_0 a}{L}; \; R = \frac{L}{W} \qquad [6.43]$$

Using equations [6.43] and [6.42], we obtain the expression of buckling load $\widehat{N}$ in out-of-phase sequence of NDNPS as:

$$\widehat{N} = \frac{\left[(m\pi)^2 + R^2(n\pi)^2\right]^2 + 2K + 2K\mu^2\left[(m\pi)^2 + R^2(n\pi)^2\right]}{(m\pi)^2\left[1 + \mu^2\left[(m\pi)^2 + R^2(n\pi)^2\right]\right]} \qquad [6.44]$$

6.6.3.2. *Both nanoplates buckling synchronously: (w_1-w_2=0)*

Here, we will consider the in-phase (synchronous) sequence of buckling of the double-nanoplate-system under uniaxial compression. The schematic illustration is shown in Figure 6.16, which is the first synchronous-type buckling. For the present nanoplate system, the relative displacements between the two nanoplates are absent, i.e. $w_1 - w_2 = 0$. The nanoplates are buckled in the same direction (synchronous). In synchronous buckling state, the double nanoplate system can be considered to be as one of the nanoplates (i.e. nanobeam-2).

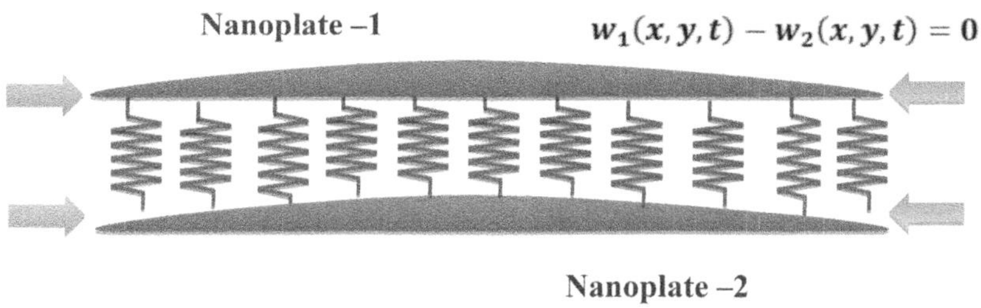

Figure 6.16. *In-phase-type buckling behavior of the double-nanoplate-system*

Here, we solve equation [6.40] for the synchronous sequence of buckling. We apply the same procedure as earlier for solving equation [6.39]. The buckling load of the non-local double-nanoplate-system is evaluated as:

$$\widehat{N} = \frac{\left[(m\pi)^2+R^2(n\pi)^2\right]^2}{(m\pi)^2\left[1+\mu^2[(m\pi)^2+R^2(n\pi)^2]\right]} \qquad [6.45]$$

Here, we see for this case, the buckling phenomenon in the double-nanoplate-system is independent of the stiffness of the connecting springs and therefore the double-nanoplate-system can be effectively treated as a single nanoplate.

6.6.3.3. *One nanoplate is fixed in double-nanoplate-system ($w_2 = 0$)*

Consider the case of double-nanoplate-system when one of the two nanoplates (i.e. nanoplate-2) is stationary ($w_2 = 0$). The schematic configuration of the NDNPS is shown in Figure 6.17.

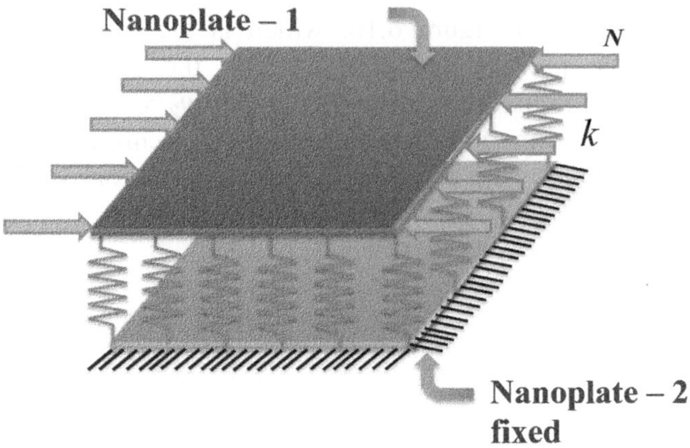

Figure 6.17. *Buckling behavior of uniaxially compressed NDNPS with one nanoplate stationary*

We utilize the equations from non-local elasticity and the governing equation for the NDNBS in this case reduces to:

$$D\nabla^4 w(x,y,t) + kw(x,y,t) - k(e_0a)^2\nabla^2 w(x,y,t)$$
$$-Nw''(x,y,t) + N(e_0a)^2\nabla^2\, w''(x,y,t) = 0 \qquad [6.46]$$

Here, it is worth noting that in this case, the double-nanoplate-system behaves as a nanoplate embedded or supported on an elastic medium or by other forces at the nanoscale (Figure 6.17). The elastic medium can be modeled as a Winkler elastic foundation. The stiffness of the elastic medium is denoted by k. By following the same procedure as solution of equation [6.39], the explicit non-local buckling load of double nanoplate can be easily obtained. The buckling load is evaluated as:

$$\widehat{N} = \frac{\left[(m\pi)^2+R^2(n\pi)^2\right]^2+K+K\mu^2\left[(m\pi)^2+R^2(n\pi)^2\right]}{(m\pi)^2\left[1+\mu^2[(m\pi)^2+R^2(n\pi)^2]\right]} \qquad [6.47]$$

In fact, when one of the nanoplates (i.e. nanoplate-2) in NDNPS is fixed ($w_2 = 0$), the double-nanoplate-systems behave as nanoplate on an elastic medium.

6.7. Results and discussion

6.7.1. *Coupled double-graphene-sheet-system*

As an illustration, the properties of the nanoplates are considered that of a single-walled GS. The two GSs are coupled by the embedded polymer matrix (Figure 6.18). The bonded double-graphene-sheet-system is uniaxially compressed. The Young's modulus of the GS is considered as $E = 1.06$TPa, the Poisson ratio $\vartheta = 0.25$. The thickness of the GS is taken as $h = 0.34$ nm.

The non-local double-plate theory for NDNPS illustrated here is a generalized theory and can be applied for the buckling analysis of coupled GSs (multiple-walled), gold nanoplates, etc. The reliability of non-local elasticity theory in the analysis of nanostructures (nanotubes and GSs) can be observed in various earlier works [ERI 83, LU 06, YAN 10, MUR 09, MUR 09, MUR 10, WAN 08, RED 08, PED 03, WAN 06, LU 07, AYD 09, HEI 08, AYD 09, WAN 09, YAN 10, KHA 10, RED 07, WAN 08, NIU 09, SIM 10, SIM 11, HEI 08a, HEI 08b, TOU 09a, TOU 09b, LIM 07, PRA 09, AYD 09, FIL 10, AKS 11, WAN 10, NAR 11a, NAR 11b].

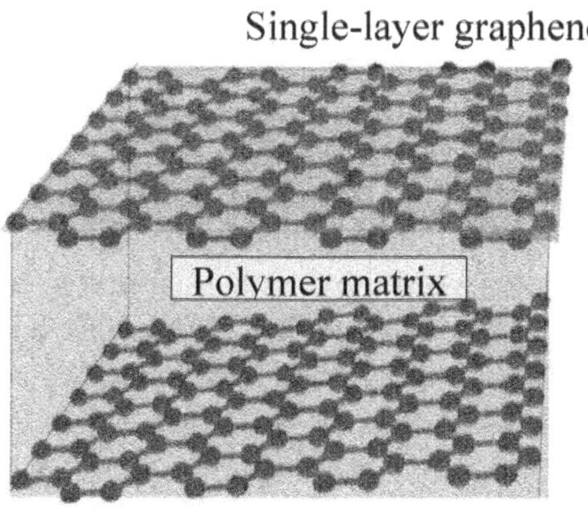

Figure 6.18. *Uniaxially compressed bonded double-graphene-sheet-system*

The buckling results of the NDNPS are presented in terms of the buckling load parameters (equation [6.43]). The non-local parameter and the stiffness of the springs are computed as given in same equation [6.43]. Different values of spring parameters, K, are considered. Spring stiffness represents the stiffness of the enclosing elastic medium. Both high and large stiffness of springs are assumed. Values of K range from 10 to 100. Both the GSs (GS-1 and GS-2) are assumed to have the same geometrical and material properties. In general, the non-local parameters may be taken as $e_0 = 0.39$ [ERI 83] and $a = 0.142$ nm (distance between carbon-carbon atoms). For CNTs and graphenes, the range of $e_0 a = 0 - 2.0$ nm has been widely used. In the present study, we take the scale coefficient μ or non-local parameter in the similar range as $\mu = 0 - 1$ [AYD 09].

6.7.2. *Effect of small scale on NDNPS undergoing compression*

To see the influence of small scales on the natural buckling load of the coupled-GS-systems, curves have been plotted for buckling load parameter and scale coefficient (non-local parameter, μ).

To express the small-scale effect, we introduce a parameter buckling load reduction percent (BLRP). BLRP is defined as:

$$\text{BLRP} = 100 \times \left(\frac{\widehat{N}_{\text{Local}} - \widehat{N}_{\text{Nonocal}}}{\widehat{N}_{\text{Local}}}\right) \qquad [6.48]$$

Figure 6.19 shows the variation of the BLRP with the scale coefficient for different cases of NDNPS. The results for the load reduction parameter are in the dimensionless form as in equation [6.48]. The stiffness parameter of the coupling springs between the GSs is assumed to be constant ($K = 10$). From Figure 6.19, it can be observed that as the scale coefficient μ increases the BLRP increases. This implies that for increasing scale coefficient, the value of load parameter $\widehat{N}$ decreases. The reduction in load parameter is due to the assimilation of small-scale effects in the NDNPS in the material properties of the GSs. The small-scale effect reduces the stiffness of the material and hence the comparative lower buckling load. Therefore, by the non-local elastic model, the size effects are reflected in the NDNPS.

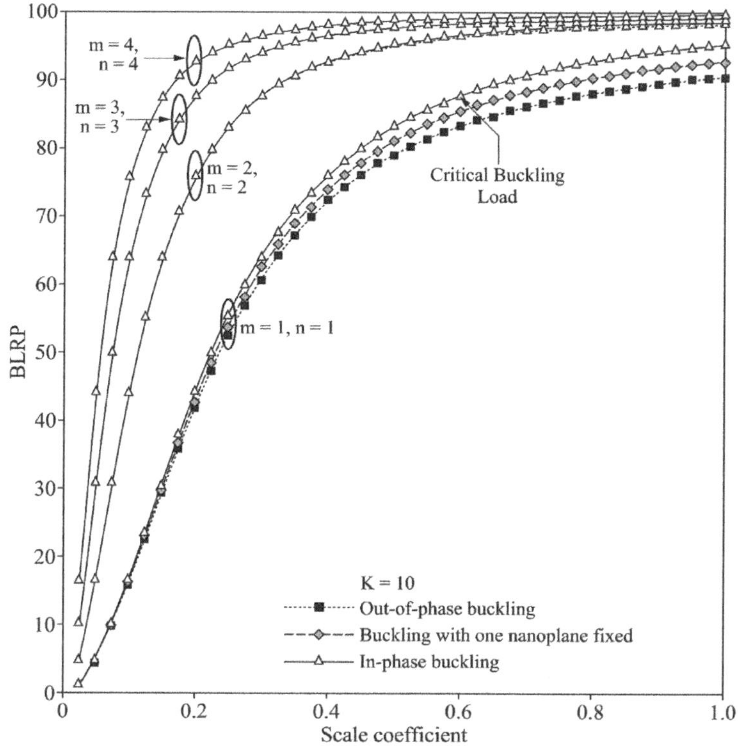

Figure 6.19. *Effect of scale coefficient on buckling load reduction percent (BLRP)*

Three different cases of NDNPS are considered. Case 1, case 2 and case 3 depict the conditions (i) when both the GSs buckle in the out-of-phase (asynchronous) sequence $(w_1 - w_2 \neq 0)$, (ii) when one of the GSs in NDNPS is stationary $(w_2 = 0)$ and (iii) when both the GSs buckle with in-phase (synchronous) sequence $(w_1 - w_2 = 0)$, respectively. Comparing the three cases of coupled-graphene-sheet-system, we observe that the BLRP for case 3 (in-phase buckling behavior) is larger than the BLRP for case 1 (out-of-phase buckling behavior) and case 2 (one GS fixed). In other words, the scale coefficient significantly reduces the in-phase buckling load (thus higher BLRP) compared to other cases considered. The relative higher BLRP in case 3 is due to the absence of coupling effect of the spring and the two nanoplates (GSs).

In addition, it can be seen that the values of the BLRP for case 2 (one GS fixed) are larger than the values of the BLRP for case 1 (out-of-phase buckling behavior). For case 2, the coupled-graphene-sheet-system becomes similar to the buckling behavior characteristic of the single GS with the effect of elastic medium. In case 3 (in-phase buckling behavior), the load reduction parameter is relatively less because the NDNPS becomes independent of the effect of the spring stiffness. For case 3, the NDNPS becomes similar to the buckling response of the single GS without the effect of elastic medium. In other words, the whole NDNPS can be treated as a buckled single GS and the coupling internal structure is effectless.

In general, it is worth noticing that the small-scale effects in NDNPS are higher with increasing non-local parameter in the in-phase buckling behavior than in the out-of-phase buckling behavior. This is because the stiffness of the springs in out-of-phase buckling behavior reduces the non-local effects.

Higher modes of buckling in the NDNPS (i.e. $m = 1, n = 1$; $m = 2, n = 2$; $m = 3, n = 3$; and $m = 4, n = 4$) are also depicted in Figure 6.19. It is observed that small-scale effects are higher for higher modes of buckling $m = 1, n = 1 < m = 2, n = 2 < m = 3, n = 3 < m = 4, n = 4$. Further, it is important to note that with increasing modes of buckling, the difference among the (i) out-of-phase (asynchronous) sequence ($w_1 - w_2 \neq 0$), (ii) in-phase (synchronous) sequence ($w_1 - w_2 = 0$) of buckling and (iii) one nanoplate stationary reduces. It should be noted that the modes for case 1 and case 3 can be referred to as submodes of original modes.

6.7.3. *Effect of stiffness of coupling springs in NDNPS*

To illustrate the influence of stiffness of the springs on the buckling load of the coupled-GS-systems, curves have been plotted for the BLRP against the scale coefficient. Spring stiffness represents the stiffness of the enclosing elastic medium (polymer resin). Different values of stiffness parameter of the coupling springs are considered. Figures 6.20–6.22 depict the stiffness of the springs on the BLRP of coupled systems. The stiffness parameters of the coupling springs are taken as $K = 0$, 50 and 100, respectively. Aspect ratio (L/W) is taken as unity. From the figure, it is noticed that as the stiffness parameter of the coupling springs increases, the BLRP decreases.

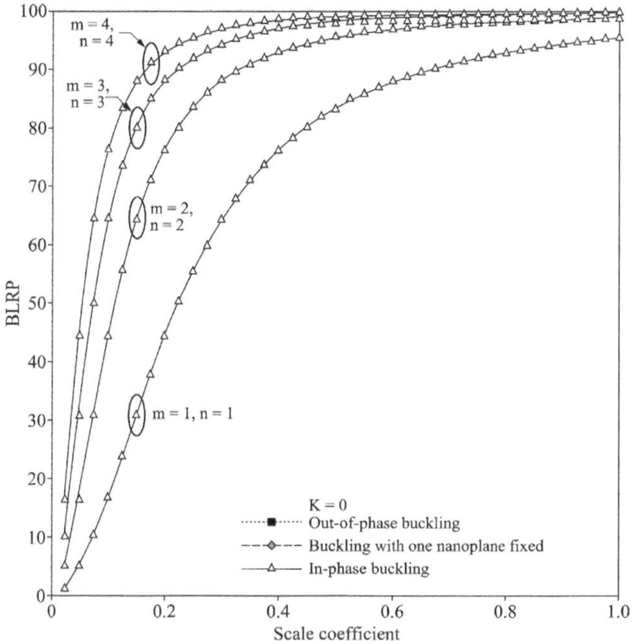

Figure 6.20. *Effect of scale coefficient on BLRP of NDNPS for K = 0*

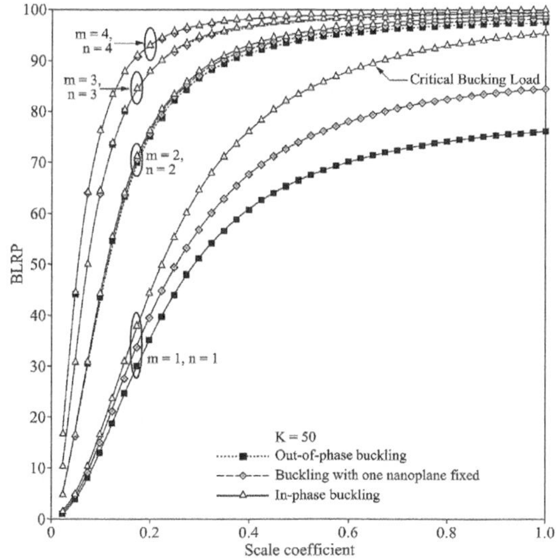

Figure 6.21. *Effect of scale coefficient on BLRP of NDNPS for K = 50*

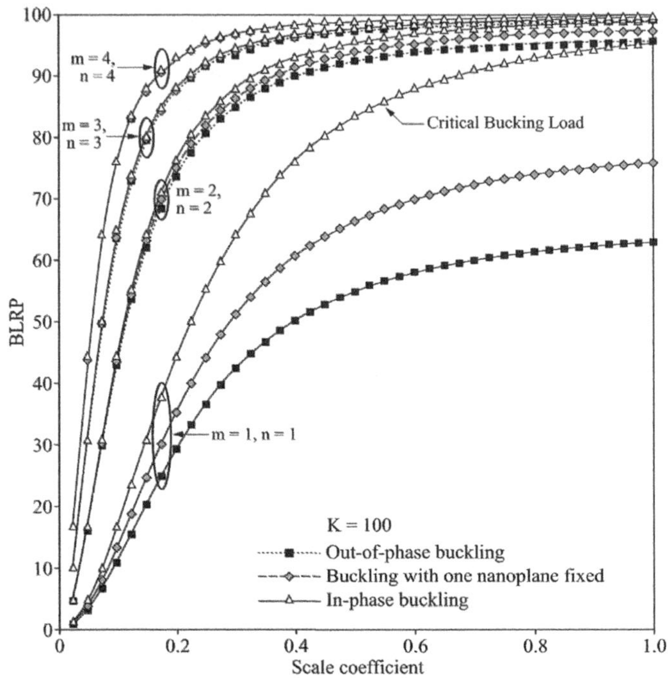

Figure 6.22. *Effect of scale coefficient on BLRP of NDNPS for K = 100*

Considering all values of the stiffness parameter; and comparing the three cases of coupled-GS system, it is noticed that the BLRP for case 3 (in-phase buckling) is larger than the BLRP for case 1 (out-of-phase buckling) and case 3 (one GS fixed). These different changes of BLRP with the increasing scale coefficient for the three different cases are more amplified as the stiffness parameter of the springs increases. For case 1 (out-of-phase buckling) and case 2 (one GS fixed), the BLRP reduces with increasing values of stiffness parameter. This observation implies that case 1 (out-of-phase buckling) and case 2 (one GS fixed) are less affected by scale effects. Comparing the case 1 and case 2, it can be seen the BLRP is less for out-of-phase buckling than for buckling in case 2. Thus, the out-of-phase buckling is less affected by the small scale or non-local effects. This out-of-phase buckling can be attributed to the fact that the coupling springs in the vibrating system dampen the non-local effects. In-phase buckling of coupled-system is unchangeable with increasing stiffness of springs. This is due to the in-phase buckling mode of behavior. For in-phase buckling, the coupled system behaves as if a single graphene without the effect of internal elastic medium. In other words, the whole coupled system can be treated as a single nanoelement and the coupling internal structure is effectless. In summary, it should be noted that the in-phase buckling of coupled-system is more affected by small-scale effects compared to out-of-phase buckling.

6.7.4. *Effect of aspect ratio on NDNPS*

Next, we illustrate the influence of aspect ratios (L/W) of the nanoplates (GS) on the buckling load of the coupled-GS-systems. The GSs are assumed to be coupled by a polymer matrix of stiffness $K = 100$. Curves have been plotted for the BLRP against the scale coefficient for different aspect ratios. Different values of aspect ratios of the NDNPS are considered. Figures 6.23–6.25 depict the effect of aspect ratio on the BLRP of coupled systems. The aspect ratios are taken as $L/W = 0.2$, 0.4 and 0.8.

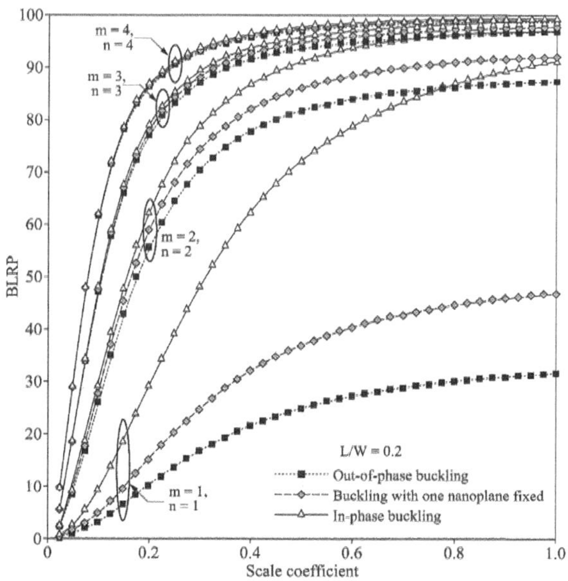

Figure 6.23. *Effect of scale coefficient on BLRP of NDNPS for L/W = 0.2*

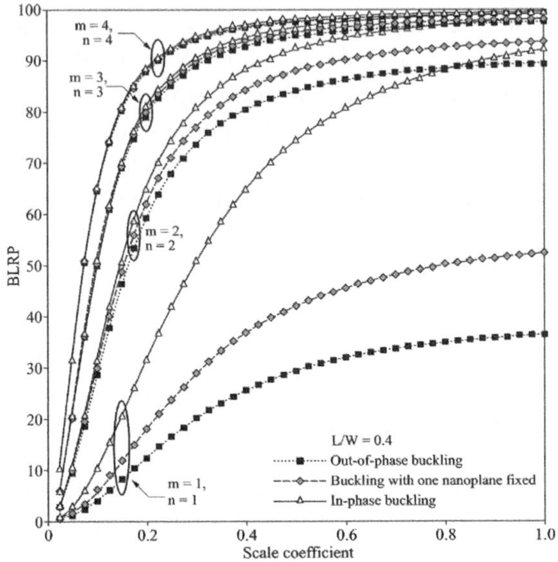

Figure 6.24. *Effect of scale coefficient on BLRP of NDNPS for L/W = 0.4*

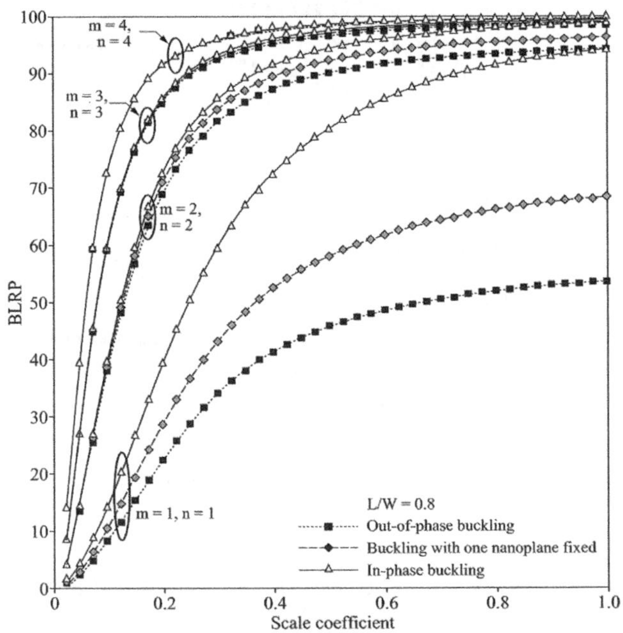

Figure 6.25. *Effect of scale coefficient on BLRP of NDNPS for L/W = 0.8*

From Figures 6.23–6.25, we see that with the increase in aspect ratios (*L*/*W*) of NDNPS, the BLRP for all cases of buckling increases. We note that the difference between in-phase buckling, out-of-phase buckling and buckling with one GS fixed becomes less for increasing aspect ratios (*L*/*W*) of NDNPS. Thus, it can be concluded that although the small-scale effects are more in higher aspect ratios (*L*/*W*) of NDNPS, the effects of stiffness of coupling springs are reduced in higher aspect ratios of NDNPS. And thus, less difference in curves between in-phase buckling, out-of-phase buckling and buckling with one GS fixed becomes less for higher aspect ratios of NDNPS.

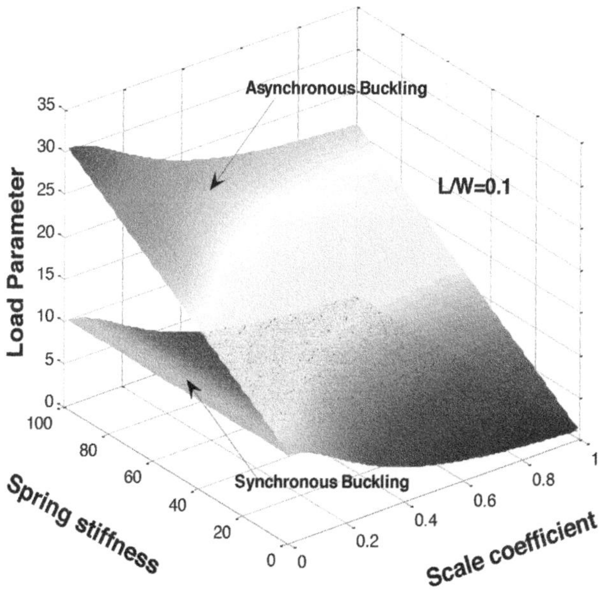

Figure 6.26. *The variation of the buckling load parameter for the asynchronous and synchronous buckling of NDNPS as a function of the spring stiffness (K) and the scale-coefficient (μ) with L/W = 0.1. For a color version of the figure, see www.iste.co.uk/murmu/non-local.zip*

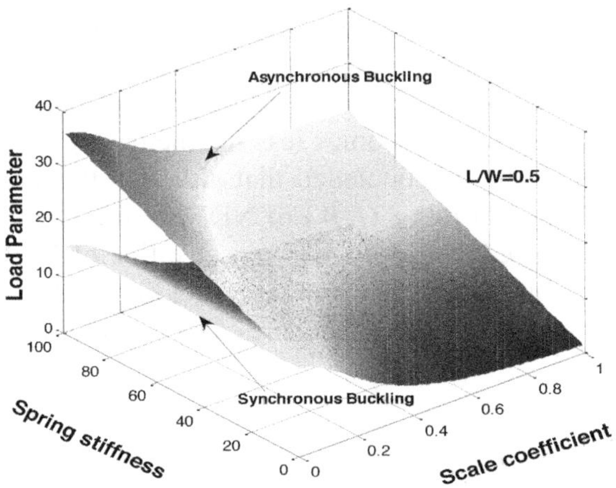

Figure 6.27. *The variation of the buckling load parameter for the asynchronous and synchronous buckling of NDNPS as a function of the spring stiffness (K) and the scale-coefficient (μ) with L/W = 0.5. For a color version of the figure, see www.iste.co.uk/murmu/non-local.zip*

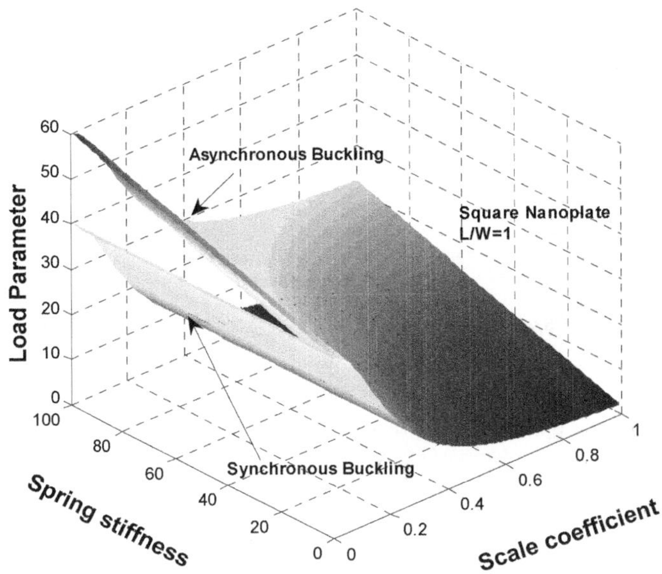

Figure 6.28. *The variation of the buckling load parameter for the asynchronous and synchronous buckling of NDNPS as a function of the spring stiffness (K) and the scale-coefficient (μ) with L/W = 1.0 (square plate). For a color version of the figure, see www.iste.co.uk/murmu/non-local.zip*

Finally, for the comprehension of the buckling phenomenon of NDNPS with small-scale effects, the buckling load parameters $\widehat{N}$ of the double nanoplate system for different non-local parameters (scale coefficient) and spring stiffness parameter (K) are plotted in a three-dimensional graph. Three different aspect ratios of the NDNPS are considered. The stiffness parameters K of the coupling springs are considered in the range of 0–100. Figures 6.26–6.28 show the plots for the aspect ratio of L/W = 0.1, 0.5 and 1.0, respectively. The non-local parameter or scale coefficient μ is varied from 0 to 1. From Figure 6.26, it is observed that the increase in the scale coefficient has a reducing effect on the non-local buckling load of the NDNPS. The stiffness of the springs has a reducing effect on the small-scale effects of the NDNPS. It should be noted that for in-phase buckling, the loads are independent of coupling springs. It is also concluded from the figure that the difference between the asynchronous and synchronous buckling decreases with increasing aspect ratios (Figures 6.27 and 6.28).

6.8. Summary

In this section, expressions for buckling of double-nanoplate-systems are established within the framework of Eringen's non-local elasticity. An analytical method is introduced for determining buckling load of non-local double-nanoplates-system (NDNPS). Explicit closed-form expressions for buckling load are derived for the case when all four ends are simply supported. Two single-layered GSs coupled by polymer matrix are considered for the study. The study highlights that the small-scale effects considerably influence the buckling load of NDNPS. The small-scale effects in NDNPS are higher with increasing values of non-local parameter for the case of synchronous modes of buckling than in the asynchronous modes of buckling. The increase in the stiffness of the coupling springs in NDNPS reduces the small-scale effects during the asynchronous modes of buckling.

This chapter may provide an analytical scale-based non-local approach which could serve as the starting point for further investigation of more complex n-nanoplates systems undergoing vibration and instability. So far only single nanostrures were considered. In the next few chapters, the analysis is extended to dynamics of multiple nanostructures.

7

Multiple Nanostructure Systems

In the previous chapters, we have focused on the dynamics of a single nanostructure. In this chapter, we consider multiple nanostructures.

In this chapter, we analyze the dynamical and stability behavior of complex multiple nanostructure systems (CMNS), which are composed of the arbitrary number of nanostructured elements such as nanorods, nanobeams and nanoplates, embedded into an elastic medium. Understanding the vibration behavior of such nanosystems is an important area of investigation in the design of modern nanodevices application. A large amount of research work was focused on developing the proper theoretical framework for describing the different physical effects appearing in such structures on the nanoscale level. Great step in this direction was the successful application of non-local continuum field theory of Eringen. By using non-local continuum models, we are able to introduce into the constitutive equations a material parameter describing the forces between the atoms and the internal length-scale. Thus, Eringen's non-local elasticity theory has been widely used by various authors to describe the mechanical behavior of nanostructures. Based on D'Alembert's principles and the non-local constitutive relation, the system of governing equations of motion in such particular case of the multiple nanostructure system is derived and solved by analytical methods. An asymptotic analysis is conducted for the case when the number of nanostructure elements tends to the infinity. The obtained analytical solutions for the non-local natural frequency and buckling load are compared with the results obtained through numerical methods and an excellent agreement is confirmed. For each of these cases numerical examples will be given to show the effects of different physical parameters on the free vibration and buckling response of such nanosystems.

7.1. Longitudinal vibration of a multi-nanorod system

Nanorods are one-dimensional structures [LAO 12] which can be grown by using several methods of synthesizing. Nanorods growing from a nanowire core can be viewed as a multi-nanorod system (see Figure 7.1). As reported in the literature, dimensions of nanorods are in the range from 1 nm to several micrometers. Theoretical analysis of such systems can be of great importance for their application and comprehension since experiments on nano-scale level cannot be well controlled. Deterministic numerical methods as molecular dynamics (MD), based on classical multi-body dynamics, non-deterministic methods as Monte-Carlo and modern multi-scale methods as a mixture of MD and continuum methods can be computationally prohibitive when nanoscale systems are composed of multiple structures with large number of molecules and atoms. Hereby, the size-dependent non-local continuum theory provides numerical solutions much faster than the previously mentioned methods and, for some cases, it is even possible to obtain exact closed-form solutions. The value of small-scale parameters or so-called non-local parameters used in non-local continuum models is usually obtained by fitting the results from non-local nanostructured models with the results from MD simulations.

The longitudinal vibration of non-local nanorod-based systems has been the research topic in many papers. Particularly, the vibration of complex systems based on nanorods coupled with elastic, viscoelastic or other type of layers was examined for various boundary conditions, aspect ratios and small-scale parameters. Danesh *et al.* [DAN 12] accounted for the small-scale effect in the longitudinal vibration of tapered nanorods using the non-local elasticity theory. The authors used the differential quadrature method to solve motion equations for three different boundary conditions. Aydogdu [AYD 12] analyzed the axial vibration of carbon nanotube (CNT) embedded into an elastic medium using the non-local model of a nanorod surrounded with the elastic layers. The author compared the longitudinal frequencies of the non-local and classical continuum models. Kiani [KIA 13] performed an extensive study of the non-local longitudinal, transversal and torsional vibration and instability of Single Walled Carbon Nanotube (SWCNT). Şimşek [SIM 12] used a Galerkin approach to obtain the natural frequencies of the longitudinal vibration of axially functionally graded tapered nanorods. The analysis was carried out for nanorods of variable cross-sections, different tapered ratios, material properties and boundary conditions. Huang [HUA 12] examined the longitudinal vibration of a nanorod considering the non-local long-range interactions. Chang [CHA 12] investigated the non-local axial vibration of elastic nanorods by using the differential quadrature method to find the solution of motion equation. Murmu and Adhikari [MUR 10] analyzed the longitudinal vibration of a double-nanorod system (DNRS) coupled with an elastic layer. The non-local elasticity theory was applied for nanorods to consider small-scale effects on the longitudinal vibration of the DNRS. In addition, a detailed parametric study was carried out.

Understanding the vibration behavior of such systems, particularly a complex multiple nanorod system (MNRS), is the key step for the application in nano-electromechanical resonators, which can be externally excited and exhibit axial vibration. Further, the resonant frequencies are in the gigahertz range, which allow us to use them in high-frequency systems. It is assumed that the system under consideration is composed of a set of m parallel and identical nanorods coupled by elastic layers. The equations of motion of non-local MNRS are described by a set of m partial differential equations derived by using D'Alembert's principle. The exact solutions for natural frequencies are derived by applying the method of separation of variables on the governing equations that are reduced to the homogeneous system of algebraic equation. Afterwards, by applying the trigonometric method, we obtain natural frequencies in the closed form for both chain systems, clamped-chain and free-chain. The accuracy of the results for natural frequencies obtained with the trigonometric method is confirmed by comparison with other results obtained using the numerical methods to solve frequency equations. In the given numerical examples, the influence of the non-local parameter and stiffness of the elastic layers on natural frequencies of MNRS for different number of nanorods and different boundary conditions are considered.

Figure 7.1. *Scanning electron microscopy images of ZnO nanorods in the multi-nanorod formation growing on nanowire core (reproduced from [LAO 12])*

7.1.1. *The governing equations of motion*

The mechanical model of MNRS is considered for two types of boundary conditions, i.e. clamped–clamped (C-C) and clamped-free (C-F), and two chain systems, "clamped-chain" and "free-chain" systems, such as shown in Figure 7.2. The main distinction between the two chain systems is in the way of coupling the first and last nanorod in an MNRS with a fixed base. The MNRS is shown in Figures 7.2(a) and (b) and is coupled in the "clamped-chain" systems, in which the first and last nanorod are connected with a fixed base by an elastic layer ($k_0 \neq 0$ and $k_m \neq 0$). The MNRS presented in Figures 7.2(c) and (d) is coupled in the "free-chain" system, where the first and last nanorod are not connected with a fixed base ($k_0 = 0$ and $k_m = 0$). Presented models consists of a set of m identical, parallel and uniform nanorods connected by an axially distributed elastic layer with stiffness per length denoted as $k_1 = k_2 = \cdots = k_i = \cdots = k_{m-1} = k$. Nanorods in an MNRS are referred to as nanorod 1, nanorod 2 and so on until the m-th nanorod. All nanorods are having the elastic modulus E, mass density ρ, uniform cross-section area A and length L. Figure 7.3 shows the i-th nanorod and a differential element of length dx. The cross-sectional area A of differential elements of nanorods is constant and the longitudinal displacements of elements in the x-direction are given by $u_i(x,t)$. We presume that a nanorod is under the dynamically varying stress field $\sigma_{xx}(x,t)$, so that adjacent sections are subjected to varying stresses. The external forces $F_i(x,t)$ and $F_{i-1}(x,t)$ per unit length from elastic interaction are also considered. The equation of motion in the $x-$ direction then becomes:

$$-N_i + (N_i + dN_i) + F_i dx - F_{i-1} dx = \ddot{u}_i dm, \qquad [7.1]$$

where $dm = \rho A dx$ is the mass of the infinitesimal element, $N_i(x,t)$ is the stress resultant of the axial stresses σ_{xx} acting internally on the cross-sectional area A of i-th nanorod and $dN_i = \frac{\partial N_i}{\partial x} dx$ is the differential change of the axial resultant force; F_i and F_{i-1} are the external forces resulting from elastic layers.

Introducing $dm = \rho A dx$ into equation [7.1] gives:

$$\frac{\partial N_i}{\partial x} = -F_i + F_{i-1} + \rho A \ddot{u}_i, \qquad [7.2]$$

where N_i is the stress resultant defined as:

$$N_i(x,t) = \int_A \sigma_{xx}(x,t)dA, \qquad [7.3]$$

and:

$$F_i = k_i(u_{i+1} - u_i), \quad F_{i-1} = k_{i-1}(u_i - u_{i-1}), \qquad [7.4]$$

are the external forces from elastic layers.

Introducing equations [7.3] into equation [2.4], the stress resultant for the non-local theory is obtained as:

$$N_i - (e_0 a)^2 \frac{\partial^2 N_i}{\partial x^2} = EA \frac{\partial u_i}{\partial x}. \qquad [7.5]$$

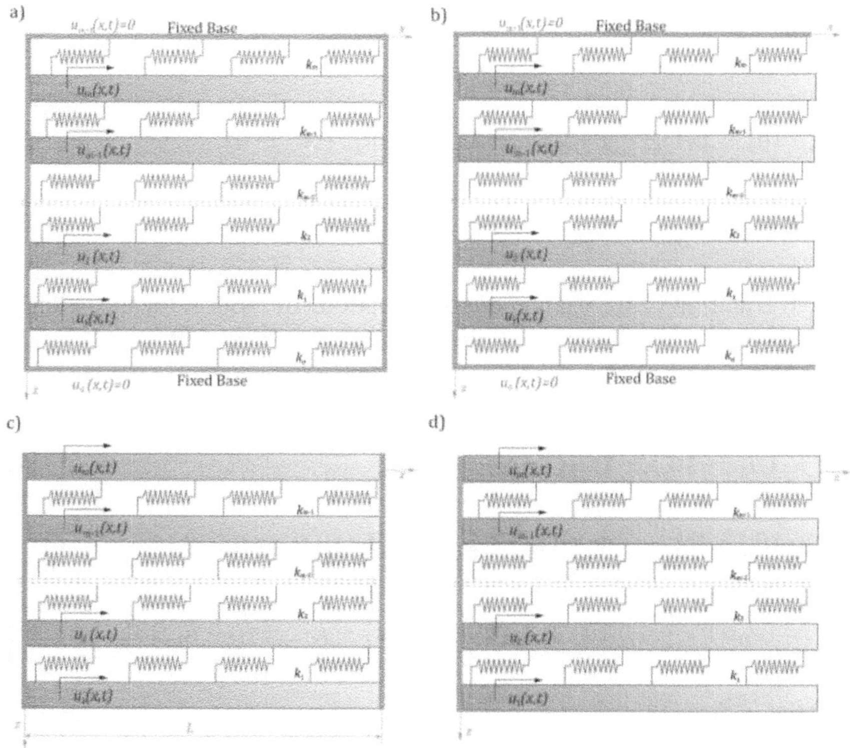

Figure 7.2. *The MNRS for different boundary conditions: a) clamped-chain system with C-C boundary conditions, b) clamped-chain system with C–F boundary conditions, c) free–chain system with C–C boundary conditions and d) free–chain system with C–F boundary conditions*

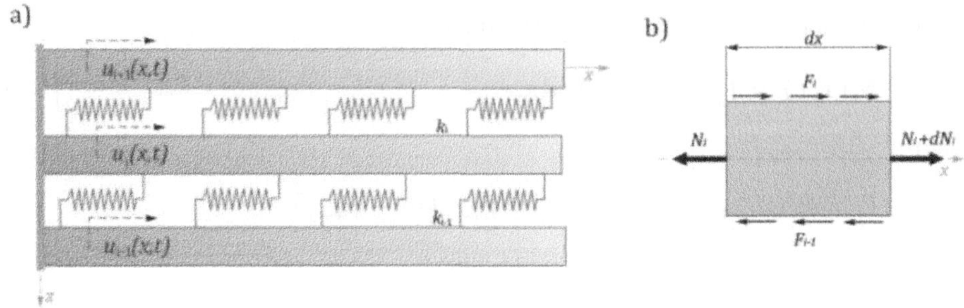

Figure 7.3. *a) The i-th nanorod of MNRS and b) differential element of the i-th nanorod*

By substituting equations [7.2] and [7.4] into equation [7.5], we get the following equation of motion expressed in terms of the displacement $u_i(x,t)$ for non-local elastic constitutive relation:

$$\rho A\ddot{u}_i - EA\frac{\partial^2 u_i}{\partial x^2} + k_i(u_i - u_{i+1}) + k_{i-1}(u_i - u_{i-1})$$

$$= (e_0 a)^2 \frac{\partial^2}{\partial x^2}[\rho A\ddot{u}_i + k_i(u_i - u_{i+1}) + k_{i-1}(u_i - u_{i-1})], \qquad [7.6]$$

From equation [7.6], we can obtain equations of motion for the "clamped-chain" system ($k_0 = k_m = k$), Figures 7.2(a) and (b) as follows:

$$\rho A\ddot{u}_1 - EA\frac{\partial^2 u_1}{\partial x^2} + k_1(u_1 - u_2) + k_0 u_1 = (e_0 a)^2 \frac{\partial^2}{\partial x^2}[\rho A\ddot{u}_1 + k_1(u_1 - u_2) + k_0 u_1], \qquad [7.7a]$$

$$\rho A\ddot{u}_i - EA\frac{\partial^2 u_i}{\partial x^2} + k_i(u_i - u_{i+1}) + k_{i-1}(u_i - u_{i-1})$$

$$= (e_0 a)^2 \frac{\partial^2}{\partial x^2}[\rho A\ddot{u}_i + k_i(u_i - u_{i+1}) + k_{i-1}(u_i - u_{i-1})], \qquad [7.7b]$$

$i = 2,3, \ldots, m-1,$

$$\rho A\ddot{u}_m - EA\frac{\partial^2 u_m}{\partial x^2} + k_m u_m + k_{m-1}(u_m - u_{m-1})$$

$$= (e_0 a)^2 \frac{\partial^2}{\partial x^2}[\rho A\ddot{u}_m + k_m u_m + k_{m-1}(u_m - u_{m-1})]. \qquad [7.7c]$$

From equation [7.6], we can obtain equations of motion for the "free-chain" system ($k_0 = K_m = 0$), Figures 7.2(c) and (d) in the following form:

$$\rho A\ddot{u}_1 - EA\frac{\partial^2 u_1}{\partial x^2} + k_1(u_1 - u_2) = (e_0 a)^2 \frac{\partial^2}{\partial x^2}[\rho A\ddot{u}_1 + K_1(u_1 - u_2)], \quad [7.8a]$$

$$\rho A\ddot{u}_i - EA\frac{\partial^2 u_i}{\partial x^2} + k_i(u_i - u_{i+1}) + k_{i-1}(u_i - u_{i-1})$$

$$= (e_0 a)^2 \frac{\partial^2}{\partial x^2}[\rho A\ddot{u} + k_i(u_i - u_{i+1}) + k_{i-1}(u_i - u_{i-1})], \quad [7.8b]$$

$$i = 2,3, \ldots, m-1,$$

$$\rho A\ddot{u}_m - EA\frac{\partial^2 u_m}{\partial x^2} + k_{m-1}(u_m - u_{m-1}) = (e_0 a)^2 \frac{\partial^2}{\partial x^2}[\rho A\ddot{u}_m + k_{m-1}(u_m - u_{m-1})]. \quad [7.8c]$$

7.1.2. *Exact solution*

Suppose that the ends of nanorods are clamped-clamped (see Figures 7.2(a) and (c)) and clamped-free (see Figures 7.2(b) and (d)), the boundary conditions are given by:

Clamped-Clamped:

$$u_i(0,t) = u_i(L,t) = 0, \quad [7.9]$$

Clamped-Free:

$$u_i(0,t) = N_i(L,t) = 0, \quad [7.10]$$

where $N_i(L,t), (i = 1,2,..,m)$ are the stress resultants on the right side of a set of m nanorods.

By substituting equation [7.6] into equation [7.5], we get the stress resultant force N_i as follows:

$$N_i(x,t) = (e_0 a)^2 \frac{\partial}{\partial x}[\rho A\ddot{u}_i + k_i(u_i - u_{i+1}) + k_{i-1}(u_i - u_{i-1})] + EA\frac{\partial u_i}{\partial x}. \quad [7.11]$$

Introducing equation (10) into the boundary conditions (19), yields:

$$N_i(L,t) = (e_0a)^2 \frac{\partial}{\partial x}\left[\rho A\ddot{u}_i(L,t) + k_i\left(u_i(L,t) - u_{i+1}(L,t)\right)\right.$$

$$\left.+k_{i-1}\left(u_i(L,t) - u_{i-1}(L,t)\right)\right] + EA\frac{\partial u_i(L,t)}{\partial x} = 0. \qquad [7.12]$$

The equations of motion equation [7.6] with boundary conditions equations [7.19] and [7.10] can be solved by using the method of separations of variables and assuming the solutions in the following form:

$$u_i(x,t) = \sum_{n=1}^{\infty} U_{ni} \sin \alpha_n \, e^{i\omega_n t}, \qquad [7.13]$$

where for the *clamped–clamped* boundary conditions [GRA 75], we have:

$$\alpha_n = \frac{n\pi}{L}, n = 1,2,\dots,\infty, \qquad [7.14]$$

and for the *clamped–free* boundary conditions [GRA 75], we have:

$$\alpha_n = \frac{(2n-1)\pi}{2L}, n = 1,2,\dots,\infty, \qquad [7.15]$$

where $i = \sqrt{-1}$, U_{ni} is the amplitude and ω_n is the natural frequency in *n*-th mode of vibration. Introducing the assumed solutions [7.13] into equation [7.6], we obtain the system of *m* algebraic equations as:

$$-v_{ni-1}U_{ni-1} + S_{ni}U_{ni} - v_{ni}U_{ni+1} = 0, \quad i = 1,2,3,\dots,m, \qquad [7.16]$$

where:

$$S_{ni} = EA\alpha_n^2 - \rho A\omega_n^2[1 + (e_0a)^2\alpha_n^2] + k_i[1 + (e_0a)^2\alpha_n^2] + k_{i-1}[1 + (e_0a)^2\alpha_n^2], \qquad [7.17a]$$

$$v_{ni} = k_i[1 + (e_0a)^2\alpha_n^2], \qquad [7.17b]$$

$$v_{ni-1} = k_{i-1}[1 + (e_0a)^2\alpha_n^2]. \qquad [7.17c]$$

7.1.2.1. *Clamped-chain system*

In this section, we will consider the free longitudinal vibration of MNRS coupled in the clamped–chain system for C-C and C-F boundary conditions.

In addition, it is assumed that all nanorods have the same physical characteristics and are joined by the elastic layers of the same stiffness($k_0 = k_1 = \cdots = k_{m-1} = k_m = k$). Assuming the solutions equation [7.16] of the set of m partial differential equations [7.7], we obtain the system of homogeneous algebraic equations in matrix form as:

$$\begin{bmatrix} S_n & -v_n & 0 & 0 & 0 & 0 & 0 & 0 & 0 \\ -v_n & S_n & -v_n & 0 & 0 & 0 & 0 & 0 & 0 \\ \cdots & \cdots & \cdots & \cdots & \cdots & \cdots & \cdots & \cdots & \cdots \\ 0 & 0 & 0 & S_n & -v_n & 0 & 0 & 0 & 0 \\ 0 & 0 & 0 & -v_n & S_n & -v_n & 0 & 0 & 0 \\ 0 & 0 & 0 & 0 & -v_n & S_n & 0 & 0 & 0 \\ \cdots & \cdots & \cdots & \cdots & \cdots & \cdots & \cdots & \cdots & \cdots \\ 0 & 0 & 0 & 0 & 0 & 0 & -v_n & S_n & -v_n \\ 0 & 0 & 0 & 0 & 0 & 0 & 0 & -v_n & S_n \end{bmatrix} \begin{Bmatrix} U_{n1} \\ U_{n2} \\ U_{n3} \\ \cdots \\ U_{ni-1} \\ U_{ni} \\ U_{ni+1} \\ \cdots \\ U_{nm-2} \\ U_{nm-1} \\ U_{nm} \end{Bmatrix} = \begin{Bmatrix} 0 \\ 0 \\ 0 \\ \cdots \\ 0 \\ 0 \\ 0 \\ \cdots \\ 0 \\ 0 \\ 0 \end{Bmatrix} \quad [7.18]$$

where:

$$S_n = \alpha_n^2 EA - \rho A \omega_n^2 [1 + (e_0 a)^2 \alpha_n^2] + 2k[1 + (e_0 a)^2 \alpha_n^2], \quad [7.19a]$$

$$v_n = k[1 + (e_0 a)^2 \alpha_n^2]. \quad [7.19b]$$

The natural frequency ω_n is determined when the frequency determinant from the system [7.18] is equal to zero. The numerical solutions will be used to confirm the solutions obtained by the analytical method. For the homogeneous system of algebraic equations [7.18], closed-form solution for natural frequencies of MNRS can be determined by utilizing the trigonometric method [RAS 53, STO 13, KAR 14].

According to the paper by Rašković [RAS 53], Stojanović *et al.* [STO 13] and Karličić *et al.* [KAR 14], solution of the *i*-th algebraic equation is assumed in the following form:

$$U_{ni} = V cos(i\, \varphi_{cc}) + M\, sin(i\, \varphi_{cc}),\ i = 1,2,3, \ldots, m. \quad [7.20]$$

By substituting equation [7.20] into the i-th algebraic equation of the system [7.18], we get two trigonometric equations, assuming that the constants V and N are not simultaneously equal to zero:

$$V\{-v_n \cos[(i-1)\varphi_{cc}] + S_n \cos(i\,\varphi_{cc}) - v_n \cos[(i+1)\varphi_{cc}]\} = 0,$$

$$i = 2,3,\dots,m-1, \qquad [7.21a]$$

$$M\{-v_n \sin[(i-1)\varphi_{cc}] + S_n \sin(i\,\varphi_{cc}) - v_n \sin[(i+1)\varphi_{cc}]\} = 0,$$

$$i = 2,3,\dots,m-1. \qquad [7.21b]$$

After some algebra, equation [7.21] reduces to:

$$(S_n - 2v_n \cos\varphi_{cc}) V \cos(i\,\varphi_{cc}) = 0, \qquad [7.22a]$$

$$(S_n - 2v_n \cos\varphi_{cc}) M \sin(i\,\varphi_{cc}) = 0. \qquad [7.22b]$$

From equation [7.22], we can say that system will have an oscillatory behavior under the following conditions $V \neq 0$ and $\cos(i\,\varphi_{cc}) \neq 0$ or $M \neq 0$ and $\sin(i\,\varphi_{cc}) \neq$, for $i = 2,3,\dots,m-1$. Thus, we obtain the frequency equation as:

$$S_n = 2v_n \cos\varphi_{cc}. \qquad [7.23]$$

Now, from equation [7.23], it is possible to obtain analytical expressions for the natural frequencies and critical buckling loads, but it is necessary to find the value of unknown φ_{cc}. The value of φ_{cc} is determined from the upper and lower conditions of the chain system, i.e. equation [7.20] must satisfy the first and last equation of the system [7.18]. If we introduce expressions $U_{1n} = V\cos\varphi_{cc} + M\sin\varphi_{cc}$ and $U_{2n} = V\cos(2\varphi_{cc}) + M\sin(2\varphi_{cc})$ into the first equation and $U_{m-1n} = V\cos[(m-1)\varphi_{cc}] + M\sin[(m-1)\varphi_{cc}]$ and $U_{m-1n} = V\cos(m\varphi_{cc}) + M\sin(m\varphi_{cc})$ into the last equation of the system [7.18], we obtain the system of algebraic equations:

$$V[S_n \cos\varphi_{cc} - v_n \cos(2\varphi_{cc})] + M[S_n \sin\varphi_{cc} - v_n \sin(2\varphi_{cc})] = 0, \qquad [7.24a]$$

$$V\big[S_n \cos(m\varphi_{cc}) - v_n \cos[(m-1)\varphi_{cc}]\big] + M\big[S_n \sin(m\varphi_{cc}) - v_n \sin[(m-1)\varphi_{cc}]\big] = 0. \qquad [7.24b]$$

Non-trivial solutions for the constants N and M can be obtained, which yields the following trigonometric equation:

$$\begin{vmatrix} 1 & 0 \\ cos[(m+1)\varphi_{cc}] & sin[(m+1)\varphi_{cc}] \end{vmatrix} = 0 \Rightarrow sin[(m+1)\varphi_{cc}] = 0, \quad [7.25]$$

from which, we can obtain solutions for unknown $\varphi_{cc,s}$ as:

$$\varphi_{cc,s} = \frac{s\pi}{m+1}, \qquad s = 1,2,\dots,m. \qquad [7.26]$$

Introducing parameter $\varphi_{cc,s}$ and equations [7.19] into equation [7.26], the frequency equation is obtained as:

$$EA\alpha_n^2 - \rho A\omega_n^2[1+(e_0a)^2\alpha_n^2] + 2v_n(1-cos\varphi_{cc,s}) = 0. \qquad [7.27]$$

Thus, we obtain closed-form solution for natural frequencies of MNRS coupled in the "clamped–chain" system, as:

$$\omega_{ncc,s} = \sqrt{\frac{EA\alpha_n^2+2v_n(1-cos\varphi_{cc,s})}{\rho A[1+(e_0a)^2\alpha_n^2]}}, \qquad s = 1,2,\dots,m. \qquad [7.28a]$$

or in the dimensionless form:

$$\Omega_{ncc,s} = \sqrt{\frac{(n\pi)^2+2K[1+\xi^2(n\pi)^2](1-cos\varphi_{cc,s})}{[1+\xi^2(n\pi)^2]}}, \qquad s = 0,1,\dots,m-1, \qquad [7.28b]$$

where:

$$\Omega_{ncc,s} = \omega_{ncc,s}L\sqrt{\frac{\rho}{E}}, \quad K = \frac{kL^2}{EA}, \quad \xi = \frac{e_0a}{L},$$

are the dimensionless parameters.

7.1.2.2. *Free-chain system*

Here, we assume that the stiffness $k_0 = k$ and $k_m = k$ are equal to zero, so that there is no coupling of the first and last nanorod with a fixed base. Also, it is assumed that all nanorods have the same physical characteristics and are joined by elastic layers of the same stiffness. Introducing the assumed solutions [7.13] into the set of *m* partial differential equations [7.8], we obtain the system of algebraic equations in the matrix form as:

$$\begin{bmatrix} S_n - v_n & -v_n & 0 & 0 & 0 & 0 & 0 & 0 & 0 \\ -v_n & S_n & -v_n & 0 & 0 & 0 & 0 & 0 & 0 \\ \cdots & \cdots & \cdots & \cdots & \cdots & \cdots & \cdots & \cdots & \cdots \\ 0 & 0 & 0 & S_n & -v_n & 0 & 0 & 0 & 0 \\ 0 & 0 & 0 & -v_n & S_n & -v_n & 0 & 0 & 0 \\ 0 & 0 & 0 & 0 & -v_n & S_n & 0 & 0 & 0 \\ \cdots & \cdots & \cdots & \cdots & \cdots & \cdots & \cdots & \cdots & \cdots \\ 0 & 0 & 0 & 0 & 0 & 0 & -v_n & S_n & -v_n \\ 0 & 0 & 0 & 0 & 0 & 0 & 0 & -v_n & S_n - v_n \end{bmatrix} \begin{Bmatrix} U_{n1} \\ U_{n2} \\ U_{n3} \\ \cdots \\ U_{ni-1} \\ U_{ni} \\ U_{ni+1} \\ \cdots \\ U_{nm-2} \\ U_{nm-1} \\ U_{nm} \end{Bmatrix} = \begin{Bmatrix} 0 \\ 0 \\ 0 \\ \cdots \\ 0 \\ 0 \\ 0 \\ \cdots \\ 0 \\ 0 \\ 0 \end{Bmatrix}, \quad [7.29]$$

where expressions for S_n and v_n are defined in equations [7.19].

Following the same procedures as in the previous case for the "clamped–chain" system, assuming the solution given by equation [7.20] and entering into the $i-th$ algebraic equation of the system [7.29], we get the frequency equation as:

$$S_n = 2v_n cos\varphi_{fc}, \quad [7.30]$$

where unknown φ_{cf} can be determined based on the conditions that the assumed solutions [7.20] must satisfy the first and last equation of the system [7.29]. Introducing the assumed solutions $U_{1n} = Vcos\varphi_{fc} + M\, sin\varphi_{fc}$ and $U_{2n} = Vcos(2\varphi_{fc}) + M\, sin(2\varphi_{fc})$ into the first equation and $U_{m-1n} = Vcos[(m-1)\varphi_{fc}] + M\, sin[(m-1)\varphi_{fc}]$ and $U_{m-1n} = Vcos(m\varphi_{fc}) + M\, sin(m\varphi_{fc})$ into the last algebraic equation of the system [7.29], after some algebra we have the system of algebraic equations:

$$V[(S_n - v_n)cos\varphi_{fc} - v_n cos(2\varphi_{fc})] + M[(S_n - v_n)sin\varphi_{fc} - v_n sin(2\varphi_{fc})] = 0, \quad [7.31]$$

$$V\left[(S_n - v_n)cos(m\varphi_{fc}) - v_n cos[(m-1)\varphi_{fc}]\right]$$

$$+M\left[(S_n - v_n)sin(m\varphi_{fc}) - v_n sin[(m-1)\varphi_{fc}]\right] = 0. \qquad [7.32]$$

Non-trivial solutions for the constants V and M can be obtained, which yields a trigonometric equation in the following form:

$$\begin{vmatrix} 1 - cos\varphi_{fc} & -sin\varphi_{fc} \\ cos[(m+1)\varphi_{cc}] - cos(m\varphi_{fc}) & sin[(m+1)\varphi_{cc}] - sin(m\varphi_{fc}) \end{vmatrix} = 0 \;\Rightarrow$$

$$sin(m\varphi_{fc}) = 0, \qquad [7.33]$$

from which we can obtain the solutions for unknown $\varphi_{fc,s}$ as:

$$\varphi_{fc,s} = \frac{s\pi}{m}, \qquad s = 0,2,\ldots,m-1. \qquad [7.34]$$

By substituting expression for $\varphi_{fc,s}$ and equations [7.19] into equation [7.30], the frequency equation is obtained in the following form:

$$EA\alpha_n^2 - \rho A\omega_n^2(1 + \mu\alpha_n^2) + 2v_n(1 - cos\varphi_{fc,s}) = 0. \qquad [7.35]$$

The natural frequencies of MNRS coupled in the "free-chain" system are:

$$\omega_{nfc,s} = \sqrt{\frac{EA\alpha_n^2 + 2v_n(1 - cos\varphi_{fc,s})}{\rho A[1 + (e_0 a)^2\alpha_n^2]}}, \qquad s = 0,1,\ldots,m-1. \qquad [7.36a]$$

or in the dimensionless form:

$$\Omega_{nfc,s} = \sqrt{\frac{(n\pi)^2 + 2K[1 + \xi^2(n\pi)^2](1 - cos\varphi_{fc,s})}{[1 + \xi^2(n\pi)^2]}}, \qquad s = 0,1,\ldots,m-1, \qquad [7.36b]$$

where:

$$\Omega_{nfc,s} = \omega_{nfc,s}L\sqrt{\frac{\rho}{E}}, \quad K = \frac{kL^2}{EA}, \quad \xi = \frac{e_0 a}{L},$$

are the dimensionless parameters.

7.1.3. *Asymptotic analysis*

Assuming that a number of nanorods tend to infinity, i.e. introducing $m \to \infty$ into equation [7.28] or equation [7.36], we obtain the asymptotic value of the natural frequency of MNRS in the following form:

$$\omega_{n,m\to\infty} = \sqrt{\frac{EA\alpha_n^2}{\rho A[1+(e_0a)^2\alpha_n^2]}}. \quad [7.37]$$

Expression [7.37] is representing the lowest natural frequency of the system when the number of nanorods tends to the infinity, and it is the same for both "clamped-chain" and "free-chain" systems. Now, consider the case when the number of modes and the number of nanorods tends to infinity, i.e. introducing $n\to\infty$ and $m\to\infty$ into equation [7.28] or equation [7.36], which is previously divided by α_n^2. Thus, we get the asymptotic natural frequency:

$$\omega_{n\to\infty,m\to\infty} = \sqrt{\frac{EA}{\rho A(e_0a)^2}}. \quad [7.38]$$

This implies that the asymptotic natural frequencies are independent of the boundary conditions.

7.1.4. *Numerical examples and discussions*

In this section, we present a comparative study of the analytical results derived in this chapter with the results obtained numerically in the software package Wolfram Mathematica. In addition, we present several numerical examples for some particular cases of MNRS, when the system consists of three, five and seven coupled nanorods. To examine the influence of non-local effects on the longitudinal vibration of MNRS, natural frequencies are analyzed for different (1) values of non-local parameters, (2) number of nanorods and (3) the stiffness coefficient of the elastic layers.

7.1.4.1. *Validation of the proposed method*

In order to justify the results obtained for arbitrary number of nanorods, in MNRS we compared the obtained analytical solutions from equations [7.28] and [7.36] with numerical solutions of the homogeneous system of algebraic equations given in equations [7.18] and [7.29]. The following values are used for the comparative study: $L = 1$ [nm], $\rho A = 10^{-9}$ [kg/m], $EA = 1$ [nN], $e_0a = 0 - 2$ [nm], stiffness coefficient $k = 8$ [N/nm]. Table 7.1 shows the natural frequencies of MNRS with three, five and seven coupled nanorods for "clamped-chain" and "free-chain" systems. For the case of the "clamped-chain" system, natural frequencies are calculated numerically from equation [7.18] and analytically from equation [7.28] representing the numerical and analytical solution, respectively. The natural frequencies of the "free-chain" system are calculated numerically from equation [7.29] and analytically from equation [7.36] also representing the numerical and analytical solution, respectively. Based on the presented results, we can observe that the fundamental natural frequencies of the system decrease for an increase in the number of nanorods in MNRS. Further, it can be concluded that the fundamental natural frequency ($s = 0$) of the "free-chain" system is independent of the number of nanorods, while the remaining natural frequencies decrease for an increase in the number of nanorods in MNRS. The next lowest frequency is considerably higher than the fundamental one. In the case of natural frequencies of the "clamped-chain" system presented in Table 7.2, there is no fundamental frequency and the lowest natural frequency and intervals between frequencies are also decreasing for an increase in the number of nanorods in MNRS. Finally, it can be concluded that the results obtained by analytical and numerical methods are in excellent agreement, for both boundary conditions and "chain" systems.

Coupled multiple-nanorod system for C-F boundary conditions					
Number of nanorods		"Free-Chain" system		"Clamped-Chain" system	
		N. S.* of equation [7.2])	A. S.* equation [7.36]	N. S.* of equation [7.18]	A. S.* equation [7.28]
$m = 3$	1	0.4764	0.4764	2.2165	2.2165
	2	2.8682	2.8682	4.0282	4.0282
	3	4.9220	4.9220	5.2479	5.2479
$m = 5$	1	0.4764	0.4764	1.5396	1.5396
	2	1.8118	1.8118	2.8682	2.8682
	3	3.3589	3.3589	4.0282	4.0282
	4	4.6012	4.6012	4.9220	4.9220
	5	5.4010	5.4010	5.4848	5.4848
$m = 7$	1	0.4764	0.4764	1.2020	1.2020
	2	1.3459	1.3459	2.2165	2.2165
	3	2.5002	2.5002	3.1786	3.1786
	4	3.5590	3.5590	4.0282	4.0282
	5	4.4482	4.4482	4.7275	4.7275
	6	5.1188	5.1188	5.2479	5.2479
	7	5.5355	5.5355	5.5685	5.5685

*N.S.- Numerical Solutions ; *A.S.- Analytical Solutions; *m*- number of nanorods

Table 7.1. *Comparison of the analytical and numerical solutions for natural frequencies (GHz) of MNRS with C-F boundary conditions*

Coupled multiple-nanorod system for C-C boundary conditions					
Number of nanorods		"Free-Chain" system		"Clamped-Chain" system	
		N. S.* of equation [7.29]	A. S.* equation [7.36]	N. S.* of equation [7.18]	A. S.* equation [7.28]
$m = 3$	1	0.4937	0.4937	2.2203	2.2203
	2	2.8712	2.8712	4.0303	4.0303
	3	4.9238	4.9238	5.2495	5.2495
$m = 5$	1	0.4937	0.4937	1.5451	1.5451
	2	1.8164	1.8164	2.8712	2.8712
	3	3.3614	3.3614	4.0303	4.0303
	4	4.6030	4.6030	4.9238	4.9238
	5	5.4026	5.4026	5.4863	5.4863
$m = 7$	1	0.4937	0.4937	1.2090	1.2090
	2	1.3521	1.3521	2.2203	2.2203
	3	2.5035	2.5035	3.1813	3.1813
	4	3.5613	3.5613	4.0303	4.0303
	5	4.4501	4.4501	4.7293	4.7293
	6	5.1205	5.1205	5.2495	5.2495
	7	5.5370	5.5370	5.5700	5.5700

*N.S.- Numerical Solutions; *A.S.- Analytical Solutions; *m*- number of nanorods

Table 7.2. *Comparison of the analytical and numerical solutions for natural frequencies (GHz) of MNRS with C-C boundary conditions*

7.1.4.2. *Non-local effect on natural frequency of MNRS*

Numerical results are obtained for the same system parameters as in Tables 7.1 and 7.2. Natural frequencies are determined for $m = 3, 5$ and 7 nanorods in MNRS, three different stiffness coefficients of the layer $k = 4, -12\ [N/nm]$ and the values of non-local parameter in the range$(e_0a) = 0 - 2\ [nm]$. Natural frequencies given in Figures 7.4 and 7.5 are obtained for the value of the parameter $s = 1$ and first vibration mode in both cases, "free-chain" and "clamped-chain" systems.

The main property of MNRS that can be observed from Figures 7.4 and 7.5 is that natural frequencies decrease with an increase in the non-local parameter and the number of nanorods. But, for an increase in the stiffness coefficient of elastic layers, the natural frequencies increase. A common property of MNRS in all cases is that frequencies are much higher when non-local parameter is equal to zero, which yields the solution for the local case. This means that classical continuum theories overdetermined natural frequencies, which is also stated by other authors. Further, we can state that for the lowest value of the layers stiffness coefficients, the highest number of nanorods and an increase in the non-local parameter, natural frequency is approaching to the lowest frequency of the system. However, there are some differences for MNRS with different boundary conditions. For C-F boundary conditions and the "free-chain" system (Figure 7.4(a)), the effect of the non-local parameter on natural frequency is less pronounced as the value of the parameter increases, i.e. non-local natural frequencies decrease slightly (Figure 7.4(a)). Also, in the case of an MNRS with C-F boundary conditions coupled in the "clamped-chain" system (Figure 7.4(b)) values of the natural frequencies are similar to the previous case but slightly lower for the same parameters of MNRS. For MNRS with C-C boundary conditions in both "free-chain" and "clamped-chain" systems (Figures 7.5(a) and (b)), when the non-local parameter $(e_0a) = 0$, the difference between the lowest natural frequencies of an MNRS with different number of nanorods and different values of stiffness coefficient is negligible. This is not the case when the non-local parameter is taken into account where the difference between non-local frequencies becomes more significant. Influence of the non-local parameter on natural frequencies is very strong for lower values of this parameter, but this effect is weakening as its value increases. Hence, we can conclude that in the classical theory of elasticity, the effect of an increase in a number of nanorods on natural frequencies is less pronounced than in the case when the non-local elasticity theory is employed. The main difference between C-F and C-C boundary condition cases of an MNRS can be observed from Figures 7.4 and 7.5, where the effect of the non-local parameter on natural frequencies is much more noticeable for the C-C boundary condition than for the C-F boundary condition, especially for the lower values of the non-local parameter.

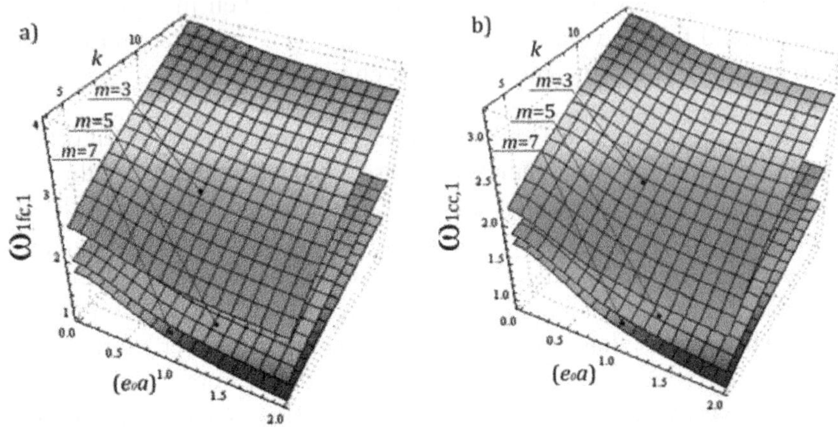

Figure 7.4. *Effect of the non-local parameter (e_0a) on natural frequencies, for various values of stiffness k, different number of nanorods m and C-F boundary conditions: a) free-chain system and b) clamped-chain system. For a color version of the figure, see www.iste.co.uk/murmu/non-local.zip*

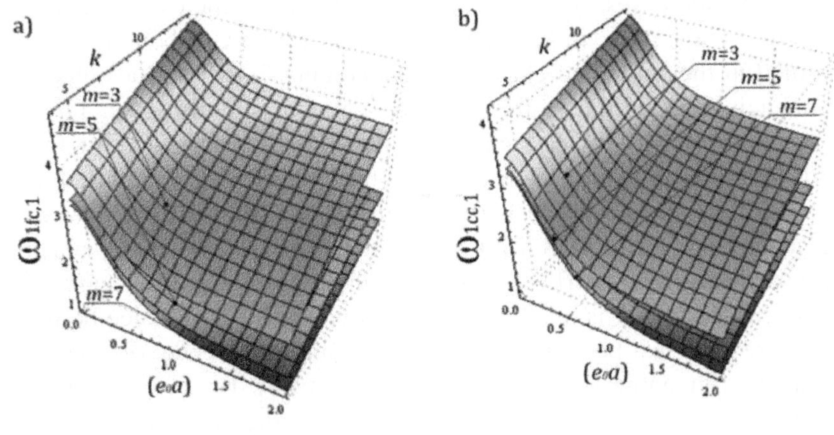

Figure 7.5. *Effect of the non-local parameter (e_0a) on natural frequencies for various values of stiffness k, different number of nanorods m and C-C boundary conditions: a) free-chain system and b) clamped-chain system. For a color version of the figure, see www.iste.co.uk/murmu/non-local.zip*

Figure 7.6 shows the effect of a significant increase in the number of nanorods in the "clamped-chain" MNRS for clamped-clamped (C-C) and clamped-free (C-F) boundary conditions, respectively. It can be observed that the lowest natural frequency of the system tends to the asymptotic value of natural frequency for significant increase in the number of nanorods, which is in line with equation [7.37]. The lowest value of natural frequency of the "free-chain" system is for s=0 and it is independent of the number of nanorods, which also indicates that it is equal to the asymptotic value of natural frequency.

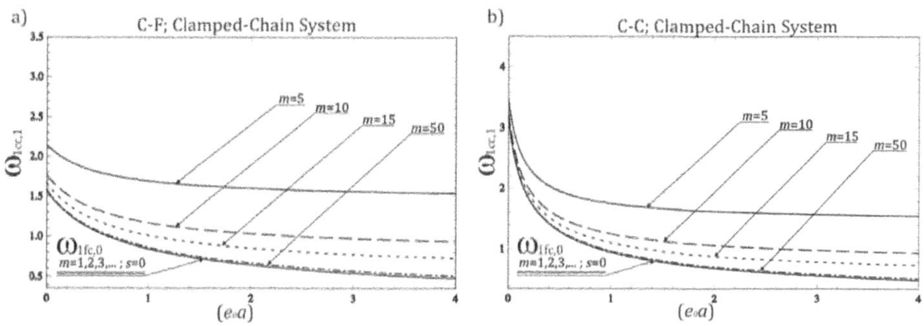

Figure 7.6. *Effect of the number of nanorods and non-local parameter* $(e_0 a)$ *on natural frequency* $\omega_{1cc,1}$ *for* $k = 8\,[N/nm]$*: a) C-F boundary conditions and b) C-C boundary conditions*

7.2. Transversal vibration and stability of a multiple-nanobeam system

The problem of the free transversal vibration and buckling analysis of complex systems with coupled nanobeams such as multiple-nanostructures has attracted much attention from the scientific community. Such systems are being increasingly used for nanodevices such as biosensors [WAN 03], mass sensors [MEH 13], nanobeam resonators [YOO 11] and nanoactuators [LEE 07]. Therefore, the issue of vibration and stability of multiple-nanobeam systems (MNBSs) can be very important for nanoengineering practice. In many papers, theoretical analyses of the mechanical behavior of nanomaterials have been proposed apart from the experimental work [GUZ 07]. CNTs [IIJ 91], boron nitride nanotubes (BNNTs) [CHO 95] and zinc oxide nanotubes (ZnO) [LIU 09] are the most common materials used for nanotechnology due to their extraordinary mechanical, chemical, electrical and thermal properties, which can be modeled as nanobeam structures.

Double-nanosolid system is the simplest model of a coupled multi-nanosolid system, which is composed of two elastic nanobeams or nanorods or nanoplates continuously joined by the Winkler elastic medium or medium with other properties. Pioneering studies of vibration and stability of double-nanobeam system using non-local continuum theory approach are papers by Murmu and Adhikari [MUR 10, MUR 11, MUR 12]. In [MUR 10], the natural frequencies of the non-local double-nanobeam model are obtained for the case of simply-supported nanobeams. The problem of axial instability of the double-nanobeam system where the material length-scale parameter is taken into account is discussed in [MUR 11]. Furthermore, Murmu and Adhikari [MUR 12] proposed an analytical method to obtain natural frequencies of the non-local double-nanobeam-system under the initial compressive prestressed conditions. Arani and Roudbari [ARA 13] presented an analytical solution for vibration of a visco-Pasternak coupled BNNT system under a moving nanoparticle with considered non-local piezoelectric surface effect. The authors have investigated the influences of the small-scale effect, electrical potential, viscose medium and dimensionless velocity parameter on the vibration behavior of the observed system. On the subject of investigation of coupled macroscopic structures, few references will be mentioned. In paper by Kelly and Srinivas [KEL 09], a general theory to determine mode shapes and natural frequencies of an axially loaded multi-beam system coupled with elastic layers has been proposed. The authors used the Euler–Bernoulli beam theory and obtained an exact solution in the general case. Also, as an alternative way to find a solution, they developed a Rayleigh–Ritz approximation method using mode shapes of the corresponding unscratched beams. Stojanović *et al.* [STO 13] suggested a general procedure of finding the natural frequencies and buckling load of a multi-beam system coupled with elastic layers and under compressive axial loading. For beams, they used Timoshenko and other higher order theories and obtained exact solutions for natural frequencies and critical buckling loads using the trigonometric method.

In this section, we apply the analytical method for finding the closed-form solutions of natural frequencies and critical buckling loads for the transverse vibration and stability of the system of coupled multiple nanobeams. It is assumed that the system is composed of m parallel and identical nanobeams coupled by the Winkler elastic medium. The nanobeams are having the same length L and boundary conditions. Mathematical expressions for the free vibration and buckling of the MNBS are obtained according to the Eringen's non-local elasticity and Euler–Bernoulli beam theory. Analytical solutions for natural frequency and critical buckling load are derived by using the classical Bernoulli–Fourier method and trigonometric method for a different number of identical nanobeams in MNBS. In addition, analytical solutions for natural frequencies and buckling loads are compared with numerical solutions of the frequency polynomial. Asymptotic values of natural frequencies and critical buckling loads are investigated in the case when the number of nanobeams tends to the infinity. Particular attention is dedicated to the evaluation of the effects of non-local parameters, number of nanobeams and stiffness coefficients of the Winkler elastic medium on the value of natural frequencies and critical buckling loads of MNBS.

7.2.1. *The governing equations of motion*

Consider the multiple straight axially loaded nanobeams, continuously joined by the Winkler elastic medium as shown in Figure 7.7. Nanobeams of MNBS are referred to as nanobeam 1, nanobeam 2 and so on until the m-th nanobeam. All nanobeams are made of the same material of elastic modulus E, mass density ρ, with uniform cross-section area A and moment of inertia I. Each nanobeam is under the influence of the same compressive axial load P. The transverse displacement of the $i-th$ nanobeam is $w_i(x,t), i = 1,2,3 \dots m$. This study is limited to the case of the non-local Euler–Bernoulli's beam theory with simply-supported nanobeams. We will also consider the two cases of coupling. In the first case, the "clamped-chain" system, the first and last nanobeam are coupled with a fixed base by the Winkler elastic medium of stiffness k_0 and k_m, respectively, Figure 7.1(a). In the second case, the "free-chain" system, it is assumed that the stiffness of the Winkler elastic medium k_0 and k_m are equal to zero, Figure 7.1(b), i.e. there is no coupling with a fixed base. Other nanobeams in the MNBS are also coupled by the continuous linear Winkler type of elastic medium with stiffness per length $k_0 = k_1 = k_2 = \dots = k_i = \dots = k_{m-1} = k_m = k$.

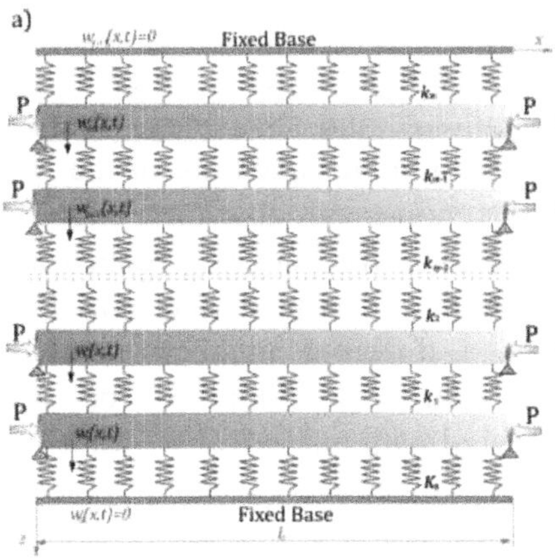

Figure 7.7. *The multi-nanobeam system coupled by the Winkler elastic medium: a) "Clamped-Chain"($k_0 \neq 0,\ k_m \neq 0$) and b) "Free-Chain" ($k_0 = k_m = 0$). For a color version of the figure, see www.iste.co.uk/murmu/non-local.zip*

Now, we analyze the equilibrium of the stress resultant forces acting on the differential element of $i - th$ nanobeam, as shown in Figure 7.8(b). The cross-sectional area is constant along the x coordinate. We observe the shear force F_T and banding moment M_f acting on the left face of a differential element. Both quantities vary along the length of nanobeam with time, $F_T = F_T(x,t)$ and $M_f = M_f(x,t)$, measured in the horizontal direction x. Figure 7.8(b) shows the external distributed force $q_i(x,t)$ and $q_{i-1}(x,t)$ of the Winkler elastic medium having dimensions of force per unit length. On the right end of differential element, we have force $F_T + dF_T$ and bending moment $M_f + dM_f$.

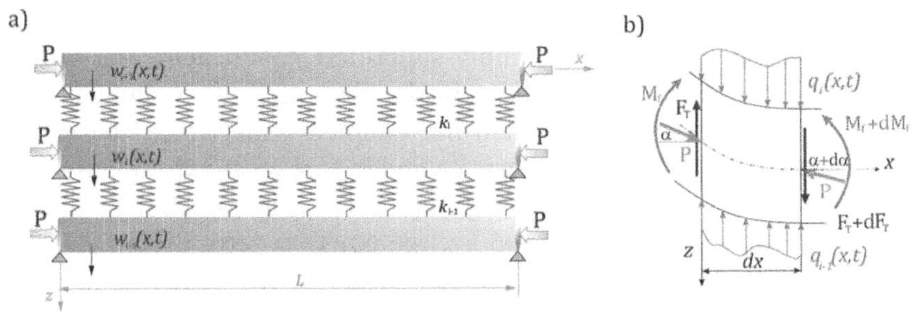

Figure 7.8. *a) $i - th$ non-local Euler–Bernoulli's nanobeam subjected to the axial load F and external loads $q_i(x,t)$ and $q_{i-1}(x,t)$, b) differential element of $i - th$ nanobeam of length dx. For a color version of the figure, see www.iste.co.uk/murmu/non-local.zip*

By summing the forces in the z-direction and moments around y-axis, we obtain the following equation of motion:

$$-\frac{\partial F_T}{\partial x} + \rho A \frac{\partial^2 w_i}{\partial t^2} + P \frac{\partial^2 w_i}{\partial x^2} - q_i + q_{i-1} = 0, \; i = 1,2,3, \dots m, \qquad [7.39a]$$

$$F_T = \frac{\partial M_f}{\partial x}, \qquad [7.39b]$$

where M_f and F_T are the stress resultants defined in Chapter 3 as:

$$M_f = \int_0^A z\sigma_{xx}\, dA, \qquad [7.40]$$

and the external loads from the Winkler elastic medium are:

$$q_i = k_i(w_{i+1} - w_i), \quad q_{i-1} = k_{i-1}(w_i - w_{i-1}). \qquad [7.41]$$

Governing equations of motion can be expressed in terms of the transversal displacement $w_i(x,t)$ using the non-local theory, by combining equations [7.39] and [7.40], and introducing them into the non-local constitutive relation equation [2.1]. This yields the following equation of motion for the $i - th$ nanobeam:

$$\rho A \frac{\partial^2 w_i}{\partial t^2} + P \frac{\partial^2 w_i}{\partial x^2} - k_i(w_{i+1} - w_i) + k_{i-1}(w_i - w_{i-1}) + EI \frac{\partial^4 w_i}{\partial x^4}$$

$$= (e_0 a)^2 \frac{\partial^2}{\partial x^2}\left[\rho A \frac{\partial^2 w_i}{\partial t^2} + P \frac{\partial^2 w_i}{\partial x^2} - k_i(w_{i+1} - w_i) + k_{i-1}(w_i - w_{i-1})\right],$$

$$i = 1,2,3, \dots, m. \qquad [7.42]$$

From equation [7.42], we obtain equations of motion for the "clamped-chain" system (Figure 7.7(a)) of the form:

$$\rho A \frac{\partial^2 w_1}{\partial t^2} + P \frac{\partial^2 w_1}{\partial x^2} + 2kw_1 - kw_2 + EI \frac{\partial^4 w_1}{\partial x^4}$$

$$= (e_0 a)^2 \frac{\partial^2}{\partial x^2}\left[\rho A \frac{\partial^2 w_1}{\partial t^2} + P \frac{\partial^2 w_1}{\partial x^2} + 2kw_1 - kw_2\right], \qquad [7.43a]$$

$$\rho A \frac{\partial^2 w_i}{\partial t^2} + P \frac{\partial^2 w_i}{\partial x^2} - k_i(w_{i+1} - w_i) + k_{i-1}(w_i - w_{i-1}) + EI \frac{\partial^4 w_i}{\partial x^4}$$

$$= (e_0 a)^2 \frac{\partial^2}{\partial x^2} \left[\rho A \frac{\partial^2 w_i}{\partial t^2} + P \frac{\partial^2 w_i}{\partial x^2} - k_i(w_{i+1} - w_i) + k_{i-1}(w_i - w_{i-1})\right],$$

$$i = 2,3, \ldots, m-1, \qquad [7.43b]$$

$$\rho A \frac{\partial^2 w_m}{\partial t^2} + P \frac{\partial^2 w_m}{\partial x^2} + 2k w_m - k w_{m-1} + EI \frac{\partial^4 w_m}{\partial x^4}$$

$$= (e_0 a)^2 \frac{\partial^2}{\partial x^2} \left[\rho A \frac{\partial^2 w_m}{\partial t^2} + P \frac{\partial^2 w_m}{\partial x^2} + 2k w_m - k w_{m-1}\right], \qquad [7.43c]$$

and for the "free-chain" system (Figure 7.7(b)) as:

$$\rho A \frac{\partial^2 w_1}{\partial t^2} + P \frac{\partial^2 w_1}{\partial x^2} + k w_1 - k w_2 + EI \frac{\partial^4 w_1}{\partial x^4}$$

$$= (e_0 a)^2 \frac{\partial^2}{\partial x^2} \left[\rho A \frac{\partial^2 w_1}{\partial t^2} + P \frac{\partial^2 w_1}{\partial x^2} + k w_1 - k w_2\right], \qquad [7.44a]$$

$$\rho A \frac{\partial^2 w_i}{\partial t^2} + P \frac{\partial^2 w_i}{\partial x^2} - k_i(w_{i+1} - w_i) + k_{i-1}(w_i - w_{i-1}) + EI \frac{\partial^4 w_i}{\partial x^4}$$

$$= (e_0 a)^2 \frac{\partial^2}{\partial x^2} \left[\rho A \frac{\partial^2 w_i}{\partial t^2} + P \frac{\partial^2 w_i}{\partial x^2} - k_i(w_{i+1} - w_i) + k_{i-1}(w_i - w_{i-1})\right],$$

$$i = 2,3, \ldots, m-1, \qquad [7.44b]$$

$$\rho A \frac{\partial^2 w_m}{\partial t^2} + P \frac{\partial^2 w_m}{\partial x^2} + k w_m - k w_{m-1} + EI \frac{\partial^4 w_m}{\partial x^4}$$

$$= (e_0 a)^2 \frac{\partial^2}{\partial x^2} \left[\rho A \frac{\partial^2 w_m}{\partial t^2} + P \frac{\partial^2 w_m}{\partial x^2} + k w_m - k w_{m-1}\right]. \qquad [7.44c]$$

7.2.2. *Exact solution*

The initial conditions in general form and boundary conditions for simply-supported non-local Euler–Bernoulli's nanobeams of the same length L are assumed as follows:

$$w_i(0,t) = w_i(L,t) = 0, \quad M_{fi}(0,t) = M_{fi}(L,t) = 0, \quad i = 1,2,3,\ldots,m, \qquad [7.45a]$$

where banding moment resultants are defined as:

$$M_{fi}(x,t) = (e_0 a)^2 \frac{\partial^2}{\partial x^2}\left(\rho A \frac{\partial^2 w_i}{\partial t^2} + P\frac{\partial^2 w_i}{\partial x^2} - q_i + q_{i-1}\right) - EI\frac{\partial^2 w_i}{\partial x^2}. \quad [7.45b]$$

By using the governing boundary conditions [7.45a], equation [7.42] can be solved by assuming the solution for generalized displacement $w_i(x,t)$ in the form of expansions as:

$$w_i(x,t) = \sum_{n=1}^{\infty} W_{in} \sin \alpha_n x \, e^{i\omega_n t}, i = 1,2,3,..,m, \quad [7.46]$$

where $i = \sqrt{-1}$, $\alpha_n = \frac{n\pi}{L}$ $(n = 1,2,3,...)$; W_{in} are the amplitudes and $\omega_n (i = 1,2,3,..,m)$ denotes the natural frequencies. Introducing the assumed solution equation [7.46] into equation [7.42], the system of m algebraic equations is obtained as:

$$-v_{i-1n}W_{i-1n} + S_{in}W_{in} - v_{in}W_{i+1n} = 0, \quad i = 1,2,3,...,m, \quad [7.47a]$$

where:

$$S_{in} = EI\alpha_n^4 - \rho A\omega_n^2[1 + (e_0 a)^2\alpha_n^2] - \alpha_n^2 P[1 + (e_0 a)^2\alpha_n^2] + v_{in} + v_{i-1n}, \quad [7.47b]$$

$$v_{in} = k_i[1 + (e_0 a)^2\alpha_n^2], \quad [7.47c]$$

$$v_{i-1n} = k_{i-1}[1 + (e_0 a)^2\alpha_n^2]. \quad [7.47d]$$

7.2.2.1. *Clamped-chain system*

Assuming that all nanobeams in MNBS have the same material and geometrical characteristics and coupled by the Winkler elastic medium of same stiffness, the system of partial differential equations [7.43] can be reduced to the system of algebraic equations as:

$$\begin{bmatrix} S_n & -v_n & 0 & \dots & 0 & 0 & 0 & \dots & 0 & 0 & 0 \\ -v_n & S_n & -v_n & \dots & 0 & 0 & 0 & \dots & 0 & 0 & 0 \\ \dots & \dots & \dots & \dots & \dots & \dots & \dots & \dots & \dots & \dots & \dots \\ 0 & 0 & 0 & \dots & S_n & -v_n & 0 & \dots & 0 & 0 & 0 \\ 0 & 0 & 0 & \dots & -v_n & S_n & -v_n & \dots & 0 & 0 & 0 \\ 0 & 0 & 0 & \dots & 0 & -v_n & S_n & \dots & 0 & 0 & 0 \\ \dots & \dots & \dots & \dots & \dots & \dots & \dots & \dots & \dots & \dots & \dots \\ 0 & 0 & 0 & \dots & 0 & 0 & 0 & \dots & 0 & S_n & -v_n \\ 0 & 0 & 0 & \dots & 0 & 0 & 0 & \dots & 0 & -v_n & S_n \end{bmatrix} \begin{Bmatrix} W_{1n} \\ W_{2n} \\ W_{3n} \\ \dots \\ W_{i-1n} \\ W_{in} \\ W_{i+1n} \\ \dots \\ W_{m-2n} \\ W_{m-1n} \\ W_{mn} \end{Bmatrix} = \begin{Bmatrix} 0 \\ 0 \\ 0 \\ \dots \\ 0 \\ 0 \\ 0 \\ \dots \\ 0 \\ 0 \\ 0 \end{Bmatrix}, \quad [7.48]$$

where:

$$S_n = EI\alpha_n^4 - \rho A\omega_n^2[1 + (e_0a)^2\alpha_n^2] - \alpha_n^2 P[1 + (e_0a)^2\alpha_n^2] + 2v_n, \quad [7.49a]$$

$$v_n = k[1 + (e_0a)^2\alpha_n^2]. \quad [7.49b]$$

The natural frequencies and buckling loads are obtained in two ways: (1) the numerical solution of determinants of the system [7.48] and (2) the explicit analytical solution using the trigonometric method, similarly to the previous section (section 7.1.2.1). Following the previously used methodology based on the trigonometric method, we obtain natural frequencies and buckling loads in the closed form. The numerical solution of the system of equations [7.48] is used to verify the proposed analytical solutions. For the non-trivial solution, $W_{1n} \neq 0, W_{2n} \neq 0, \dots, W_{m-1n} \neq 0$ and $W_{mn} \neq 0$, determinant of matrix equation [7.48] must be equal to zero. The analytical solution of determinant is possible only for the case of identical nanobeams and identical properties of coupling medium. According to Rašković [RAS 53], the assumed solution of the $i-th$ algebraic equation is given in the following form:

$$W_{in} = V\cos(i\,\varphi_{cc}) + M\sin(i\,\varphi_{cc}), \; i = 1,2,3,\dots,m. \quad [7.50]$$

From the system of algebraic equations [7.48], we can observe that the first and last equation are different from other equations. This depends on the upper and lower boundary conditions of the chain system, which are equal to $w_0 = w_{m+1} = 0$ in this case. Introducing equation [7.50] into the $i - th$ algebraic equation of the system [7.48], we get two trigonometric equations, assuming that the constants V and M are not simultaneously equal to zero. By applying similar procedure as in section 7.1.2.1, we obtain the frequency equation:

$$S_n = 2v_n cos\varphi_{cc}. \quad [7.51]$$

Now, from equation [7.51], it is possible to obtain analytical expressions for natural frequencies and buckling loads, but first it is necessary to find the value of unknown φ_{cc}. To determineφ_{cc}, we use the upper and lower boundary conditions of the chain system, i.e. equation [7.50] must satisfy the first and last equation of the system [7.48]. If we introduce expressions $W_{1n} = Vcos\varphi_{cc} + M\ sin\varphi_{cc}$ and $W_{2n} = Vcos(2\varphi_{cc}) + M\ sin(2\varphi_{cc})$ into the first equation and $W_{m-1n} = Vcos[(m-1)\varphi_{cc}] + M\ sin[(m-1)\varphi_{cc}]$ and $W_{m-1n} = Vcos(m\varphi_{cc}) + M\ sin(m\varphi_{cc})$ into the last equation of the system [7.48], we obtain a system of algebraic equations:

$$V[S_n cos\varphi_{cc} - v_n cos(2\varphi_{cc})] + M[S_n sin\varphi_{cc} - v_n sin(2\varphi_{cc})] = 0, \quad [7.52a]$$

$$V\big[S_n cos(m\varphi_{cc}) - v_n cos[(m-1)\varphi_{cc}]\big] + M\big[S_n sin(m\varphi_{cc}) - v_n sin[(m-1)\varphi_{cc}]\big] = 0. \quad [7.52b]$$

We can obtain non-trivial solutions for the constants V and M, which yields the following trigonometric equation:

$$\begin{vmatrix} 1 & 0 \\ cos[(m+1)\varphi_{cc}] & sin[(m+1)\varphi_{cc}] \end{vmatrix} = 0 \quad \Rightarrow \quad sin[(m+1)\varphi_{cc}] = 0, \quad [7.53]$$

from which, we obtain solutions for unknown $\varphi_{cc,s}$ as:

$$\varphi_{cc,s} = \frac{s\pi}{m+1}, \qquad s = 1,2,\dots,m. \quad [7.54]$$

Introducing $\varphi_{cc,s}$ and equation [7.49] into equation [7.51], we get the frequency and stability equation in the following form:

$$EI\alpha_n^4 - \rho A\omega_n^2[1 + (e_0a)^2\alpha_n^2] - \alpha_n^2 P[1 + (e_0a)^2\alpha_n^2] + 2v_n(1 - cos\varphi_{cc,s}) = 0. \quad [7.55]$$

For the case when $P = 0$, from equation [7.55] we obtain the natural frequency of MNBS as:

$$\omega_{Bncc,s} = \sqrt{\frac{EI\alpha_n^4 + 2v_n(1-\cos\varphi_{cc,s})}{\rho A[1+(e_0 a)^2\alpha_n^2]}}, \quad s = 1,2,\dots,m, \qquad [7.56a]$$

or in the dimensionless form:

$$\Omega_{Bncc,s} = \sqrt{\frac{(n\pi)^4 + 2K[1+\xi^2(n\pi)^2](1-\cos\varphi_{cc,s})}{[1+\xi^2(n\pi)^2]}}, \quad s = 1,2,\dots,m, \qquad [7.56b]$$

where:

$$\Omega_{Bncc,s} = \omega_{ncc,s}L^2\sqrt{\frac{\rho A}{EI}}, \quad K = \frac{kL^4}{EI}, \quad \xi = \frac{e_0 a}{L},$$

are the dimensionless parameters.

When an axial compressive load P applied to each nanobeam of MNBS reaches a certain critical value, MNBS becomes unstable and we can consider that the system begins to buckle. Introducing $\omega_{Bn} = 0$ into equation [7.55] yields:

$$P_{Bcc,s}^b = \frac{EI\alpha_n^4 + 2v_n(1-\cos\varphi_{cc,s})}{\alpha_n^2[1+(e_0 a)^2\alpha_n^2]}, \quad s = 1,2,\dots,m, \qquad [7.57a]$$

or in the dimensionless form:

$$N_{cc,s}^b = \frac{(n\pi)^4 + 2K[1+\xi^2(n\pi)^2](1-\cos\varphi_{cc,s})}{(n\pi)^2[1+\xi^2(n\pi)^2]}, \quad s = 1,2,\dots,m, \qquad [7.57b]$$

where:

$$N_{cc,s}^b = \frac{P_{Bcc,s}^b L^2}{EI}, \quad K = \frac{kL^4}{EI}, \quad \xi = \frac{e_0 a}{L},$$

are the dimensionless parameters.

From equation [7.57a], the value of critical buckling load corresponding to the minimum mode n of MNBS can be obtained by:

$$P_{Bcc,s}^b = \min_{\substack{n=1,2,3,\dots \\ s=1,2,3,\dots m}} \frac{EI\alpha_n^4 + 2v_n(1-\cos\varphi_{cc,s})}{\alpha_n^2[1+(e_0 a)^2\alpha_n^2]}, \quad s = 1,2,\dots,m. \qquad [7.57c]$$

7.2.2.2. *Free-chain system*

Here, we consider MNBS coupled in the "free-chain" system, which differs from the previous case only in a manner of coupling of the first and last nanobeam with a fixed base. In this case, stiffness k_0 and k_m are equal to zero. It is assumed that all nanobeams are identical and coupled by the elastic layers of same stiffness. The system of partial differential equations [7.44] can be reduced to the system of algebraic equation as:

$$\begin{bmatrix} S_n - v_n & -v_n & 0 & \dots & 0 & 0 & 0 & \dots & 0 & 0 & 0 \\ -v_n & S_n & -v_n & \dots & 0 & 0 & 0 & \dots & 0 & 0 & 0 \\ \dots & \dots & \dots & \dots & \dots & \dots & \dots & \dots & \dots & \dots & \dots \\ 0 & 0 & 0 & \dots & S_n & -v_n & 0 & \dots & 0 & 0 & 0 \\ 0 & 0 & 0 & \dots & -v_n & S_n & -v_n & \dots & 0 & 0 & 0 \\ 0 & 0 & 0 & \dots & 0 & -v_n & S_n & \dots & 0 & 0 & 0 \\ \dots & \dots & \dots & \dots & \dots & \dots & \dots & \dots & \dots & \dots & \dots \\ 0 & 0 & 0 & \dots & 0 & 0 & 0 & \dots & 0 & S_n & -v_n \\ 0 & 0 & 0 & \dots & 0 & 0 & 0 & \dots & 0 & -v_n & S_n - v_n \end{bmatrix} \begin{Bmatrix} W_{1n} \\ W_{2n} \\ W_{3n} \\ \dots \\ W_{i-1n} \\ W_{in} \\ W_{i+1n} \\ \dots \\ W_{m-2n} \\ W_{m-1n} \\ W_{mn} \end{Bmatrix} = \begin{Bmatrix} 0 \\ 0 \\ 0 \\ \dots \\ 0 \\ 0 \\ 0 \\ \dots \\ 0 \\ 0 \\ 0 \end{Bmatrix}, \qquad [7.58]$$

where expressions S_n and v_n are defined in equations [7.49].

Based on the methodology developed in section 7.1.2.2, where the assumed solution given by equation [7.50] is entered into the $i-th$ algebraic equation of the system [7.58] yields the frequency equation:

$$S_n = 2v_n cos\varphi_{fc}, \qquad [7.59]$$

where unknown φ_{cf} can be determined based on the condition that assumed solutions [7.50] must satisfy the first and last equation of the system [7.58]. Introducing the assumed solutions $W_{1n} = Vcos\varphi_{fc} + M\,sin\varphi_{fc}$ and $W_{2n} = Vcos(2\varphi_{fc}) + M\,sin(2\varphi_{fc})$ into the first equation and $W_{m-1n} = Vcos[(m-1)\varphi_{fc}] + M\,sin[(m-1)\varphi_{fc}]$ and $W_{m-1n} = Vcos(m\varphi_{fc}) + M\,sin(m\varphi_{fc})$ into the last algebraic equation of the system [7.58], after some algebra we obtain the non-trivial solutions for the constants V and M in the following form:

$$\begin{vmatrix} 1-cos\varphi_{fc} & -sin\varphi_{fc} \\ cos[(m+1)\varphi_{cc}]-cos(m\varphi_{fc}) & sin[(m+1)\varphi_{cc}]-sin(m\varphi_{fc}) \end{vmatrix} = 0 \Rightarrow sin(m\varphi_{fc}) = 0, \quad [7.60]$$

from which, we obtain solutions for unknown $\varphi_{fc,s}$ as:

$$\varphi_{fc,s} = \frac{s\pi}{m}, \qquad s = 0,1,\dots,m-1. \quad [7.61]$$

By substituting equations [7.49] and [7.61] into equation [7.59], the frequency and stability of equation are obtained in the following form:

$$EI\alpha_n^4 - \rho A\omega_n^2[1+(e_0a)^2\alpha_n^2] - \alpha_n^2 P[1+(e_0a)^2\alpha_n^2] + 2v_n(1-cos\varphi_{fc,s}) = 0. \quad [7.62]$$

In the case when $P = 0$, the natural frequency of MNBS is:

$$\omega_{Bnfc,s} = \sqrt{\frac{EI\alpha_n^4+2v_n(1-cos\varphi_{fc,s})}{\rho A[1+(e_0a)^2\alpha_n^2]}}, \quad s = 0,1\dots,m-1, \quad [7.63a]$$

or in the dimensionless form:

$$\Omega_{Bnfc,s} = \sqrt{\frac{(n\pi)^4+2K[1+\xi^2(n\pi)^2](1-cos\varphi_{fc,s})}{[1+\xi^2(n\pi)^2]}}, \quad s = 0,1\dots,m-1, \quad [7.63b]$$

where:

$$\Omega_{nfc,s} = \omega_{nfc,s}L^2\sqrt{\frac{\rho A}{EI}}, \quad K = \frac{kL^4}{EI}, \quad \xi = \frac{e_0a}{L},$$

are the dimensionless parameters.

By setting the natural frequency $\omega_n = 0$ in equation [7.62] gives the critical buckling load as:

$$P_{Bnfc,s}^b = \frac{EI\alpha_n^4+2v_n(1-cos\varphi_{fc,s})}{\alpha_n^2[1+(e_0a)^2\alpha_n^2]}, \quad s = 0,1\dots,m-1, \quad [7.64a]$$

or in the dimensionless form:

$$N^b_{fc,s} = \frac{(n\pi)^4 + 2K[1+\xi^2(n\pi)^2](1-\cos\varphi_{fc,s})}{(n\pi)^2[1+\xi^2(n\pi)^2]}, \qquad s = 0,1\,...,m-1, \qquad [7.64b]$$

where:

$$N^b_{fc,s} = \frac{F^b_{fc,s}L^2}{EI}, \quad K = \frac{kL^4}{EI}, \quad \xi = \frac{e_0 a}{L},$$

are the dimensionless parameters.

From equation [7.64a] and assuming that the value of critical buckling load corresponds to the minimum mode n of MNBS, we can obtain:

$$F^b_{nfc,s} = \min_{\substack{n=1,2,3,....\\ s=1,2,.,m-1}} \frac{EI\alpha_n^4 + 2v_n(1-\cos\varphi_{fc,s})}{\alpha_n^2[1+(e_0 a)^2\alpha_n^2]}, \qquad s = 0,...,m-1, \quad [7.64c]$$

7.2.3. *Asymptotic analysis*

If the number of nanobeams tends to the infinity, i.e. introducing $m \longrightarrow \infty$ into equations [7.56b] and [7.57b] or equations [7.63b] and [7.64b], we get the asymptotic values of natural frequency and buckling load, respectively, as:

$$\bar{\Omega}_{n,m\to\infty} = \sqrt{\frac{(n\pi)^4}{[1+\xi^2(n\pi)^2]}}, \quad \bar{N}^b_{n,m\to\infty} = \frac{(n\pi)^2}{[1+\xi^2(n\pi)^2]}, \quad n = 1,2,3\,.... \qquad [7.65]$$

The asymptotic relations [7.65] are the lowest natural frequency and buckling load of MNBS when the number of nanobeams tends to infinity. The asymptotic values are for both cases "clamped-chain" and "free-chain" systems. Moreover, from these expressions, we can conclude that they are independent of the chain conditions in MNBS.

7.2.4. *Numerical examples and discussions*

Here, we validate the results obtained from explicit analytical expressions for natural frequencies and buckling loads of MNBS by comparing them with the corresponding results obtained with numeric methods available in the software package Wolfram Mathematica. In addition, few numerical examples are given to perform the parametric study that shows the influence of the non-local parameter, number of nanobeams and stiffness coefficients of the Winkler elastic medium on values of the natural frequencies and critical buckling loads. Both chain systems are considered and some conclusions are drawn.

7.2.4.1. *Validation of the proposed method*

To validate the analytical solutions for natural frequencies and buckling loads determined with the trigonometric method for systems composed of three or more coupled nanobeams ($m > 2$), we compare them with solutions obtained by using the numerical method available in Wolfram Mathematica to solve the system of algebraic equations. The obtained results are given in Tables 7.1 and 7.2. For this purpose, we consider the system with $m = 3, 5, 7$ CNTs represented by nanobeam models and coupled by the Winkler's type of elastic medium. Consider the single-walled carbon nanotube (SWCNT) with the following properties, the armchair type (5, 5) with thickness($h = 0.34\, nm$), Poisson's ratio ($\nu = 0.3$), mass density ($\rho = 2300\, kg/m^3$), length ($L = 20\, nm$) and Young's modulus ($E = 0.971\, TPa$). In the present analysis, the non-local parameter is taken in the range $e_0 a = 0 - 2\, nm$. The stiffness of the Winkler's elastic medium is represented by the stiffness of the springs $k = 0.01\, E$.

Tables 7.1 and 7.2 show the values of natural frequencies and buckling loads obtained from numerical simulations and those obtained via analytical method for the systems with three, five and seven ($m = 3, 5, 7$) coupled nanobeams. The results predicted by the presented trigonometric method are found to be in excellent agreement with the results obtained by numerical methods, which indicates the potential of this analytical method for the exact prediction of natural frequencies and buckling loads of MNBS. From Tables 7.3 and 7.4, it can be observed that values of the lowest natural frequency and buckling load decrease for an increase in the number of nanobeams in MNBS. Further, it can be noted that for the "clamped-chain" system these values are approaching to the asymptotic values of natural frequency and buckling load. For the "free-chain" system, it can be noted that the lowest natural frequency and buckling load of the system remain unchanged regardless of an increase in the number of nanobeams in MNBS. It should be noted that the fundamental frequency of the "free-chain" system for $s = 0$ represents at the same time the asymptotic value of frequency, which is independent of the number of nanobeams in MNBS.

The natural frequencies of the MNBS					
		"Free-Chain" system		"Clamped-Chain" system	
Number of nanobeams		N. S.* of equation [7.58] ($P = 0$)	A. S.* equation [7.63a]	N. S.* of equation [7.48] ($P = 0$)	A. S.* equation [7.56a]
$m = 3$	1	0.12903698278022	0.12903698278022	1.74764121801651	1.74764121801652
	2	2.28082372820514	2.28082372820514	3.22298979448620	3.22298979448620
	3	3.94628553852617	3.94628553852617	4.20964091150470	4.20964091150471
$m = 5$	1	0.12903698278022	0.12903698278022	1.18579204671867	1.18579204671867
	2	1.41327198885515	1.41327198885515	2.28082372820514	2.28082372820514
	3	2.68008284399964	2.68008284399963	3.22298979448622	3.22298979448620
	4	3.68679839156187	3.68679839156188	3.94628553852613	3.94628553852617
	5	4.33335767231853	4.33335767231853	4.40104801753666	4.40104801753664
$m = 7$	1	0.12903698278022	0.12903698278022	0.89782897919498	0.89782897919498
	2	1.02161816197385	1.02161816197385	1.74764121801652	1.74764121801652
	3	1.98026326006163	1.98026326006163	2.53354464905831	2.53354464905833
	4	2.84251575007402	2.84251575007399	3.22298979448624	3.22298979448620
	5	3.56306478208091	3.56306478208097	3.78899431801537	3.78899431801531
	6	4.10534820101474	4.10534820101468	4.20964091150450	4.20964091150471
	7	4.44202912663226	4.44202912663229	4.46869439040569	4.46869439040556

*N.S. - Numerical Solutions; *A.S. - Analytical Solutions;

Table 7.3. *Comparison of the analytical and numerical solutions for natural frequencies (THz) of MNBS*

The buckling load of the MNBS					
		"Free-Chain" system		"Clamped-Chain" system	
Number of nanobeams		N. S.* of equation [7.58] ($\omega_n = 0$)	A. S.* equation [7.57a]	N. S.* of equation [7.48] ($\omega_n = 0$)	A. S.* equation [7.64a]
$m = 3$	1	1.263620624434	1.263620624434	231.789022786629	231.789022786629
	2	394.795097891274	394.795097891274	788.326575158114	788.326575158113
	3	1181.858052424954	1181.858052424954	1344.864127529597	1344.864127529598
$m = 5$	1	1.263620624434	1.263620624434	106.710062154310	106.710062154310
	2	151.579269297410	151.579269297410	394.795097891274	394.795097891274
	3	545.110746564249	545.110746564250	788.326575158113	788.326575158113
	4	1031.542403751977	1031.542403751977	1181.858052424955	1181.858052424954
	5	1425.073881018818	1425.073881018817	1469.943088161915	1469.943088161917
$m = 7$	1	1.263620624434	1.263620624434	61.175220666585	61.175220666585
	2	79.207356043971	79.207356043971	231.789022786630	231.789022786629
	3	297.600849585559	297.600849585560	487.130622229753	487.130622229756
	4	613.188591432869	613.188591432863	788.326575158121	788.326575158113
	5	963.464558883356	963.464558883364	1089.522528086411	1089.522528086471
	6	1279.052300730656	1279.052300730667	1344.864127529766	1344.864127529598
	7	1497.445794272270	1497.445794272256	1515.477929649527	1515.477929649642

*N.S. - Numerical Solutions; *A.S. - Analytical Solutions;

Table 7.4. *Comparison of the analytical and numerical solutions for critical buckling loads (nN) of MNBS*

7.2.4.2. *Numerical examples*

In order to analyze the influence of the non-local parameter and stiffness coefficient on the free transverse vibration and buckling load of MNBS, we will use the dimensionless equations for natural frequencies equations [7.56b] and [7.63b] and buckling loads equations [7.57b] and [7.64b] in both cases, the "clamped-chain" and "free-chain" systems (Figures 7.9 and 7.10). Moreover, we analyzed the effects of the number of nanobeams m on the lowest dimensionless natural frequency and buckling load of MNBS coupled in "clamped-chain" system (Figure 7.11). In the following examples, we have considered the dimensionless non-local parameter in the range $\xi = 0 - 1$ and stiffness coefficient of the elastic medium in the range $K = 20 - 100$. Numerical results for natural frequencies and critical buckling loads, given in Figures 7.9–7.11, are obtained for the value of parameter $s = 1$ in the first mode $m = 1$, for both cases, the "clamped-chain" and "free-chain" system.

Figure 7.9 shows the influence of small scale and stiffness coefficient on the dimensionless natural frequency of MNBS, for different numbers of nanobeams, $m = 3, 5, 7$. It can be noted that an increase in the small-scale coefficient causes a significant reduction in natural frequency. However, we can see that an increase in the stiffness of Winkler's elastic medium causes a reduction in the influence of small-scale coefficients on natural frequencies for both cases, the "clamped-chain" and "free-chain" systems, Figures 7.9(a) and (b). The effect of an increase in the number of nanobeams m causes a decrease in the non-dimensional natural frequency.

To explore the small-scale effects on the dimensionless buckling load, the systems with three, five and seven coupled nanobeams $m = 3, 5, 7$ are considered. The dimensionless buckling loads corresponding to different values of stiffness coefficient of Winkler'a elastic medium are plotted in Figure 7.10. Figure 7.10 shows that the influence of a small-scale coefficient in an MNBS on the dimensionless buckling load is larger than on the dimensionless natural frequency. Furthermore, it can be noticed that the dimensionless buckling load is more sensitive to the small-scale coefficient for lower values of stiffness coefficient in both cases, the "clamped-chain" and "free-chain" systems (Figures 7.10(a) and (b)). An increase in the stiffness coefficient of Winkler's elastic medium causes a reduction in the influence of small-scale coefficient on the dimensionless buckling load. Furthermore, it can be observed that an increase in the number of nanobeams in MNBS causes a decrease in the dimensionless buckling load. We can conclude that an increase in the stiffness coefficient of the elastic medium causes a reduction in the influence of the number of nanobeams on the critical buckling load.

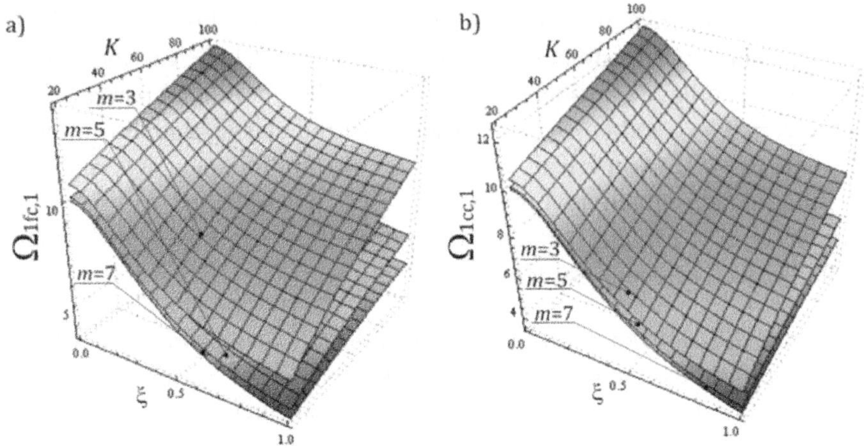

Figure 7.9. *The influence of the scale and stiffness coefficients on the dimensionless natural frequency: a) "free- chain" system and b) "clamped-chain" system, for* $s = 1$. *For a color version of the figure, see www.iste.co.uk/murmu/non-local.zip*

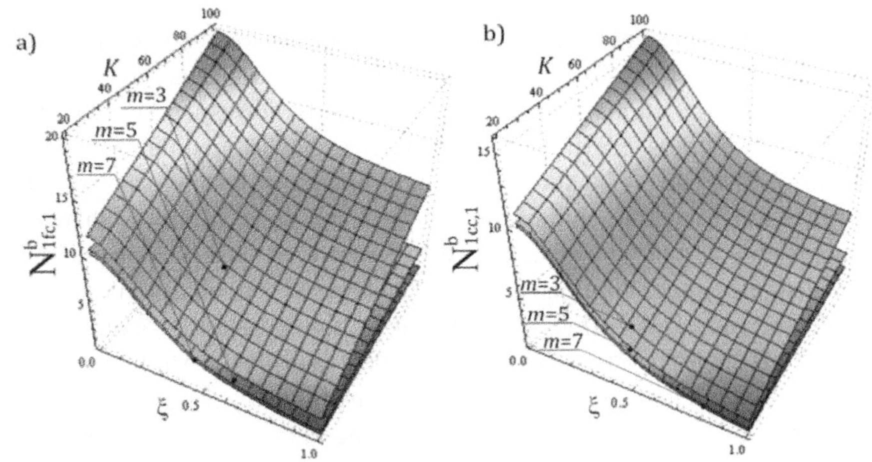

Figure 7.10. *The influence of the scale and stiffness coefficients on the dimensionless buckling load: a) "free-chain" system and b) "clamped-chain" system, for* $s = 1$. *For a color version of the figure, see www.iste.co.uk/murmu/non-local.zip*

Now we investigate the effect of the number of nanobeams at the lowest dimensionless natural frequency and buckling load for $m = 3, 5, 10, 15, 50,$ nanobeams in MNBS are shown in Figure 7.11. For this purpose, it is assumed that the stiffness coefficient of the Winkler's elastic medium is constant $K = 20$. It can be conclude that the lowest natural frequency and buckling load of MNBS are dependent on the number of nanobeams, for the case of "clamped-chain" system. In addition, these results indicate that an increase in the number of nanobeams, the values of lowest natural frequency and buckling load tend to the fundamental natural frequency and critical buckling load, equation [7.65], which are independent of the number of nanobeams. Thus, it can be concluded that the lowest natural frequency and buckling load represent the fundamental natural frequency and critical buckling load of MNBS when the number of nanobeams tends to the infinity $m \to \infty$.

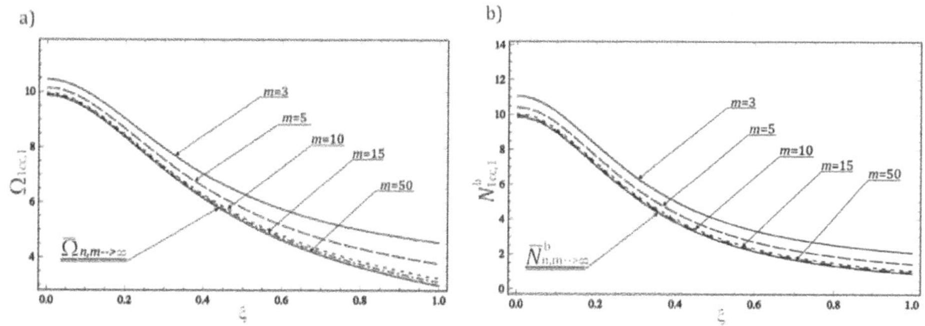

Figure 7.11. *Influence of the number of nanobeams and scale coefficient at the lowest value of: a) dimensionless natural frequency and b) dimensionless buckling load, for K=20*

7.3. Transversal vibration and buckling of the multi-nanoplate system

Analogously to the plate structures widely used in macroengineering practice, there are also two-dimensional nanostructures called nanoplates. Such nanostructures are synthesized from various types of new nanomaterials such as graphene sheets [GEI 07], gold nanoplates [XIE 07], silver nanoplates [HUA 11] and boron-nitride sheets [PAC 08]. Due to their exceptional characteristics, nanoplates are convenient for possible application in nanoelectromechanical systems (NEMS), nanooptomechanical systems (NOMS), nanocomposites, nanosensors, nanoactuators and biomedical systems [ZHU 10]. A special class of nanostructure systems are multi-nanoplate systems (MNPS) bounded with certain type of medium, which can be modeled using the same phenomenological assumptions as in the case of classical plate systems [ONI 00]. Experimental investigations [WON 97] and atomistic simulations [HAN 11] have shown that the small-scale effects cannot be neglected in the analysis of mechanical properties of nanostructures. MD simulation is a convenient method to simulate the mechanical behavior of smaller nanostructures. However, this method can be computationally expensive for larger nanostructures and complex nanoscale systems such as nanocomposites, which are having large number of atoms. For this purpose, much effective approaches are the one that uses methods of classical continuum mechanics with necessary modifications in order to consider the effects appearing at nanoscale level. In [KIT 05, HE 05], authors have used continuum-based models of single and multi-layered graphene sheets taking into account van der Waals (vdW) interactions between sheets and using different potential models. Another way to consider small-scale effects and atomic forces in the analysis of nanostructures is application of the non-local elasticity theory of Eringen [ERI 83]. This theory has been used in numerous of studies of nanostructures for reliable and fast analysis.

Complex nanoplate systems bonded by a certain type of medium are important from both practical and theoretical point of view. Pradhan and Phadicar [PRA 09a, PRA 09b] show important results on the vibration of a single, double and multi-layer graphene sheet embedded into an elastic medium, with vdW forces considered. The authors determined natural frequencies analytically for a single and double-layer graphene sheet system, whereas for a multi-layer system the frequencies are determined numerically. Recently, Murmu and Adhikari [MUR 11] and Pouresmaeeli *et al.* [POU 12] used the non-local theory for vibration analysis of double-nanoplate systems bonded with elastic medium. For the same model, Murmu *et al.* [MUR 11] performed buckling analysis of the system under uniaxial compression and later in [MUR 13] for the system under biaxial compression.

In the following section, we analyze the free transverse vibration and buckling of MNPS by analytical method. A system of partial differential equations of motion is derived based on the Eringen's non-local elasticity theory and modification of the Kirchhoff's plate theory. Simply-supported boundary conditions are employed to all four ends of nanoplates. Further, nanoplates in MNPS are coupled with a fixed base in two manners, "clamped-chain" and "free-chain" systems. In addition, procedures of solution described in the sections 7.1.2.1 and 7.1.2.2 are used to obtain analytical solutions, i.e. method of separation variables and trigonometric method are employed to find closed-form solutions for natural frequencies and buckling loads of elastically coupled MNPS. The obtained analytical results are compared with the results obtained from numerical methods. At the end of this section, we examine the effects of the non-local parameter, stiffness of the elastic medium and number of nanoplates on natural frequencies and buckling loads of MNPS.

7.3.1. *The governing equations of motion*

Consider a set of m isotropic elastic nanoplates embedded into an elastic medium, as shown in Figure 7.1. We assume that all nanoplates are made of a same material and with same dimensions such as uniform cross-section area A, thickness h, length a, width b, same elastic modulus E, Poisson coefficient ϑ, shear modulus G and mass density ρ. In addition, we assume that elastic matrix, which is located between nanoplates, modeled with continuously distributed linear elastic springs, i.e. Winkler's type of elastic medium. The elastic layers representing the medium are of the stiffness k_i , $i = 1,2,3 \dots m$. Each nanoplate is under the influence of biaxial compression loads N_x and N_y in x and y directions (see Figures 7.12(b) and (c)). The transverse displacements of m-coupled nanoplates are denoted by $w_i(x, y, t)$, $(i = 1,2,3 \dots m)$. For all nanoplates in MNPS, we consider simply-supported boundary conditions and two types of chain systems. In the case of "clamped-chain" system, the first and last nanoplate are coupled with a fixed base by the elastic layers of stiffness k_0 and k_m, respectively, as shown in Figure 7.12(b). In the second case of "free-chain" system, the first and last nanoplate are not coupled with a fixed base and stiffness k_0 and k_m are equal to zero (see Figure 7.12(c)). The x and y are the coordinates in the directions of width and length of nanoplates, respectively, while z is the coordinate in the direction of the thickness.

According to the Kirchhoff–Love plate theory, displacement components (u_{ix}, v_{iy}, w_{iz}) are defined for an arbitrary point of the middle surface of i-th nanoplate as:

$$u_{ix} = -z\frac{\partial w_i}{\partial x}, \quad v_{iy} = -z\frac{\partial w_i}{\partial y}, \quad w_{iz} = w_i(x, y, t), \qquad [7.66]$$

where u_{ix} and v_{iy} are the in-plane rotation displacements of nanoplates and w_{iz} is the transverse displacement. Based on the given displacement field and assuming small deformations, we can obtain non-zero strain–displacement relations as:

$$\varepsilon_{xx} = -z\frac{\partial^2 w_i}{\partial x^2}, \quad \varepsilon_{yy} = -z\frac{\partial^2 w_i}{\partial y^2}, \quad \gamma_{xy} = -2z\frac{\partial^2 w_i}{\partial x \partial y}, \qquad [7.67]$$

where ε_{xx} and ε_{yy} are the normal strains and γ_{xy} is the shear strain. Using the D' Alembert's principle, we get the equilibrium equations expressed through stress resultant terms in the following form:

$$q + N_x\frac{\partial^2 w_i}{\partial x^2} + N_y\frac{\partial^2 w_i}{\partial y^2} + \frac{\partial Q_x}{\partial x} + \frac{\partial Q_y}{\partial y} = \rho h\frac{\partial^2 w_i}{\partial t^2}, \qquad [7.68a]$$

$$\frac{\partial M_x}{\partial x} + \frac{\partial M_{xy}}{\partial y} = Q_x, \qquad [7.68b]$$

$$\frac{\partial M_y}{\partial y} + \frac{\partial M_{xy}}{\partial x} = Q_y. \qquad [7.68c]$$

Introducing expressions [7.68a] and [7.68b] into the first equilibrium equation [7.68a] leads to the governing equation of motion of i-th nanoplate in the form:

$$q - N_x\frac{\partial^2 w_i}{\partial x^2} - N_y\frac{\partial^2 w_i}{\partial y^2} + \frac{\partial^2 M_x}{\partial x^2} + \frac{\partial^2 M_y}{\partial y^2} + 2\frac{\partial^2 M_{xy}}{\partial x \partial y} = \rho h\frac{\partial^2 w_i}{\partial t^2}, \qquad [7.69]$$

where M_x and M_y are the bending moments and M_{xy} is the twisting moment:

$$(M_x, M_y, M_{xy}) = \int_{-\frac{h}{2}}^{\frac{h}{2}}(\sigma_{xx}, \sigma_{yy}, \tau_{xy})\, z dz. \qquad [7.70]$$

External load from Winkler elastic medium is defined as:

$$q = F_{ki} - F_{ki-1} = k_i(w_{i+1} - w_i) - k_{i-1}(w_i - w_{i-1}). \qquad [7.71]$$

By substituting equation [7.67] into the non-local constitutive relation defined in Chapter 4, equation [4.3], and then using expression [7.70] leads to:

$$[1 - (e_0a)^2\,\Delta]M_x = -D\left(\frac{\partial^2 w_i}{\partial x^2} + \vartheta\frac{\partial^2 w_i}{\partial y^2}\right), \qquad [7.72a]$$

$$[1 - (e_0a)^2\,\Delta]M_y = -D\left(\frac{\partial^2 w_i}{\partial y^2} + \vartheta\frac{\partial^2 w_i}{\partial x^2}\right), \qquad [7.72b]$$

$$[1 - (e_0a)^2\,\Delta]M_{xy} = -D(1 - \vartheta)\frac{\partial^2 w_i}{\partial x \partial y}, \qquad [7.72c]$$

where $D = Eh^3/[12(1 - \vartheta^2)]$ is the bending rigidity of a nanoplate.

Finally, using equations [7.69] and [7.72], we obtain the partial differential equations of motion in terms of transversal displacements as:

$$\rho h\frac{\partial^2 w_i}{\partial t^2} + k_i(w_i - w_{i+1}) + k_{i-1}(w_i - w_{i-1})$$

$$+D\left(\frac{\partial^4 w_i}{\partial x^4} + \frac{\partial^4 w_i}{\partial y^4} + 2\frac{\partial^2 w_i}{\partial x^2 \partial y^2}\right) + N_x\frac{\partial^2 w_i}{\partial x^2} + N_y\frac{\partial^2 w_i}{\partial y^2}$$

$$= (e_0a)^2\frac{\partial^2}{\partial x^2}\left(\rho h\frac{\partial^2 w_i}{\partial t^2} + k_i(w_i - w_{i+1}) + k_{i-1}(w_i - w_{i-1}) + N_x\frac{\partial^2 w_i}{\partial x^2} + N_y\frac{\partial^2 w_i}{\partial y^2}\right)$$

$$+(e_0a)^2\frac{\partial^2}{\partial y^2}\left(\rho h\frac{\partial^2 w_i}{\partial t^2} + k_i(w_i - w_{i+1}) + k_{i-1}(w_i - w_{i-1}) + N_x\frac{\partial^2 w_i}{\partial x^2} + N_y\frac{\partial^2 w_i}{\partial y^2}\right), \qquad [7.73]$$

for $i = 1,2,3 \ldots, m$.

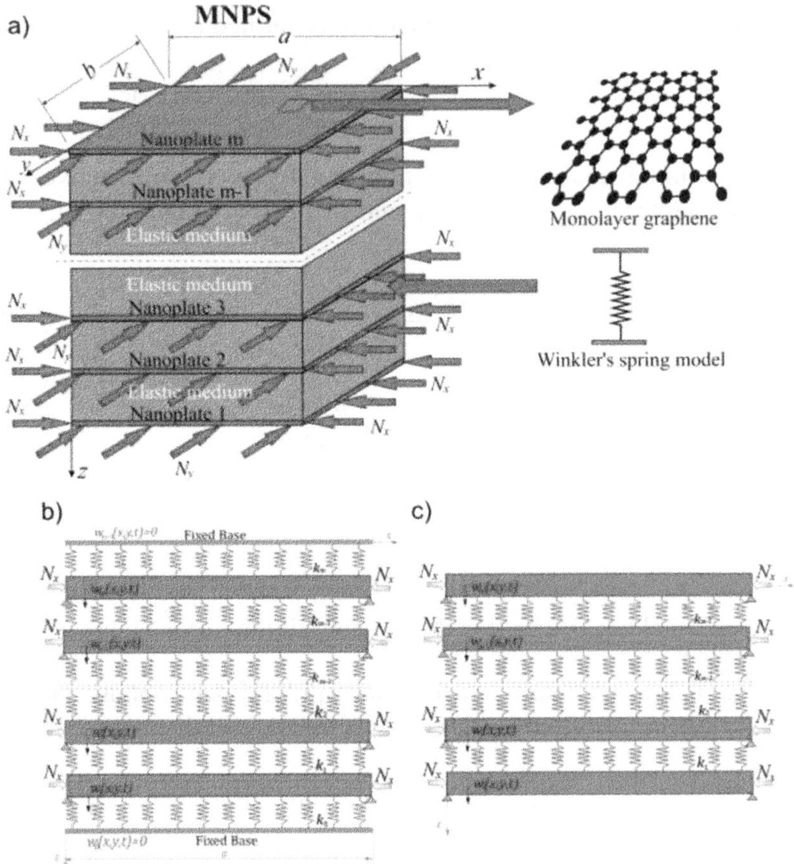

Figure 7.12. *Multi-nanoplate system: a) schematic diagram of graphene sheets embedded into an elastic medium; b) clamped-chain system and; c) free-chain system. For a color version of the figure, see www.iste.co.uk/murmu/non-local.zip*

Assuming that all rectangular nanoplates in MNPS are simply supported, we can write the following:

$$w_i(x,0,t) = w_i(x,b,t) = 0, \quad w_i(0,y,t) = w_i(a,y,t) = 0, \quad i = 1,2,3,..,m. \qquad [7.74a]$$

$$M_{xi}(0,y,t) = M_{xi}(a,y,t) = 0, \quad M_{yi}(x,0,t) = M_{yi}(x,b,t) = 0. \qquad [7.74b]$$

7.3.2. *Exact solutions*

An analytical solution for MNPS with simply-supported nanoplates can be obtained by introducing double trigonometric infinite series known as Navier's solution (see [IIJ 91]). In order to solve the equations of motion of MNPS, we assume the solution by the following expansions for generalized displacements $w_i(x,t)$:

$$w_i(x,y,t) = \sum_{r=1}^{\infty}\sum_{n=1}^{\infty} W_{irn}\, sin\, \alpha_r x\, sin\, \beta_n y\, e^{j\omega_{rn}t}\,,$$
$$i = 1,2,3,..,m, \qquad [7.75]$$

where $j = \sqrt{-1}, \alpha_r = \frac{r\pi}{a}; \beta_n = \frac{n\pi}{b}$ $(r,n = 1,2,3,\ldots)$; W_{irn} are the amplitude functions and ω_{rn} denotes the natural frequencies. Substituting the expressions for the assumed solution [7.75] into the equation of motion [7.73], we obtain the system of *m* algebraic equations of the form:

$$-v_{i-1rn}W_{i-1rn} + S_{irn}W_{irn} - v_{irn}W_{i+1rn} = 0, \quad i = 1,2,3,\ldots,m, \qquad [7.76]$$

where:

$$S_{irn} = -\rho h\omega_{rn}^2\eta_{rn} + D(\alpha_r^2 + \beta_n^2)^2 - N_x(\alpha_r^2 + \delta\beta_n^2)\eta_{rn} + v_{irn} + v_{i-1rn}, \qquad [7.77a]$$

$$v_{irn} = k_i\eta_{rn}, \qquad [7.77b]$$

$$v_{i-1rn} = k_{i-1}\eta_{rn}, \qquad [7.77c]$$

$$\eta_{rn} = 1 + (e_0a)^2(\alpha_r^2 + \beta_n^2). \qquad [7.77d]$$

in which $\delta = N_y/N_x$ defines the relation between biaxial compression loads N_x and N_y.

7.3.2.1. *Clamped-chain system*

Let us first consider MNPS with *m* identical nanoplates embedded into the Winkler's elastic medium and coupled in the "clamped-chain" system, where the first and last nanoplate i are elastically connected with a fixed base. Mathematical expressions for coupling conditions of the "clamped-chain" system are:

$$k_0 \neq 0, k_m \neq 0 \quad \text{and} \quad W_{1rn} = W_{m+1rn} = 0. \qquad [7.78]$$

Introducing expression [7.78] into equation [7.76] and assuming that stiffness of elastic layers between nanoplates are the same $k_0 = k_1 = \dots = k_m = k$, we obtain the system of algebraic equations in the matrix form:

$$\begin{bmatrix} S_{rn} & -v_{rn} & 0 & \dots & 0 & 0 & 0 & \dots & 0 & 0 & 0 \\ -v_{rn} & S_{rn} & -v_{rn} & \dots & 0 & 0 & 0 & \dots & 0 & 0 & 0 \\ \dots & \dots & \dots & \dots & \dots & \dots & \dots & \dots & \dots & \dots & \dots \\ 0 & 0 & 0 & \dots & S_{rn} & -v_{rn} & 0 & \dots & 0 & 0 & 0 \\ 0 & 0 & 0 & \dots & -v_{rn} & S_{rn} & -v_{rn} & \dots & 0 & 0 & 0 \\ 0 & 0 & 0 & \dots & 0 & -v_{rn} & S_{rn} & \dots & 0 & 0 & 0 \\ \dots & \dots & \dots & \dots & \dots & \dots & \dots & \dots & \dots & \dots & \dots \\ 0 & 0 & 0 & \dots & 0 & 0 & 0 & \dots & 0 & S_{rn} & -v_{rn} \\ 0 & 0 & 0 & \dots & 0 & 0 & 0 & \dots & 0 & -v_{rn} & S_{rn} \end{bmatrix} \begin{Bmatrix} W_{1rn} \\ W_{2rn} \\ W_{3rn} \\ \dots \\ W_{i-1rn} \\ W_{irn} \\ W_{i+1rn} \\ \dots \\ W_{m-2rn} \\ W_{m-1rn} \\ W_{mrn} \end{Bmatrix} = \begin{Bmatrix} 0 \\ 0 \\ 0 \\ \dots \\ 0 \\ 0 \\ 0 \\ \dots \\ 0 \\ 0 \\ 0 \end{Bmatrix}, \quad [7.79]$$

where:

$$S_{rn} = -\rho h \omega_{rn}^2 \eta_{rn} + D(\alpha_r^2 + \beta_n^2)^2 - N_x(\alpha_r^2 + \delta\beta_n^2)\eta_{rn} + 2v_{rn}, \quad [7.80a]$$

$$v_{rn} = k\eta_{rn}, \quad [7.80b]$$

The closed-form solutions for natural frequencies and buckling loads are obtained by using the trigonometric method as shown in section 7.1.21. In addition, it should be noted that the analytical solution of the homogeneous system of algebraic equations [7.79] is possible only under the assumption that the system is composed of a set of *m* identical nanoplates and identical elastic layers. Based on the methodology presented in section 7.1.2.1, we assumed the solution of $i-th$ algebraic equation in the form:

$$W_{irn} = V\cos(i\varphi_{cc}) + M\sin(i\varphi_{cc}),\ i = 1,2,3,\dots,m. \quad [7.81]$$

By substituting the assumed solution [7.81] into the $i-th$ algebraic equation of system [7.79], where the constants V and M are not simultaneously equal to zero, we obtain the system of two algebraic equations:

$$V\{-v_{rn}cos[(i-1)\varphi_{cc}]+S_{rn}cos(i\varphi_{cc})-v_{rn}cos[(i+1)\varphi_{cc}]\}=0,$$

$$i=2,3,\dots,m-1, \quad [7.82a]$$

$$M\{-v_{rn}sin[(i-1)\varphi_{cc}]+S_{rn}sin(i\varphi_{cc})-v_{rn}sin[(i+1)\varphi_{cc}]\}=0,$$

$$i=2,3,\dots,m-1, \quad [7.82b]$$

where after some simplifications, we obtain frequency and stability equation in the form:

$$S_{rn}=2v_{rn}cos\varphi_{cc}. \quad [7.83]$$

In order to have an oscillatory behavior of the system, we have the following conditions $V\neq 0$ and $cos(i\varphi_{cc})\neq 0$ or $M\neq 0$ and $sin(i\varphi_{cc})\neq 0$, for $i=2,3,\dots,m-1$. The unknown parameter φ_{cc} is determined from the first and last equation of the system of algebraic equations [7.79], i.e. from the chain conditions. Introducing assumed solutions equation [7.81], i.e. $W_{1rn}=Vcos\varphi_{cc}+Msin\varphi_{cc}$ and $W_{2rn}=Vcos(2\varphi_{cc})+Msin(2\varphi_{cc})$ into the first equation and $W_{m-1rn}=Vcos[(m-1)\varphi_{cc}]+Msin[(m-1)\varphi_{cc}]$ and $W_{m-1rn}=Vcos(m\varphi_{cc})+Msin(m\varphi_{cc})$ into the last equation of the system [7.79], we can obtain unknown φ_{cc} from non-trivial solutions of the determinant:

$$\begin{vmatrix} 1 & 0 \\ cos[(m+1)\varphi_{cc}] & sin[(m+1)\varphi_{cc}] \end{vmatrix}=0 \;\Rightarrow\; sin[(m+1)\varphi_{cc}]=0, \quad [7.84]$$

which yields $\varphi_{cc,s}$ in the following form:

$$\varphi_{cc,s}=\frac{s\pi}{m+1}, \qquad s=1,2,\dots,m. \quad [7.85]$$

Introducing expression [7.85] and equation [7.80] into equation [7.83], we get the frequency and stability equation as:

$$-\rho h\omega_{rn}^2\eta_{rn}+D(\alpha_r^2+\beta_n^2)^2-N_x(\alpha_r^2+\delta\beta_n^2)\eta_{rn}$$
$$+2v_{rn}\left(1-cos\varphi_{cc,s}\right)=0. \quad [7.86]$$

For the case when $N_x = 0$, from equation [7.86] we obtain the natural frequency of elastic MNPS as:

$$\omega_{rncc,s} = \sqrt{\frac{D\left(\alpha_r^2+\beta_n^2\right)^2+2v_{rn}\left(1-\cos\varphi_{cc,s}\right)}{\rho h\eta_{rn}}}, \quad s = 1,2,\dots,m, \qquad [7.87]$$

When a biaxial load applied to each nanoplate of the MNPS reaches a certain critical value, the MNPS becomes unstable and we can consider that system begins to buckle. Introducing $\omega_{rn} = 0$ into equation [7.86] yields:

$$N_{x,rncc,s} = \frac{D\left(\alpha_r^2+\beta_n^2\right)^2+2v_{rn}\left(1-\cos\varphi_{cc,s}\right)}{\left(\alpha_r^2+\delta\beta_n^2\right)\eta_{rn}}, \qquad s = 1,2,\dots,m, \qquad [7.88]$$

Introducing dimensionless parameters:

$$\Omega_{rn} = \omega_{rn}a^2\sqrt{\frac{\rho h}{D}}, \quad \widehat{N} = N_x\frac{a^2}{D}, \quad R = \frac{a}{b}, \qquad K_i = k_i\frac{a^4}{D}, \qquad \xi = \frac{(e_0 a)}{a}, \qquad [7.89]$$

we can rewrite equations [7.87] and [7.88] in dimensionless form as:

$$\Omega_{rncc,s} = \sqrt{\frac{[(r\pi)^2+R^2(n\pi)^2]^2+2K\left(1+\xi^2(r\pi)^2+\xi^2R^2(n\pi)^2\right)\left(1-\cos\varphi_{cc,s}\right)}{1+\xi^2(r\pi)^2+\xi^2R^2(n\pi)^2}},$$

$$s = 1,2,\dots,m, \qquad [7.90]$$

and

$$\widehat{N}_{rncc,s} = \frac{\left[(r\pi)^2+R^2(n\pi)^2\right]^2+2K\left(1+\xi^2(r\pi)^2+\xi^2R^2(n\pi)^2\right)\left(1-\cos\varphi_{cc,s}\right)}{[(r\pi)^2+\delta R^2(n\pi)^2](1+\xi^2(r\pi)^2+\xi^2R^2(n\pi)^2)},$$

$$s = 1,2,\dots,m, \qquad [7.91]$$

7.3.2.2. *Free-chain system*

Here, we consider the system where the first and last nanoplate in MNPS are not coupled with a fixed base. The coupling conditions of MNPS coupled in "free-chain" system are defined as:

$$k_0 = 0 \quad \text{and} \, k_m = 0. \qquad [7.92]$$

Introducing coupling conditions equation [7.92] into the system of algebraic equations [7.76] and assuming that $k_0 = k_1 = \cdots = k_m = k$ yields:

$$\begin{bmatrix} S_{rn}-v_{rn} & -v_{rn} & 0 & \dots & 0 & 0 & 0 & \dots & 0 & 0 & 0 \\ -v_{rn} & S_{rn} & -v_{rn} & \dots & 0 & 0 & 0 & \dots & 0 & 0 & 0 \\ \dots & \dots & \dots & \dots & \dots & \dots & \dots & \dots & \dots & \dots & \dots \\ 0 & 0 & 0 & \dots & S_{rn} & -v_{rn} & 0 & \dots & 0 & 0 & 0 \\ 0 & 0 & 0 & \dots & -v_{rn} & S_{rn} & -v_{rn} & \dots & 0 & 0 & 0 \\ 0 & 0 & 0 & \dots & 0 & -v_{rn} & S_{rn} & \dots & 0 & 0 & 0 \\ \dots & \dots & \dots & \dots & \dots & \dots & \dots & \dots & \dots & \dots & .. \\ 0 & 0 & 0 & \dots & 0 & 0 & 0 & \dots & 0 & S_{rn} & -v_{rn} \\ 0 & 0 & 0 & \dots & 0 & 0 & 0 & \dots & 0 & -v_{rn} & S_{rn}-v_{rn} \end{bmatrix} \begin{Bmatrix} W_{1rn} \\ W_{2rn} \\ W_{3rn} \\ \dots \\ W_{i-1rn} \\ W_{irn} \\ W_{i+1rn} \\ \dots \\ W_{m-2rn} \\ W_{m-1rn} \\ W_{mrn} \end{Bmatrix} = \begin{Bmatrix} 0 \\ 0 \\ 0 \\ \dots \\ 0 \\ 0 \\ 0 \\ \dots \\ 0 \\ 0 \\ 0 \end{Bmatrix}, \qquad [7.93]$$

where expressions S_{rn} and v_{rn} are defined in equation [7.80].

Substituting equation [7.81] into the $i-th$ algebraic equation of system [7.93], we get the frequency and stability equation as:

$$S_{rn} = 2v_{rn}cos\varphi_{fc}, \qquad [7.94]$$

where φ_{fc} is an unknown parameter, which is determined from the first and last equation of the system of algebraic equations [7.93], i.e. from chain conditions of the "free-chain" system. Introducing assumed solutions for the first and second amplitude, $W_{1rn} = Vcos\varphi_{fc} + Msin\varphi_{fc}$ and $W_{2rn} = Vcos(2\varphi_{fc}) + Msin(2\varphi_{fc})$, into the first equation and $W_{m-1n} = Vcos[(m-1)\varphi_{fc}] + Msin[(m-1)\varphi_{fc}]$ and $W_{m-1n} = Vcos(m\varphi_{fc}) + Msin(m\varphi_{fc})$ into the last equation of the system [7.93], after some transformations we obtain trigonometric equations in the following form:

$$\begin{vmatrix} 1-cos\varphi_{fc} & -sin\varphi_{fc} \\ cos[(m+1)\varphi_{cc}] - cos(m\varphi_{fc}) & sin[(m+1)\varphi_{cc}] - sin(m\varphi_{fc}) \end{vmatrix} = 0 \Rightarrow$$
$$sin(m\varphi_{fc}) = 0, \qquad [7.95]$$

where unknown $\varphi_{fc,s}$ is equal to:

$$\varphi_{fc,s} = \frac{s\,\pi}{m}, \qquad s = 0,1,\ldots,m-1. \qquad [7.96]$$

Introducing the expression for parameter $\varphi_{fc,s}$ and equation [7.80] into equation [7.94], we obtain the frequency and stability equation in the following form:

$$-\rho h\omega_{rn}^2\eta_{rn} + D(\alpha_r^2+\beta_n^2)^2 - N_x(\alpha_r^2+\delta\beta_n^2)\eta_{rn} + 2v_{rn}(1-\cos\varphi_{fc,s}) = 0. \qquad [7.97]$$

For the case when $N_x = 0$, from equation [7.97] we obtain the natural frequency of elastic MNPS as:

$$\omega_{rnfc,s} = \sqrt{\frac{D(\alpha_r^2+\beta_n^2)^2+2v_{rn}(1-\cos\varphi_{fc,s})}{\rho h\eta_{rn}}}, \quad s = 1,2,\ldots,m, \qquad [7.98]$$

Introducing $\omega_{rn} = 0$ into equation [7.97] gives:

$$N_{x,rnfc,s} = \frac{D(\alpha_r^2+\beta_n^2)^2+2v_{rn}(1-\cos\varphi_{fc,s})}{(\alpha_r^2+\delta\beta_n^2)\eta_{rn}}, \qquad s = 1,2,\ldots,m, \qquad [7.99]$$

Introducing dimensionless parameters defined in equation [7.89], we can rewrite equations [7.98] and [7.99] in dimensionless form as:

$$\Omega_{rnfc,s} = \sqrt{\frac{[(r\pi)^2+R^2(n\pi)^2]^2+2K(1+\xi^2(r\pi)^2+\xi^2R^2(n\pi)^2)(1-\cos\varphi_{fc,s})}{1+\xi^2(r\pi)^2+\xi^2R^2(n\pi)^2}},$$

$$s = 1,2,\ldots,m, \qquad [7.100]$$

and

$$\hat{N}_{rnfc,s} = \frac{[(r\pi)^2+R^2(n\pi)^2]^2+2K(1+\xi^2(r\pi)^2+\xi^2R^2(n\pi)^2)(1-\cos\varphi_{fc,s})}{[(r\pi)^2+\delta R^2(n\pi)^2](1+\xi^2(r\pi)^2+\xi^2R^2(n\pi)^2)},$$

$$s = 1,2,\ldots,m. \qquad [7.101]$$

7.3.3 *Asymptotic analysis*

When the number of nanoplates in MNPS tends to the infinity, i.e. entering $m \longrightarrow \infty$ into equations [7.90] and [7.100] or equations [7.91] and [7.100], we obtain the critical natural frequency and critical buckling load, respectively, in the form:

$$\Omega_{\substack{rn \\ m\to\infty}} = \sqrt{\frac{[(r\pi)^2+R^2(n\pi)^2]^2}{1+\xi^2(r\pi)^2+\xi^2R^2(n\pi)^2}}, \qquad [7.102a]$$

$$\hat{N}_{\substack{rn \\ m\to\infty}} = \frac{\left[(r\pi)^2+R^2(n\pi)^2\right]^2}{[(r\pi)^2+\delta R^2(n\pi)^2](1+\xi^2(r\pi)^2+\xi^2R^2(n\pi)^2)}, \quad (r,n=1,2,\dots). \qquad [7.102b]$$

Finally, we can conclude that the critical natural frequency and critical buckling load represented by the expressions [7.102] are the lowest natural frequency and critical buckling load of the system when the number of nanoplates tends to the infinity. In addition, it is observed that these critical values are independent of the chain conditions and are the same for cases, "clamped-chain" and "free-chain" systems.

It can be ascertained that obtained theoretical models represent a generalization of the vibration and stability problem to the systems with multiple coupled nanoplates. In addition, this model can be extended to consider the systems with nanoplates embedded into different types of medium.

7.3.4. *Numerical results and discussions*

For numerical examples, we consider MNPS as a system of multiple graphene sheets embedded into a polymer elastic matrix. The following material properties of graphene sheets similar to those in [MUR 11] are assumed: Young's modulus $E = 1.06$ TPa, Poisson's ratio $\vartheta = 0.25$, density $\rho = 2250$ kg/m^3 and thickness $h = 0.34$ nm. Natural frequencies and critical buckling loads of MNPS obtained as closed-form analytical solutions are computed according to equations [7.90]–[7.101].

7.3.4.1. *Validation study*

To justify the proposed trigonometric solution for natural frequencies and buckling loads, we compared analytical solutions of the homogeneous system of algebraic equations against the numerical methods of solution available in Wolfram Mathematica for the same system of equations. We should note that all plotted quantities are dimensionless. For the comparison, we considered the mode $r = 1$, $n = 1$ with aspect ratio $R = 1$, axial force ratio $\delta = 0.5$, non-local parameter $\eta = 0.5$ and stiffness coefficient $K = 100$. As shown in Table 7.5 for natural frequencies and in Table 7.6 for buckling loads, compared values obtained from analytical and numerical methods are in excellent agreement. It can be noted that the lowest natural frequency of MNPS does not depend on the number of nanoplates in the free-chain system ($s = 0$). In the case of clamped-chain system, the natural frequency decreases toward the lowest natural frequency of the system for an increase in the number of nanoplates in MNPS. The obtained analytical results confirm that the trigonometric method is a reliable solution for analysis of natural frequencies and buckling load of the MNPS.

The natural frequencies of coupled MNPS					
Number of nanoplates		"Free-Chain" system		"Clamped-Chain" system	
		N. S.* of equation (31) ($\widehat{N}_{rn} = 0$)	A. S.* equation (37)	N. S.* of equation (19) ($\widehat{N}_{rn} = 0$)	A. S.* equation (29)
$m = 2$	1	8.102641359754337	8.10264135975433	12.870617584436328	12.870617584436328
	2	16.298858763876737	16.29885876387674	19.122050020978442	19.122050020978440
$m = 5$	1	8.102641359754331	8.102641359754337	9.614973543796884	9.614973543796877
	2	10.190652487932876	10.190652487932857	12.870617584436213	12.870617584436328
	3	14.277583763711931	14.277583763711986	16.298858763876922	16.298858763876740
	4	18.095750768614074	18.095750768613915	19.122050020977852	19.122050020978440
	5	20.675013806036212	20.675013806036286	20.948935003042514	20.948935003042262
$m = 10$	1	8.102641359756117	8.102641359754337	8.588026681485388	8.588026681485225
	2	8.685706289377503	8.685706289402773	9.869249740408922	9.869249740409114
	3	10.190652487940827	10.190652487932857	11.605199275155108	11.605199275141490
	4	12.169459583146475	12.169459583165843	13.511839045938950	13.511839045978320
	5	14.277583763760147	14.277583763711986	15.400968455233263	15.400968454942843
	6	16.298858763828850	16.298858763876740	17.149803632265010	17.149803633262350
	7	18.095750768947923	18.095750768613915	18.674469194094154	18.674469192059487
	8	19.575746408604815	19.575746408842146	19.915444852405102	19.915444855535080
	9	20.675013806069206	20.675013806036286	20.830350539684872	20.830350538842062
	10	21.350974222653280	21.350974222827220	21.390450946077390	21.390450947273200

*N.S.- Numerical Solutions ; *A.S.- Analytical Solutions;

Table 7.5. *Comparison of the analytical and numerical solutions for natural frequencies of elastic MNPS*

The critical buckling load of coupled MNPS					
Number of nanoplates		"Free-Chain" system		"Clamped-Chain" system	
		N. S.* of equation (31) ($\Omega_{rn} = 0$)	A. S.* equation (38)	N. S.* of equation (19) ($\Omega_{rn} = 0$)	A. S.* equation (30)
$m = 2$	1	4.434679401304418	4.434679401304419	11.189424977460272	11.189424977460272
	2	17.944170553616118	17.94417055361612	24.698916129771966	24.698916129771973
$m = 5$	1	4.4346794013044395	4.434679401304419	6.244608023513074	6.244608023513079
	2	7.014762626037857	7.014762626037962	11.189424977460298	11.189424977460272
	3	13.769508202194016	13.769508202193814	17.9441705536164	17.94417055361612
	4	22.118832905038456	22.118832905038428	24.698916129771334	24.698916129771973
	5	28.87357848119441	28.873578481194283	29.64373308371962	29.64373308371917
m = 10	1	4.43467940130375	4.434679401304419	4.981908715865823	4.981908715865823
	2	5.095880961381969	5.095880961378354	6.579263394981978	6.579263394982316
	3	7.01476262603017	7.014762626037962	9.097335262368473	9.097335262385945
	4	10.00349088834649	10.003490888311653	12.332125110984935	12.332125110929688
	5	13.769508202105083	13.769508202193814	16.02156950497541	16.021569505120507
	6	17.94417055361673	17.94417055361612	19.86677160331807	19.866771602111736
	7	22.11883290508726	22.118832905038428	23.55621599538792	23.556215996302555
	8	25.88485021879722	25.88485021892059	26.791005847718235	26.7910058448463
	9	28.873578481444355	28.873578481194283	29.309077707986063	29.309077712249927
	10	30.7924601454757	30.792460145853894	30.906432392673373	30.906432391366423

*N.S. - Numerical Solutions; *A.S. - Analytical Solutions;

Table 7.6. *Comparison of the analytical and numerical solutions for critical buckling load of elastic MNPS*

7.3.4.2. *Numerical examples*

The following plots are obtained for the next values of dimensionless parameters for nanoplates in an MNPS: aspect ratio $R = 1$, axial force ratio $\delta = 0.5$, non-local parameter in the range $\eta = 0 - 1$, stiffness coefficient in the range $K = 0 - 100$ and$s = 1$ and mode numbers $r = 1,\ n = 1$. Figure 7.13(a) illustrates the changes in dimensionless natural frequencies $\Omega_{rn,s}$of MNPS for a "clamped-chain" system due to changes in non-local parameter and stiffness of the elastic medium and different numbers of nanoplates in the system. It can be observed that an increase in the non-local parameter causes a significant decrease in natural frequency, while an increase in dimensionless stiffness coefficient causes a slight increase in frequency. When non-local parameter is equal to zero, we have natural frequencies of the local system, i.e. classical continuum case, which yields overdetermined values of natural frequency. Furthermore, an increase in the number of nanoplates in an MNPS gives values of natural frequency that approach the fundamental frequency of the MNPS obtained from asymptotic analysis. The effect of the non-local parameter is much more pronounced and we have significant decrease in natural frequencies for an increase in the parameter. In addition, an increase in the elastic medium stiffness leads to an increase in the natural frequency, which is attributed to an increase in overall stiffness of MNPS. Figure 7.13(b) illustrates the effects of changes in non-local parameter, stiffness coefficient and number of nanoplates on natural frequencies of the "free-chain" MNPS. It is obvious that natural frequencies significantly decrease for an increase in the non-local parameter and slightly increase for an increase in the stiffness coefficient. An increase in the number of nanoplates causes a decrease in natural frequencies, but this effect is less pronounced for higher values of the elastic medium stiffness.

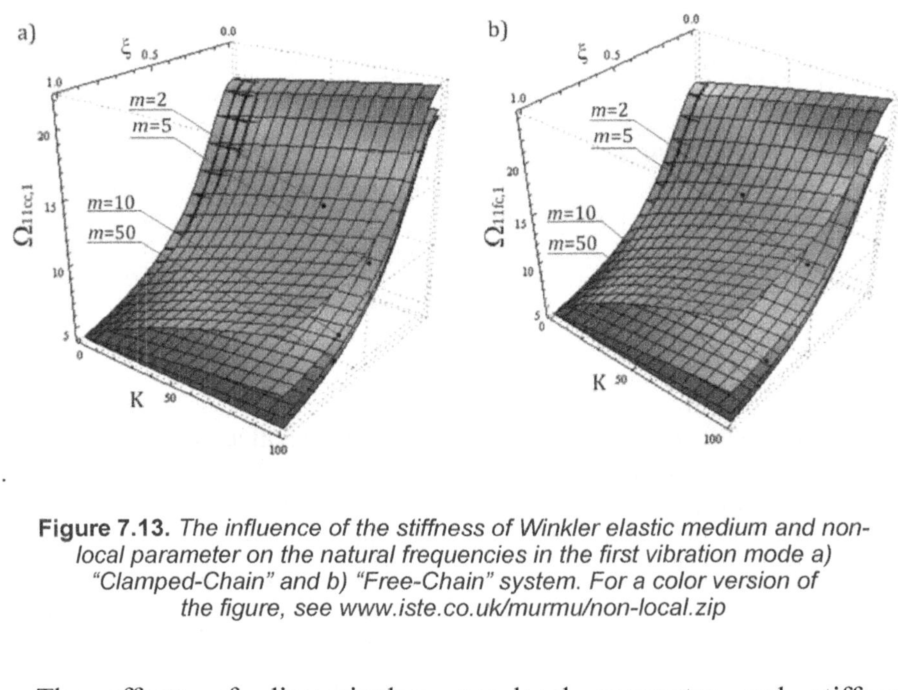

Figure 7.13. *The influence of the stiffness of Winkler elastic medium and non-local parameter on the natural frequencies in the first vibration mode a) "Clamped-Chain" and b) "Free-Chain" system. For a color version of the figure, see www.iste.co.uk/murmu/non-local.zip*

The effects of dimensionless non-local parameter and stiffness coefficients on values of critical buckling load of an MNPS are shown in Figure 7.14(a) for "clamped-chain" and in Figure 7.14(b) for the "free-chain" system. The effect of change of the number of nanoplates in MNPS is also demonstrated. We analyze only the first buckling mode ($r = 1, n = 1$). Obviously, the buckling load is significantly influenced by the dimensionless non-local parameter, i.e. an increase in the non-local parameter decreases the value of critical buckling load. This effect is more pronounced for higher values of the stiffness coefficient of elastic medium. An increase in the stiffness of medium in the range 0–100 causes a slight increase in critical buckling load. This increase is more pronounced in the case of "free-chain" MNPS (Figure 7.3(b)) than for the "clamped-chain" MNPS (Figure 7.14(a)).

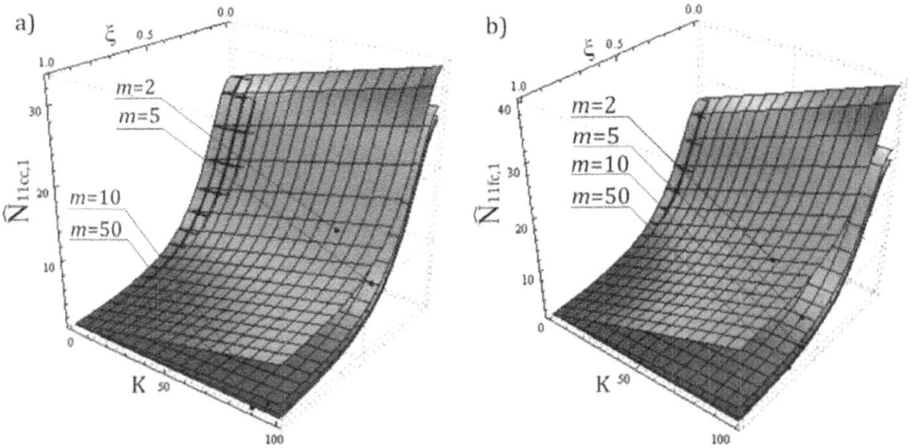

Figure 7.14. *The influence of the stiffness of the Winkler elastic medium and non-local parameter on the buckling load in the first buckling mode a) "clamped-chain" and b) "free-chain" system. For a color version of the figure, see www.iste.co.uk/murmu/non-local.zip*

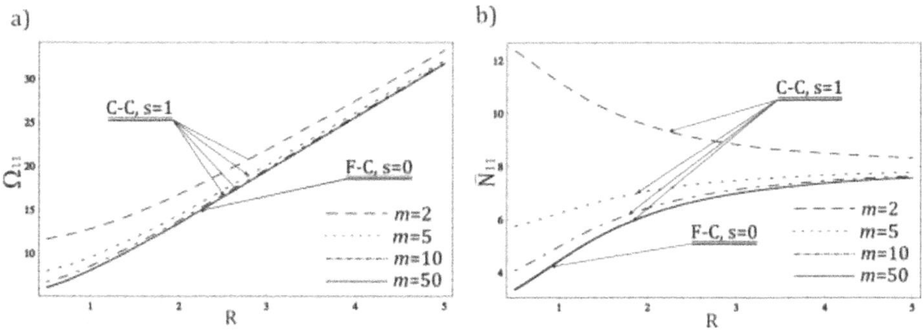

Figure 7.15. *The influence of the aspect ratio (R) in the first mode and different numbers of nanoplates on a) natural frequencies and b) critical buckling load*

From Figures 7.15(a) and (b), we can observe the influence of an increase in aspect ratio on natural frequencies and critical buckling load for different numbers of nanoplates in MNPS and both "clamped-chain" and "free-chain" systems. In both figures, curves are plotted for the same values of parameters as in Tables 7.5 and 7.6. It is obvious that there is a significant difference between "clamped-chain" and "free-chain" systems when MNPS is composed of a small number of nanoplates. The natural frequency and critical buckling load plotted for s=0 represent a fundamental natural frequency and critical buckling load in the case of "free-chain" system, which are independent of a number of nanoplates in MNPS. The previous conclusion is in line with the results presented in Tables 7.5 and 7.6 for the "free-chain" system. In the case of "clamped-chain" system, the lowest natural frequency and critical buckling load are plotted for $s = 1$. It can be observed that an increase in a number of nanoplates in an MNPS causes a decrease in natural frequencies and critical buckling loads toward the value of the lowest natural frequency and the lowest critical buckling load of the "clamped-chain" system, i.e. these two values tend to the fundamental natural frequency and critical buckling load of the "free-chain" systems. This behavior is also proved in the asymptotic analysis (equation [7.102]). If we increase the aspect ratio (R) this lead to an increase in natural frequencies of the MNPS. This is in line with other results in the literature where an increase in any dimension of a structure leads to a decrease in the influence of small-scale parameter. From Figure 7.15(b), it can be seen that for the lower numbers of nanoplates in "clamped-chain" system the value of critical buckling load decreases for an increase in the aspect ratio. Hence, for higher numbers of nanoplates in an MNPS, the value of critical buckling load slightly increases for an increase in the aspect ratio.

7.4. Summary

An analytical approach for studying the non-local vibration and stability of complex multi-nanostructures systems (CMNSSs) embedded into an elastic medium is presented. The non-local theory applied for CMNSSs is a generalized theory and can be used for the vibration and stability analysis of different complex structure systems at nanoscales such as MNRSs, MNBSs and MNPSs. The mathematical model of CMNSS is consisting of a set of m homogeneous partial differential equations that have been derived by using D'Alembert's principle and non-local elastic constitutive relations. Explicit closed-form expressions for the natural frequencies are derived successfully by using the method of separations of variables and the trigonometric method for clamped-clamped (C-C) and clamped-free (C-F) boundary conditions of the MNRS and simply supported for MNBS and MNPS. The influence of coupling conditions of the "clamped-chain" and "free-chain" system is analyzed for all presented CMNSSs. The obtained analytical results are validated by comparison with the results for natural frequencies obtained by using numerical method for cases with two and more coupled nanostructures elements in a CMNSS. Novel analytical expressions for critical values of natural frequencies and critical buckling load are obtained by asymptotic analysis for the case when the number of nanostructure elements in a CMNSS tends to infinity. Based on the numerical results presented in Tables 7.1 – 7.6, it can be concluded that the results obtained by analytical and numerical methods are in excellent agreement. Comparing the obtained values of natural frequencies for two chain systems, it is found that the lowest natural frequency predicted by the "free-c hain" system represents the fundamental frequency of a CMNSS, which is independent of the number of nanostructures elements in a CMNSS. The effects of non-local parameters, stiffness coefficients and the number of nanostructure elements in CMNSSs on natural frequencies and buckling load are also analyzed. It has been found that an increase in the non-local parameter (e_0a) and a number of nanostructures elements m reduce the system natural frequency. Also, it is noted that an increase in the stiffness coefficient causes an increase in the natural frequency. Similar conclusions can be drawn for the influence on the buckling load of the MNBS and MNPS. Furthermore, it is found that the effect of change of nanoplate aspect ratio is more evident on natural frequencies than on the critical buckling load in the MNPS. This theoretical study may be useful in practical design of nanodevices such as nanoresonators and for future investigations on other multiple-nanostructure systems embedded into different types of medium.

So far, continuum analytical theory has been used. In the next chapter, a more general finite element approach has been developed.

8

Finite Element Method for Dynamics of Non-local Systems

The scope of our analysis has been restricted to continuum-based analytical approaches so far in this book. Here, the finite element approach is developed to analyze more general nanostructures.

This chapter introduces the idea of non-local normal modes arising in the dynamic analysis of nanoscale structures. A non-local finite element approach is used for the axial vibration of nanorods, bending vibration of nanobeams and transverse vibration of nanoplates. Explicit expressions of the element mass and stiffness matrices are derived in closed-form as functions of a length-scale parameter. In general, the mass matrix can be expressed as a sum of the classical local mass matrix and a non-local part. The non-local part of the mass matrix is scale-dependent and vanishes for systems with larger length-scales. New analytical methods are developed to understand the dynamic behavior of discrete non-local systems in light of classical local systems. The conditions for the existence of classical normal modes for undamped and damped non-local systems are established. Closed-form approximate expressions of non-local natural frequencies, modes and frequency response functions are derived. The results derived in this chapter are illustrated using examples of axial and bending vibration of nanotubes and transverse vibration of graphene sheets.

8.1. Introduction

Nanoscale systems, such as those fabricated from simple and complex nanorods, nanobeams [WON 97] and nanoplates, have attracted keen interest among scientists and engineers. Examples of one-dimensional nanoscale objects include (nanorod and nanobeam) carbon nanotubes [IIJ 93], zinc oxide (ZnO) nanowires and boron nitride (BN) nanotubes, while two-dimensional nanoscale objects include graphene sheets [WAR 09] and BN nanosheets [PAC 08]. These nanoscale entities or nanostructures are found to have exciting mechanical, chemical, electrical, optical and electronic properties. Nanostructures are being used in the field of nanoelectronics, nanodevices, nanosensors, nano-oscillators, nanoactuators, nanobearings, and micromechanical resonators, transporters of drugs, hydrogen storage, electrical batteries, solar cells, nanocomposites and nano-optomechanical systems (NOMS). Understanding the dynamics of nanostructures is crucial for the development of future generation applications in these areas.

Experiments at the nanoscale can be difficult as many parameters need to be taken care of. However, atomistic computation methods such as molecular dynamic (MD) simulations [BRO 07] are computationally prohibitive for nanostructures with large numbers of atoms. Thus, continuum mechanics is an important tool for modeling, understanding and predicting physical behavior of nanostructures. Although continuum models based on classical elasticity are able to predict the general behavior of nanostructures, they lack the accountability of effects arising from the small scale. At small scales, the theory and laws of classical elasticity may not hold. Consequently, for accurate predictions, the employability of the classical continuum models has been questioned in the analysis of nanostructures and nanoscale systems. To address this, size-dependent continuum-based methods [AKG 11, AKG 12, JOM 11, KAH 10] are gaining popularity in the modeling of small-sized structures as they offer much faster solutions than MD simulations for various nanoengineering problems. Currently, research efforts are underway to bring in size effects within the formulation by modifying the traditional classical mechanics. One popular size-dependent theory is the non-local elasticity theory pioneered by Eringen [ERI 83], and applied to nanotechnology by Peddieson *et al.* [PED 03]. The theory of non-local elasticity (non-local continuum mechanics) is being increasingly used for efficient analysis of nanostructures, e.g. nanorods [AYD 09, AYD 12], nanobeams [MUR 12a], nanoplates [AKS 11, BAB 11], nanorings [WAN 08], carbon nanotubes [ART 11, AYD 11], graphenes [ANS 11, MUR 09], nanoswitches [YAN 08] and microtubules [HEI 10]. Non-local elasticity accounts for the small-scale effects at the atomistic level. At nanometer scales, size effects often become prominent. Both experimental and atomistic simulation results have shown a significant size effect in the mechanical properties when the dimensions of these structures become small [KIA 98, TAN 08]. In non-local elasticity theory, the small-scale effects are captured by assuming the stress at a point as a function of the strains at all points in the domain. Non-local theory considers long-range interatomic interaction and yields results dependent on the size of a body [ERI 83]. Some of the drawbacks of the classical continuum theory could be efficiently avoided and size-dependent phenomena can be explained by the non-local elasticity theory. A good review on non-local elasticity and application to nanostructures can be found in [ARA 12].

Several researchers have used non-local theory for dynamic analysis of continuum systems such as nanorods, nanobeams and nanoplates. Nanorods have found application in energy harvesting, light-emitting devices and microelectromechanical systems (MEMS). Using non-local elasticity, various works on mechanical behavior of nanorods [AYD 09, AYD 12, MUR 11b, MUR 10, NAR 11a] were reported. Numerous works can be found in the literature regarding analysis (mainly structural) of nanobeams using non-local elasticity [ARA 12] and coupled nanobeams [MUR 12a]. The work on nanobeams is related to carbon nanotubes, BN nanotubes and ZnO nanowires. Nanoplate models have been used to represent two-dimensional nanostructures such as graphene sheets and BN sheets. Several works on dynamics of nanoplates using non-local theory are available in the literature [PHA 10, MUR 11a].

From the literature, it is clear that significant research efforts have taken place in the analysis of nanostructures modeled as a continuum. While the results have given significant insights, the analysis is normally restricted to single-structure (e.g. a beam or a plate) with simple boundary conditions and no damping. In the future, complex nanoscale structures will be used for next-generation nanoelectromechanical systems. Therefore, it is necessary to have the ability for design and analysis of damped built-up structures. The finite element approach for nanoscale structures can provide this generality.

Work on non-local finite elements is only in the early stages. Pisano *et al.* [PIS 09] reported a finite element procedure for non-local integral elasticity. Chang [CHA 12] studied the small-scale effects on axial vibration of non-uniform and non-homogeneous nanorods by using the theory of non-local elasticity and the finite element method. Narendar and Gopalakrishnan [NAR 11b] used the concept of non-local elasticity and applied it for the development of a spectral finite element (SFE) for analysis of nanorods. Recently, Adhikari *et al.* [ADH 13] reported the free and forced axial vibrations of damped non-local rods using dynamic non-local finite element analysis. Similarly, not many works have been published on the non-local finite element formulation of nanobeams (carbon nanotubes). Phadikar and Pradhan [PHA 10] have proposed the basic finite element formulations for a non-local elastic Euler–Bernoulli beam using the Galerkin technique. Studies were carried out for bending, free vibration and buckling for non-local beams with four classical boundary conditions. Pradhan [PRA 12] updated the work of non-local finite element to Timoshenko beam theory and applied it to carbon nanotubes. With the finite element analysis, bending, buckling and vibration for non-local beams with clamped–clamped, hinged–hinged, clamped–hinged and clamped–free boundary conditions were illustrated. The basic non-local finite elements of undamped two-dimensional nanoplates (such as graphene sheets) were reported by Phadikar and Pradhan [PHA 10]. Recently, Ansari *et al.* [ANS 10] developed a non-local finite element model for vibration of embedded multi-layered graphene sheets. The proposed finite elements were based on the Mindlin-type equations of motion coupled together through the Van der Waals interaction. Vibrational characteristics of multi-layered graphene sheets with different boundary conditions embedded into an elastic medium were considered.

The majority of the reported works on non-local finite element analysis consider free vibration studies where the effect of nonlocality on the undamped eigensolutions has been studied. Damped non-local systems and forced vibration response analysis have received little attention. However, a significant body of literature is available [MEI 97, GER 97, PET 98] on finite element analysis of local dynamical systems. It is necessary to extend the idea of local modal analysis to non-local systems to gain qualitative as well as quantitative understanding. This way, the dynamic behavior of general non-local discretized systems can be explained in the light of well-known and established theories of discrete local systems. The purpose of this chapter is to make essential contributions in this open area.

The chapter is organized as follows. In section 8.2, we introduce the non-local finite element formulation for the axial vibration of rods, bending vibration of beams and transverse vibration of plates. Explicit expressions of element mass and stiffness matrices for the three systems are derived. Modal analysis of discrete non-local dynamical systems is discussed in section 8.3. The conditions for the existence of classical normal modes, approximations for non-local frequencies and modes are proposed. In section 8.4, dynamics of damped non-local systems and approximation to the frequency response function are discussed. Analytical results, including the approximations of the non-local natural frequencies and modes, are numerically illustrated for the three systems in section 8.5. In section 8.6, some conclusions are drawn based on the theoretical and numerical results obtained in this chapter.

8.2. Finite element modeling of non-local dynamic systems

8.2.1. *Axial vibration of nanorods*

The equation of motion of axial vibration for a damped non-local rod can be expressed as:

$$
\begin{aligned}
&EA\frac{\partial^2 U(x,t)}{\partial x^2} + \widehat{c}_1\left(1-(e_0a)_1^2\frac{\partial^2}{\partial x^2}\right)\frac{\partial^3 U(x,t)}{\partial x^2\partial t}\\
&= \widehat{c}_2\left(1-(e_0a)_2^2\frac{\partial^2}{\partial x^2}\right)\frac{\partial U(x,t)}{\partial t} + \left(1-(e_0a)^2\frac{\partial^2}{\partial x^2}\right)\\
&\qquad\times\left\{m\frac{\partial^2 U(x,t)}{\partial t^2} + F(x,t)\right\} \qquad [8.1]
\end{aligned}
$$

In this equation, EA is the axial rigidity, m is the mass per unit length, e_0a is the non-local parameter [ERI 83], $U(x,t)$ is the axial displacement, $F(x,t)$ is the applied force, x is the spatial variable and t is the time. The constant $\widehat{c}_1$ is the strain-rate-dependent viscous damping coefficient and $\widehat{c}_2$ is the velocity-dependent viscous damping coefficient. The parameters $(e_0a)_1$ and $(e_0a)_2$ are non-local parameters related to the two damping terms respectively. For simplicity, the non-local effect in damping is ignored in this chapter, that is we consider $(e_0a)_1 = (e_0a)_2 = 0$. We consider an element of length ℓ_e with axial stiffness EA and mass per unit length m. An element of the axially vibrating rod is shown in Figure 8.1. This element has two degrees of freedom and there are two shape functions $N_1(x)$ and $N_2(x)$. The shape function matrix for the axial deformation [PET 98] can be given by:

$$\mathbf{N}(x) = [N_1(x), N_2(x)]^T = [1 - x/\ell_e, x/\ell_e]^T \quad [8.2]$$

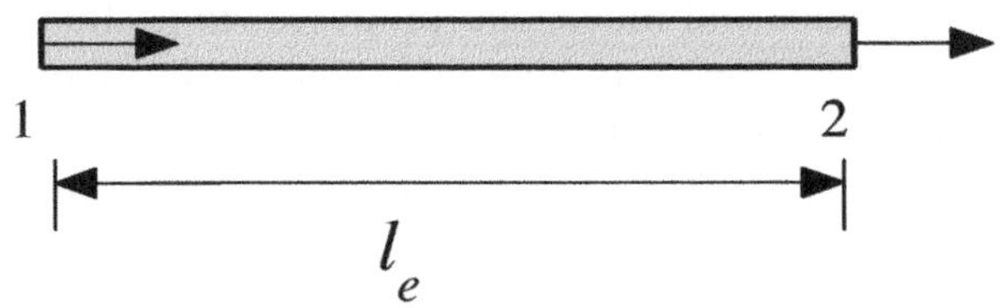

Figure 8.1. *A non-local element for the axially vibrating rod with two nodes. It has two degrees of freedom and the displacement field within the element is expressed by linear shape functions*

Using this, the stiffness matrix can be obtained using the conventional variational formulation as:

$$\mathbf{K}_e = EA \int_0^{\ell_e} \frac{d\mathbf{N}(x)}{dx} \frac{d\mathbf{N}^T(x)}{dx} \mathrm{d}x = \frac{EA}{\ell_e} \begin{bmatrix} 1 & -1 \\ -1 & 1 \end{bmatrix} \quad [8.3]$$

The mass matrix for the non-local element can be obtained as:

$$\begin{aligned} \mathbf{M}_e &= m \int_0^{\ell_e} \mathbf{N}(x)\mathbf{N}^T(x)\mathrm{d}x \\ &+ m(e_0 a)^2 \int_0^{\ell_e} \frac{d\mathbf{N}(x)}{dx} \frac{d\mathbf{N}^T(x)}{dx} \mathrm{d}x \\ &= \frac{m\ell_e}{6} \begin{bmatrix} 2 & 1 \\ 1 & 2 \end{bmatrix} + \left(\frac{e_0 a}{\ell_e}\right)^2 m\ell_e \begin{bmatrix} 1 & -1 \\ -1 & 1 \end{bmatrix} \end{aligned} \quad [8.4]$$

For the special case when the rod is local, the mass matrix derived above reduces to the classical mass matrix [PET 98, DAW 84] as $e_0 a = 0$. Therefore, for a non-local rod, the element stiffness matrix is identical to that of a classical local rod but the element mass has an additive term which is dependent on the non-local parameter.

8.2.2. *Bending vibration of nanobeams*

For the bending vibration of a non-local damped beam, the equation of motion can be expressed by:

$$
\begin{aligned}
&EI\frac{\partial^4 V(x,t)}{\partial x^4} + m\left(1-(e_0a)^2\frac{\partial^2}{\partial x^2}\right)\left\{\frac{\partial^2 V(x,t)}{\partial t^2}\right\} \\
&+\widehat{c}_1\left(1-(e_0a)_1^2\frac{\partial^2}{\partial x^2}\right)\frac{\partial^5 V(x,t)}{\partial x^4 \partial t} + \widehat{c}_2\left(1-(e_0a)_2^2\frac{\partial^2}{\partial x^2}\right)\frac{\partial V(x,t)}{\partial t} \\
&= \left(1-(e_0a)^2\frac{\partial^2}{\partial x^2}\right)\{F(x,t)\} \qquad [8.5]
\end{aligned}
$$

In this equation, EI is the bending rigidity, m is the mass per unit length, e_0a is the non-local parameter, $V(x,t)$ is the transverse displacement and $F(x,t)$ is the applied force. The constant $\widehat{c}_1$ is the strain-rate-dependent viscous damping coefficient and $\widehat{c}_2$ is the velocity-dependent viscous damping coefficient. The damping non-local parameters are assumed to be zero for simplicity. We consider an element of length ℓ_e with bending stiffness EI and mass per unit length m. An element of the beam is shown in Figure 8.2.

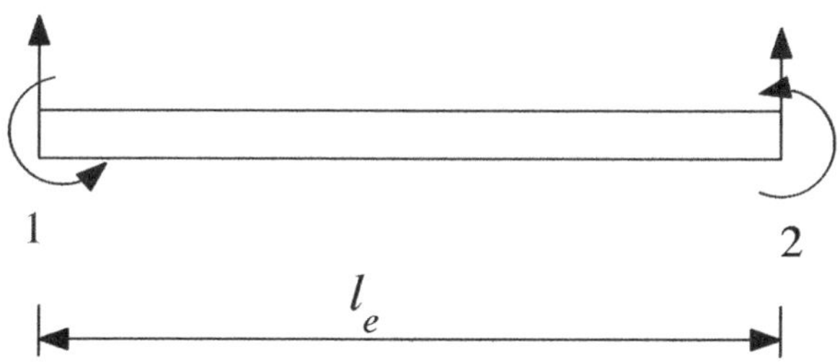

Figure 8.2. *A non-local element for the bending vibration of a beam. It has two nodes and four degrees of freedom. The displacement field within the element is expressed by cubic shape functions*

This element has four degrees of freedom and there are four shape functions. The shape function matrix for the bending deformation [PET 98] can be given by:

$$\mathbf{N}(x) = [N_1(x), N_2(x), N_3(x), N_4(x)]^T \qquad [8.6]$$

where:

$$N_1(x) = 1 - 3\frac{x^2}{\ell_e^2} + 2\frac{x^3}{\ell_e^3}, \qquad N_2(x) = x - 2\frac{x^2}{\ell_e} + \frac{x^3}{\ell_e^2},$$
$$N_3(x) = 3\frac{x^2}{\ell_e^2} - 2\frac{x^3}{\ell_e^3}, \qquad N_4(x) = -\frac{x^2}{\ell_e} + \frac{x^3}{\ell_e^2} \qquad [8.7]$$

Using this, the stiffness matrix can be obtained using the conventional variational formulation [DAW 84] as:

$$\mathbf{K}_e = EI\int_0^{\ell_e} \frac{d^2\mathbf{N}(x)}{dx^2}\frac{d^2\mathbf{N}^T(x)}{dx^2}\mathrm{d}x = \frac{EI}{\ell_e^3}\begin{bmatrix} 12 & 6\ell_e & -12 & 6\ell_e \\ 6\ell_e & 4\ell_e^2 & -6\ell_e & 2\ell_e^2 \\ -12 & -6\ell_e & 12 & -6\ell_e^2 \\ 6\ell_e & 2\ell_e^2 & -6\ell_e & 4\ell_e^2 \end{bmatrix} \qquad [8.8]$$

The mass matrix for the non-local element can be obtained as:

$$\mathbf{M}_e = m\int_0^{\ell_e} \mathbf{N}(x)\mathbf{N}^T(x)\mathrm{d}x + m(e_0a)^2\int_0^{\ell_e} \frac{d\mathbf{N}(x)}{dx}\frac{d\mathbf{N}^T(x)}{dx}\mathrm{d}x$$
$$= \frac{m\ell_e}{420}\begin{bmatrix} 156 & 22\ell_e & 54 & -13\ell_e \\ 22\ell_e & 4\ell_e^2 & 13\ell_e & -3\ell_e^2 \\ 54 & 13\ell_e & 156 & -22\ell_e \\ -13\ell_e & -3\ell_e^2 & -22\ell_e & 4\ell_e^2 \end{bmatrix} \qquad [8.9]$$
$$+ \left(\frac{e_0a}{\ell_e}\right)^2 \frac{m\ell_e}{30}\begin{bmatrix} 36 & 3\ell_e & -36 & 3\ell_e \\ 3\ell_e & 4\ell_e^2 & -3\ell_e & -\ell_e^2 \\ -36 & -3\ell_e & 36 & -3\ell_e \\ 3\ell_e & -\ell_e^2 & -3\ell_e & 4\ell_e^2 \end{bmatrix}$$

For the special case when the beam is local, the mass matrix derived above reduces to the classical mass matrix [PET 98, DAW 84] as $e_0a = 0$.

8.2.3. *Transverse vibration of nanoplates*

For the transverse bending vibration of a non-local damped thin plate, the equation of motion can be expressed by:

$$\begin{aligned} &D\nabla^4 V(x,y,t) + m\left(1-(e_0a)^2\nabla^2\right)\left\{\frac{\partial^2 V(x,y,t)}{\partial t^2}\right\} \\ &+\widehat{c_1}\left(1-(e_0a)_1^2\nabla^2\right)\nabla^4\frac{\partial V(x,y,t)}{\partial x^4\partial t} \\ &\quad+\widehat{c_2}\left(1-(e_0a)_2^2\nabla^2\right)\frac{\partial V(x,y,t)}{\partial t} = \left(1-(e_0a)^2\nabla^2\right)\{F(x,y,t)\} \end{aligned} \quad [8.10]$$

In the above equation, $\nabla^2 = \left(\frac{\partial^2}{\partial x^2} + \frac{\partial^2}{\partial x^2}\right)$ is the differential operator, $D = \frac{Eh^3}{12(1-\nu^2)}$ is the bending rigidity, h is the thickness, ν is Poisson's ratio, m is the mass per unit area, e_0a is the non-local parameter, $V(x,y,t)$ is the transverse displacement and $F(x,y,t)$ is the applied force. The constant $\widehat{c_1}$ is the strain-rate-dependent viscous damping coefficient and $\widehat{c_2}$ is the velocity-dependent viscous damping coefficient. The damping non-local parameters are assumed to be zero for simplicity as before. We consider an element of dimension $2c \times 2b$ with bending stiffness D and mass per unit area m. An element of the plate is shown in Figure 8.3 together with the local coordinate system.

The shape function matrix for the bending deformation is a 12×1 vector [DAW 84] and can be expressed as:

$$\mathbf{N}(x,y) = \mathbf{C}_e^{-1}\boldsymbol{\alpha}(x,y) \quad [8.11]$$

Here, the vector of polynomials is given by:

$$\boldsymbol{\alpha}(x,y) = \left[1\ x\ y\ x^2\ xy\ y^2\ x^3\ x^2y\ xy^2\ y^3\ x^3y\ xy^3\right]^T \quad [8.12]$$

and the coefficient matrix can be obtained as:

$$\mathbf{C}_e{}^{-1} = \frac{1}{8a^3b^3} \times$$

$$\begin{bmatrix} 2c^3b^3 & c^3b^4 & c^4b^3 & 2c^3b^3 & c^3b^4 & -b^3c^4 & 2c^3b^3 & -c^3b^4 & -c^4b^3 & 2c^3b^3 & -c^3b^4 & c^4b^3 \\ -3c^2b^3 & -c^2b^4 & -c^3b^3 & 3c^2b^3 & c^2b^4 & -c^3b^3 & 3c^2b^3 & -c^2b^4 & -c^3b^3 & -3c^2b^3 & c^2b^4 & -c^3b^3 \\ -3c^3b^2 & -c^3b^3 & -c^4b^2 & -3c^3b^2 & -c^3b^3 & c^4b^2 & 3c^3b^2 & -c^3b^3 & -c^4b^2 & 3c^3b^2 & -c^3b^3 & c^4b^2 \\ 0 & 0 & -c^2b^3 & 0 & 0 & c^2b^3 & 0 & 0 & c^2b^3 & 0 & 0 & -c^2b^3 \\ 4c^2b^2 & c^2b^3 & c^3b^2 & -4c^2b^2 & -c^2b^3 & c^3b^2 & 4c^2b^2 & -c^2b^3 & -c^3b^2 & -4c^2b^2 & c^2b^3 & -c^3b^2 \\ 0 & -c^3b^2 & 0 & 0 & -c^3b^2 & 0 & 0 & c^3b^2 & 0 & 0 & c^3b^2 & 0 \\ b^3 & 0 & cb^3 & -b^3 & 0 & cb^3 & -b^3 & 0 & cb^3 & b^3 & 0 & cb^3 \\ 0 & 0 & c^2b^2 & 0 & 0 & -c^2b^2 & 0 & 0 & c^2b^2 & 0 & 0 & -c^2b^2 \\ 0 & c^2b^2 & 0 & 0 & -c^2b^2 & 0 & 0 & c^2b^2 & 0 & 0 & -c^2b^2 & 0 \\ c^3 & c^3b & 0 & c^3 & c^3b & 0 & -c^3 & c^3b & 0 & -c^3 & c^3b & 0 \\ -b^2 & 0 & -cb^2 & b^2 & 0 & -cb^2 & -b^2 & 0 & cb^2 & b^2 & 0 & cb^2 \\ -c^2 & -c^2b & 0 & c^2 & c^2b & 0 & -c^2 & c^2b & 0 & c^2 & -c^2b & 0 \end{bmatrix} \quad [8.13]$$

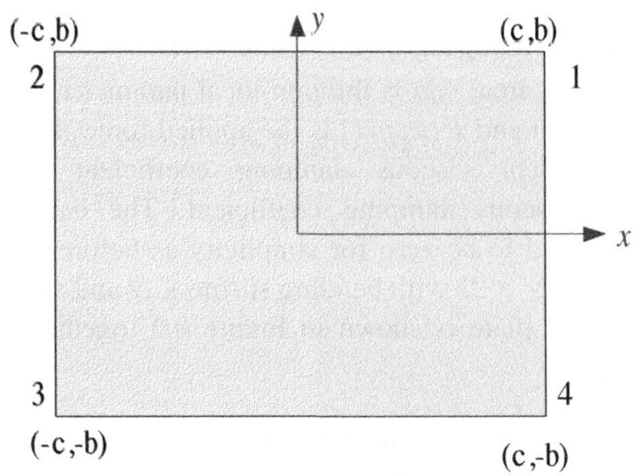

Figure 8.3. *A non-local element for the bending vibration of a plate. It has four nodes and 12 degrees of freedom. The displacement field within the element is expressed by cubic shape functions in both directions*

Using the shape functions in equation [8.11], the stiffness matrix can be obtained using the conventional variational formulation [DAW 84] as:

$$\mathbf{K}_e = \int_{A_e} \mathbf{B}^T \mathbf{E} \mathbf{B} \mathrm{d}A_e \quad [8.14]$$

In the preceding equation, **B** is the strain-displacement matrix and the matrix **E** is given by:

$$\mathbf{E} = D \begin{bmatrix} 1 & \nu & 0 \\ \nu & 1 & 0 \\ 0 & 0 & \frac{1-\nu}{2} \end{bmatrix} \qquad [8.15]$$

Evaluating the integral in equation [8.14], we can obtain the element stiffness matrix in closed-form as:

$$\mathbf{K}_e = \frac{Eh^3}{12(1-\nu^2)} \mathbf{C}^{-1^T} \mathbf{k}_e \mathbf{C}^{-1} \qquad [8.16]$$

where

$$\mathbf{k}_e = \begin{bmatrix} 0 & & & & & & & & & & & \\ 0 & 0 & & & & & & & & & & \\ 0 & 0 & 0 & & & & & & & & & \\ 0 & 0 & 0 & k_{e44} & & & & & & & & \\ 0 & 0 & 0 & 0 & k_{e55} & & & & & & & \\ 0 & 0 & 0 & k_{e64} & 0 & k_{e66} & & & & \text{symmetric} & & \\ 0 & 0 & 0 & 0 & 0 & 0 & k_{e77} & & & & & \\ 0 & 0 & 0 & 0 & 0 & 0 & 0 & k_{e88} & & & & \\ 0 & 0 & 0 & 0 & 0 & 0 & k_{e97} & 0 & k_{e99} & & & \\ 0 & 0 & 0 & 0 & 0 & 0 & 0 & k_{e10\,8} & 0 & k_{e10\,10} & & \\ 0 & 0 & 0 & 0 & k_{e11\,5} & 0 & 0 & 0 & 0 & 0 & k_{e11\,11} & \\ 0 & 0 & 0 & 0 & k_{e12\,5} & 0 & 0 & 0 & 0 & 0 & k_{e12\,11} & k_{e12\,12} \end{bmatrix} \qquad [8.17]$$

with

$$k_{e44} = 16cb \qquad k_{e55} = 8cb(1-\nu) \qquad k_{e64} = 16\nu cb$$

$$k_{e66} = 16cb \qquad k_{e77} = 48c^3 b \qquad k_{e88} = \frac{16cb\left((2-2\nu)\,c^2 + b^2\right)}{3}$$

$$k_{e97} = 16\nu c^3 b \qquad k_{e99} = \frac{16cb(c^2 + 2(1-\nu)b^2)}{3} \qquad k_{e10\,8} = 16\nu cb^2$$

$$k_{e10\,10} = 48cb^3 \qquad k_{e11\,5} = 8\,(1-\nu)\,c^3 b \qquad k_{e11\,11} = 4c^3 b(\frac{9(1-\nu)c^2}{5} + 4b^2)$$

$$k_{e12\,5} = 8\,(1-\nu)\,cb^3 \qquad k_{e12\,11} = 8(\nu+1)c^3 b^3 \qquad k_{e12\,12} = \frac{8\,cb^3\,(10c^2 + 9(1-\nu)b^2)}{5}$$

The mass matrix for the non-local element can be obtained as:

$$
\begin{aligned}
\mathbf{M}_e &= \rho h \int_{A_e} \{\mathbf{N}(x,y)\mathbf{N}^T(x,y) + (e_0 a)^2 \\
&\quad \times \left(\frac{\partial \mathbf{N}(x,y)}{\partial x} \frac{d\mathbf{N}^T(x,y)}{dx} + \frac{\partial \mathbf{N}(x,y)}{\partial x} \frac{d\mathbf{N}^T(x,y)}{dx} \right) \Bigg\} \, \mathrm{d}A_e \\
&= \mathbf{M}_{0_e} + \left(\frac{e_0 a}{c}\right)^2 \mathbf{M}_{x_e} + \left(\frac{e_0 a}{b}\right)^2 \mathbf{M}_{y_e}
\end{aligned}
\quad [8.18]
$$

The three matrices appearing in the above expression can be obtained in closed-form as:

$$
\mathbf{M}_{0_e} = \frac{\rho h c b}{3150} \times
$$

$$
\begin{bmatrix}
1727 & 461b & 461c & 613 & 199b & -274c & 197 & -116b & -116c & 613 & -274b & 199c \\
461b & 160b^2 & 126cb & 199b & 80b^2 & -84cb & 116b & -60b^2 & -56cb & 274b & -120b^2 & 84cb \\
461c & 126cb & 160c^2 & 274c & 84cb & -120c^2 & 116c & -56cb & -60c^2 & 199c & -84cb & 80c^2 \\
613 & 199b & 274c & 1727 & 461b & -461c & 613 & -274b & -199c & 197 & -116b & 116c \\
199b & 80b^2 & 84cb & 461b & 160b^2 & -126cb & 274b & -120b^2 & -84cb & 116b & -60b^2 & 56cb \\
-274c & -84cb & -120c^2 & -461c & -126cb & 160c^2 & -199c & 84cb & 80c^2 & -116c & 56cb & -60c^2 \\
197 & 116b & 116c & 613 & 274b & -199c & 1727 & -461b & -461c & 613 & -199b & 274c \\
-116b & -60b^2 & -56cb & -274b & -120b^2 & 84cb & -461b & 160b^2 & 126cb & -199b & 80b^2 & -84cb \\
-116c & -56cb & -60c^2 & -199c & -84cb & 80c^2 & -461c & 126cb & 160c^2 & -274c & 84cb & -120c^2 \\
613 & 274b & 199c & 197 & 116b & -116c & 613 & -199b & -274c & 1727 & -461b & 461c \\
-274b & -120b^2 & -84cb & -116b & -60b^2 & 56cb & -199b & 80b^2 & 84cb & -461b & 160b^2 & -126cb \\
199c & 84cb & 80c^2 & 116c & 56cb & -60c^2 & 274c & -84cb & -120c^2 & 461c & -126cb & 160c^2
\end{bmatrix}
\quad [8.19]
$$

$$
\mathbf{M}_{x_e} = \frac{\rho h c b}{630} \times
$$

$$
\begin{bmatrix}
276 & 66b & 42c & -276 & -66b & 42c & -102 & 39b & 21c & 102 & -39b & 21c \\
66b & 24b^2 & 0 & -66b & -24b^2 & 0 & -39b & 18b^2 & 0 & 39b & -18b^2 & 0 \\
42c & 0 & 112c^2 & -42c & 0 & -28c^2 & -21c & 0 & -14c^2 & 21c & 0 & 56c^2 \\
-276 & -66b & -42c & 276 & 66b & -42c & 102 & -39b & -21c & -102 & 39b & -21c \\
-66b & -24b^2 & 0 & 66b & 24b^2 & 0 & 39b & -18b^2 & 0 & -39b & 18b^2 & 0 \\
42c & 0 & -28c^2 & -42c & 0 & 112c^2 & -21c & 0 & 56c^2 & 21c & 0 & -14c^2 \\
-102 & -39b & -21c & 102 & 39b & -21c & 276 & -66b & -42c & -276 & 66b & -42c \\
39b & 18b^2 & 0 & -39b & -18b^2 & 0 & -66b & 24b^2 & 0 & 66b & -24b^2 & 0 \\
21c & 0 & -14c^2 & -21c & 0 & 56c^2 & -42c & 0 & 112c^2 & 42c & 0 & -28c^2 \\
102 & 39b & 21c & -102 & -39b & 21c & -276 & 66b & 42c & 276 & -66b & 42c \\
-39b & -18b^2 & 0 & 39b & 18b^2 & 0 & 66b & -24b^2 & 0 & -66b & 24b^2 & 0 \\
21c & 0 & 56c^2 & -21c & 0 & -14c^2 & -42c & 0 & -28c^2 & 42c & 0 & 112c^2
\end{bmatrix}
\quad [8.20]
$$

$$\mathbf{M}_{y_e} = \frac{\rho hcb}{630} \times$$

$$\begin{bmatrix} 276 & 42b & 66c & 102 & 21b & -39c & -102 & 21b & 39c & -276 & 42b & -66c \\ 42b & 112b^2 & 0 & 21b & 56b^2 & 0 & -21b & -14b^2 & 0 & -42b & -28b^2 & 0 \\ 66c & 0 & 24c^2 & 39c & 0 & -18c^2 & -39c & 0 & 18c^2 & -66c & 0 & -24c^2 \\ 102 & 21b & 39c & 276 & 42b & -66c & -276 & 42b & 66c & -102 & 21b & -39c \\ 21b & 56b^2 & 0 & 42b & 112b^2 & 0 & -42b & -28b^2 & 0 & -21b & -14b^2 & 0 \\ -39c & 0 & -18c^2 & -66c & 0 & 24c^2 & 66c & 0 & -24c^2 & 39c & 0 & 18c^2 \\ -102 & -21b & -39c & -276 & -42b & 66c & 276 & -42b & -66c & 102 & -21b & 39c \\ 21b & -14b^2 & 0 & 42b & -28b^2 & 0 & -42b & 112b^2 & 0 & -21b & 56b^2 & 0 \\ 39c & 0 & 18c^2 & 66c & 0 & -24c^2 & -66c & 0 & 24c^2 & -39c & 0 & -18c^2 \\ -276 & -42b & -66c & -102 & -21b & 39c & 102 & -21b & -39c & 276 & -42b & 66c \\ 42b & -28b^2 & 0 & 21b & -14b^2 & 0 & -21b & 56b^2 & 0 & -42b & 112b^2 & 0 \\ -66c & 0 & -24c^2 & -39c & 0 & 18c^2 & 39c & 0 & -18c^2 & 66c & 0 & 24c^2 \end{bmatrix} \quad [8.21]$$

For the special case when the plate is local, the mass matrix derived above reduces to the classical mass matrix as $e_0 a = 0$ [DAW 84].

Based on the discussions in this section for all the three systems considered here, in general the element mass matrix of a non-local dynamic system can be expressed as:

$$\mathbf{M}_e = \mathbf{M}_{0_e} + \mathbf{M}_{\mu_e} \quad [8.22]$$

Here, $\mathbf{M}_{0_e}$ is the element stiffness matrix corresponding to the underlying local system and $\mathbf{M}_{\mu_e}$ is the additional term arising due to the non-local effect.

8.3. Modal analysis of non-local dynamical systems

Using the finite element formulation, the stiffness matrix of the local and non-local system turns out to be identical to each other. The mass matrix of the non-local system is, however, different from its equivalent local counterpart. Assembling the element matrices and applying the boundary conditions, following the usual procedure of the finite element method [ZIE 91] we obtain the global mass matrix as:

$$\mathbf{M} = \mathbf{M}_0 + \mathbf{M}_\mu \quad [8.23]$$

In this equation, $\mathbf{M}_0$ is the usual global mass matrix arising in the conventional local system and $\mathbf{M}_\mu$ is the matrix arising due to non-local nature of the systems. In general, we can express this matrix by:

$$\mathbf{M}_\mu = \left(\frac{e_0 a}{L}\right)^2 \widehat{\mathbf{M}}_\mu \qquad [8.24]$$

where $\widehat{\mathbf{M}}_\mu$ is a non-negative definite matrix. The matrix $\mathbf{M}_\mu$ is, therefore, a scale-dependent matrix and its influence reduces if the length of the system L is large compared to the parameter $e_0 a$. Majority of the current finite element software and other computational tools do not explicitly consider the non-local part of the mass matrix. For the design and analysis of future generation of nanoelectromechanical systems, it is vitally important to consider the non-local influence. In this section, we are interested in understanding the impact of the difference in the mass matrix on the dynamic characteristics of the system. In particular, the following questions of fundamental interest have been addressed:

– under what condition a non-local system possess classical local normal modes?

– how the vibration modes and frequencies of a non-local system can be understood in the light of the results from classical local systems?

By addressing these questions, it would be possible to extend conventional "local" elasticity-based finite element software to analyze non-local systems arising in the modeling of complex nanoscale built-up structures.

8.3.1. *Conditions for classical normal modes*

The equation of motion of a discretized non-local damped system with n degrees of freedom can be expressed as:

$$[\mathbf{M}_0 + \mathbf{M}_\mu]\,\ddot{\mathbf{u}}(t) + \mathbf{C}\dot{\mathbf{u}}(t) + \mathbf{K}\mathbf{u}(t) = \mathbf{f}(t) \qquad [8.25]$$

Here, $\mathbf{u}(t) \in \mathbb{R}^n$ is the displacement vector, $\mathbf{f}(t) \in \mathbb{R}^n$ is the forcing vector, $\mathbf{K}, \mathbf{C} \in \mathbb{R}^{n\times n}$ are, respectively, the global stiffness and the viscous damping matrix. In general, $\mathbf{M}_0$ and $\mathbf{M}_\mu$ are the positive definite symmetric matrices, and $\mathbf{C}$ and $\mathbf{K}$ are the non-negative definite symmetric matrices. The equation of motion of corresponding local system is given by:

$$\mathbf{M}_0\ddot{\mathbf{u}}_0(t) + \mathbf{C}\dot{\mathbf{u}}_0(t) + \mathbf{K}\mathbf{u}_0(t) = \mathbf{f}(t) \qquad [8.26]$$

where $\mathbf{u}_0(t) \in \mathbb{R}^n$ is the local displacement vector. The natural frequencies ($\omega_j \in \mathbb{R}$) and the mode shapes ($\mathbf{x}_j \in \mathbb{R}^n$) of the corresponding undamped local system can be obtained by solving the matrix eigenvalue problem [MEI 97] as:

$$\mathbf{K}\mathbf{x}_j = \omega_j^2 \mathbf{M}_0 \mathbf{x}_j, \quad \forall\, j = 1, 2, \ldots, n \tag{8.27}$$

The undamped local eigenvectors satisfy an orthogonality relationship over the local mass and stiffness matrices, that is:

$$\mathbf{x}_k^T \mathbf{M}_0 \mathbf{x}_j = \delta_{kj} \tag{8.28}$$

$$\text{and} \quad \mathbf{x}_k^T \mathbf{K} \mathbf{x}_j = \omega_j^2 \delta_{kj}, \quad \forall\, k, j = 1, 2, \ldots, n \tag{8.29}$$

where δ_{kj} is the Kroneker's delta function. We construct the local modal matrix:

$$\mathbf{X} = [\mathbf{x}_1, \mathbf{x}_2, \ldots, \mathbf{x}_n] \in \mathbb{R}^n \tag{8.30}$$

The local modal matrix can be used to diagonalize the local system [8.26] provided that the damping matrix $\mathbf{C}$ is simultaneously diagonalizable with $\mathbf{M}_0$ and $\mathbf{K}$. This condition, known as the proportional damping, originally introduced by Lord Rayleigh [RAY 77] in 1877, is still widely used today. The mathematical condition for proportional damping can be obtained from the commutitative behavior of the system matrices [CAU 65]. This can be expressed as:

$$\mathbf{C}\mathbf{M}_0^{-1}\mathbf{K} = \mathbf{K}\mathbf{M}_0^{-1}\mathbf{C} \tag{8.31}$$

or, equivalently, $\mathbf{C} = \mathbf{M}_0 f(\mathbf{M}_0^{-1}\mathbf{K})$ as shown in [ADH 06].

Considering undamped non-local system and premultiplying the equation by $\mathbf{M}_0^{-1}$, we have:

$$\left(\mathbf{I}_n + \mathbf{M}_0^{-1}\mathbf{M}_\mu\right)\ddot{\mathbf{u}}(t) + \left(\mathbf{M}_0^{-1}\mathbf{K}\right)\mathbf{u}(t) = \mathbf{M}_0^{-1}\mathbf{f}(t) \tag{8.32}$$

This system can be diagonalized by a similarity transformation which also diagonalizes $\left(\mathbf{M}_0^{-1}\mathbf{K}\right)$ provided that the matrices $\left(\mathbf{M}_0^{-1}\mathbf{M}_\mu\right)$ and $\left(\mathbf{M}_0^{-1}\mathbf{K}\right)$

commute. This implies that the condition for existence of classical local normal modes is:

$$\left(\mathbf{M}_0^{-1}\mathbf{K}\right)\left(\mathbf{M}_0^{-1}\mathbf{M}_\mu\right) = \left(\mathbf{M}_0^{-1}\mathbf{M}_\mu\right)\left(\mathbf{M}_0^{-1}\mathbf{K}\right) \tag{8.33}$$

$$\text{or} \quad \mathbf{K}\mathbf{M}_0^{-1}\mathbf{M}_\mu = \mathbf{M}_\mu\mathbf{M}_0^{-1}\mathbf{K} \tag{8.34}$$

If the above condition is satisfied, then a non-local undamped system can be diagonalized by the classical local normal modes. However, it is also possible to have non-local normal modes which can diagonalize the non-local undamped system as discussed in the next section.

8.3.2. *Non-local normal modes*

Non-local normal modes can be obtained by the undamped non-local eigenvalue problem:

$$\mathbf{K}\mathbf{u}_j = \lambda_j^2\left[\mathbf{M}_0 + \mathbf{M}_\mu\right]\mathbf{u}_j, \quad \forall j = 1, 2, \ldots, n \tag{8.35}$$

Here, λ_j and $\mathbf{u}_j$ are the non-local natural frequencies and non-local normal modes of the system. We can define a non-local modal matrix as:

$$\mathbf{U} = \left[\mathbf{u}_1, \mathbf{u}_2, \ldots, \mathbf{u}_n\right] \in \mathbb{R}^n \tag{8.36}$$

which will unconditionally diagonalize the non-local undamped system. It should be remembered that in general non-local normal modes and frequencies will be different from their local counterparts.

Under certain restrictive conditions, it may be possible to diagonalize the damped non-local system using classical normal modes. Premultiplying the equation of motion [8.25] by $\mathbf{M}_0^{-1}$, the required condition is that $\left(\mathbf{M}_0^{-1}\mathbf{M}_\mu\right)$, $\left(\mathbf{M}_0^{-1}\mathbf{C}\right)$ and $\left(\mathbf{M}_0^{-1}\mathbf{K}\right)$ must commute pairwise. This implies that in addition to the two conditions given by equations [8.31] and [8.34], we also need a third condition:

$$\mathbf{C}\mathbf{M}_0^{-1}\mathbf{M}_\mu = \mathbf{M}_\mu\mathbf{M}_0^{-1}\mathbf{C} \tag{8.37}$$

If we consider the diagonalization of the non-local system by the non-local modal matrix in [8.31], then the concept of proportional damping can be applied similar to that of the local system. We can obtain the required condition similar to Caughey's condition [CAU 65] as in equation [8.31] by replacing the mass matrix with $\mathbf{M}_0 + \mathbf{M}_\mu$. If this condition is satisfied, then the equation of motion can be diagonalized by the non-local normal modes and in general not by the classical normal modes.

8.3.3. *Approximate non-local normal modes*

The majority of existing finite element software calculate the classical normal modes. However, it was shown that only under certain restrictive conditions, the classical normal modes can be used to diagonalize the system. In general, we need to use non-local normal modes to diagonalize the equation of motion [8.25], which is necessary for efficient dynamic analysis and physical understanding of the system. In this section, we aim to express non-local normal modes in terms of classical normal modes. Since the classical normal modes are well understood, this approach will allow us to develop physical understanding of the non-local normal modes.

For distinct undamped eigenvalues (ω_l^2), local eigenvectors $\mathbf{x}_l, \forall l = 1, \ldots, n$ form a complete set of vectors. For this reason, each non-local normal mode $\mathbf{u}_j$ can be expanded as a linear combination of $\mathbf{x}_l$. Thus, an expansion of the form:

$$\mathbf{u}_j = \sum_{l=1}^{n} \alpha_l^{(j)} \mathbf{x}_l \qquad [8.38]$$

may be considered. Without any loss of generality, we can assume that $\alpha_j^{(j)} = 1$ (normalization) which leaves us to determine $\alpha_l^{(j)}, \forall l \neq j$. Substituting the expansion of $\mathbf{u}_j$ into the eigenvalue equation [8.35], we obtain:

$$\left[-\lambda_j^2 \left(\mathbf{M}_0 + \mathbf{M}_\mu\right) + \mathbf{K}\right] \sum_{l=1}^{n} \alpha_l^{(j)} \mathbf{x}_l = \mathbf{0} \qquad [8.39]$$

For the case when $\alpha_l^{(j)}$ are approximate, the error involving the projection in equation [8.38] can be expressed as:

$$\varepsilon_j = \sum_{l=1}^{n} \left[-\lambda_j^2 \left(\mathbf{M}_0 + \mathbf{M}_\mu\right) + \mathbf{K}\right] \alpha_l^{(j)} \mathbf{x}_l \qquad [8.40]$$

We use a Galerkin approach to minimize this error by viewing the expansion as a projection in the basis functions $\mathbf{x}_l \in \mathbb{R}^n, \forall l = 1, 2, \ldots n$. Therefore, making the error orthogonal to the basis functions we have:

$$\varepsilon_j \perp \mathbf{x}_l \quad \text{or} \quad \mathbf{x}_k^T \varepsilon_j = 0 \quad \forall\, k = 1, 2, \ldots, n \qquad [8.41]$$

Using the orthogonality property of the undamped local modes described by equations [8.28] and [8.29], we obtain:

$$\sum_{l=1}^{n} \left[-\lambda_j^2 \left(\delta_{kl} + M'_{\mu_{kl}} \right) + \omega_k^2 \delta_{kl} \right] \alpha_l^{(j)} = 0 \qquad [8.42]$$

where $M'_{\mu_{kl}} = \mathbf{x}_k^T \mathbf{M}_\mu \mathbf{x}_l$ are the elements of the non-local part of the modal mass matrix. The j-th equation of this set is obtained by setting $k = j$ and can be written as:

$$-\lambda_j^2 \left(1 + M'_{\mu_{jj}} \right) + \omega_j^2 - \lambda_j^2 \sum_{l \neq j}^{n} \left(M'_{\mu_{jl}} \right) \alpha_l^{(j)} = 0 \qquad [8.43]$$

Assuming that the off-diagonal terms of the non-local part of the modal mass matrix are small and $\alpha_l^{(j)} \ll 1, \forall l \neq j$, approximate non-local frequencies can be obtained as:

$$\lambda_j \approx \frac{\omega_j}{\sqrt{1 + M'_{\mu_{jj}}}} \qquad [8.44]$$

This important equation gives an explicit closed-form expression relating to non-local natural frequencies λ_j and local natural frequencies ω_j. If the length-scale parameter is large, then diagonal elements of the non-local part of the modal mass matrix become smaller and consequently the non-local frequencies approach the classical local frequencies. Equation [8.44] can also be viewed as a general correction to the local frequencies due to the non-local effect arising due to small-length scale.

For the general case when $k \neq j$, from equation [8.42] we have:

$$\left[-\lambda_j^2 \left(1 + M'_{\mu_{kk}} \right) + \omega_k^2 \right] \alpha_k^{(j)} - \lambda_j^2 \sum_{l \neq k}^{n} \left(M'_{\mu_{kl}} \right) \alpha_l^{(j)} = 0 \qquad [8.45]$$

Recalling that $\alpha_j^{(j)} = 1$, this equation can be expressed as:

$$\left[-\lambda_j^2 \left(1 + M'_{\mu_{kk}}\right) + \omega_k^2\right] \alpha_k^{(j)} = \lambda_j^2 \left[M'_{\mu_{kj}} + \sum_{l \neq k \neq j}^{n} M'_{\mu_{kl}} \alpha_l^{(j)}\right] \quad [8.46]$$

Again, assuming that the off-diagonal terms of the non-local part of the modal mass matrix are small and $\alpha_l^{(j)} \ll 1, \forall l \neq j$, we can obtain:

$$\alpha_k^{(j)} \approx \frac{\lambda_j^2 M'_{\mu_{kj}}}{-\lambda_j^2 \left(1 + M'_{\mu_{kk}}\right) + \omega_k^2} = \frac{\lambda_j^2}{\left(\lambda_k^2 - \lambda_j^2\right)} \frac{M'_{\mu_{kj}}}{\left(1 + M'_{\mu_{kk}}\right)} \quad [8.47]$$

Substituting this in the original expansion [8.38], the non-local normal modes can be expressed in terms of the classical normal modes as:

$$\mathbf{u}_j \approx \mathbf{x}_j + \sum_{k \neq j}^{n} \frac{\lambda_j^2}{\left(\lambda_k^2 - \lambda_j^2\right)} \frac{M'_{\mu_{kj}}}{\left(1 + M'_{\mu_{kk}}\right)} \mathbf{x}_k \quad [8.48]$$

This equation explicitly relates to non-local normal modes with the classical normal modes. From this expression, the following insights about the non-local normal modes can be deduced:

– each non-local mode can be viewed as a sum of two principal components. One of them is parallel to the corresponding local mode and the other is orthogonal to it as all $\mathbf{x}_k$ are orthogonal to $\mathbf{x}_j$ for $j \neq k$;

– due to the term $\left(\lambda_k^2 - \lambda_j^2\right)$ in the denominator, for a given non-local mode, only few adjacent local modes contribute to the orthogonal component;

– for systems with well-separated natural frequencies, the contribution of the orthogonal component becomes smaller compared to the parallel component.

Equations [8.44] and [8.48] completely define the non-local natural frequencies and mode shapes in terms of the local natural frequencies and mode shapes. Accuracy of these expressions will be investigated through numerical examples in section 8.5. Dynamic response of non-local damped systems is considered next.

8.4. Dynamics of damped non-local systems

In this section, we consider forced response of damped non-local systems in the frequency domain. Assuming that all the initial conditions are zero and taking the Fourier transformation of the equation of motion [8.25], we have:

$$\mathbf{D}(\mathrm{i}\omega)\bar{\mathbf{u}}(\mathrm{i}\omega) = \bar{\mathbf{f}}(\mathrm{i}\omega) \tag{8.49}$$

where the non-local dynamic stiffness matrix is given by:

$$\mathbf{D}(\mathrm{i}\omega) = -\omega^2\left[\mathbf{M}_0 + \mathbf{M}_\mu\right] + \mathrm{i}\omega\mathbf{C} + \mathbf{K} \tag{8.50}$$

In equation [8.49], $\bar{\mathbf{u}}(\mathrm{i}\omega)$ and $\bar{\mathbf{f}}(\mathrm{i}\omega)$ are, respectively, the Fourier transformations of the response and the forcing vectors. Using the local modal matrix [8.30], the dynamic stiffness matrix can be transformed into the modal coordinate as:

$$\mathbf{D}'(\mathrm{i}\omega) = \mathbf{X}^T\mathbf{D}(\mathrm{i}\omega)\mathbf{X} = -\omega^2\left[\mathbf{I} + \mathbf{M}'_\mu\right] + \mathrm{i}\omega\mathbf{C}' + \boldsymbol{\Omega}^2 \tag{8.51}$$

where $\mathbf{I}$ is an n-dimensional identity matrix, $\boldsymbol{\Omega}^2$ is a diagonal matrix containing the squared local natural frequencies and $(\bullet)'$ denotes that the quantity is in the modal coordinates. Unless all the conditions derived in section 8.3.2 are satisfied, in general $\mathbf{M}'_\mu$ and $\mathbf{C}'$ are not diagonal matrices. We separate the diagonal and off-diagonal terms of these matrices and rewrite equation [8.51] as:

$$\mathbf{D}'(\mathrm{i}\omega) = \underbrace{-\omega^2\left[\mathbf{I} + \overline{\mathbf{M}}'_\mu\right] + \mathrm{i}\omega\overline{\mathbf{C}}' + \boldsymbol{\Omega}^2}_{\text{diagonal}} + \underbrace{\left(-\omega^2\Delta\mathbf{M}'_\mu + \mathrm{i}\omega\Delta\mathbf{C}'\right)}_{\text{off-diagonal}} \tag{8.52}$$

$$= \overline{\mathbf{D}}'(\mathrm{i}\omega) + \Delta\mathbf{D}'(\mathrm{i}\omega) \tag{8.53}$$

From equation [8.49], the dynamic response of the system can be obtained as:

$$\bar{\mathbf{u}}(\mathrm{i}\omega) = \mathbf{H}(\mathrm{i}\omega)\bar{\mathbf{f}}(\mathrm{i}\omega) = \left[\mathbf{X}\mathbf{D}'^{-1}(\mathrm{i}\omega)\mathbf{X}^T\right]\bar{\mathbf{f}}(\mathrm{i}\omega) \tag{8.54}$$

where the matrix $\mathbf{H}(\mathrm{i}\omega)$ is known as the transfer function matrix. From the expression of the modal dynamic stiffness matrix in equation [8.53], we have:

$$\mathbf{D}'^{-1}(\mathrm{i}\omega) = \left[\overline{\mathbf{D}}'(\mathrm{i}\omega)\left(\mathbf{I} + \overline{\mathbf{D}}'^{-1}(\mathrm{i}\omega)\Delta\mathbf{D}'(\mathrm{i}\omega)\right)\right]^{-1} \tag{8.55}$$

$$\approx \overline{\mathbf{D}}'^{-1}(\mathrm{i}\omega) - \overline{\mathbf{D}}'^{-1}(\mathrm{i}\omega)\Delta\mathbf{D}'(\mathrm{i}\omega)\overline{\mathbf{D}}'^{-1}(\mathrm{i}\omega) \tag{8.56}$$

In the above equation, the diagonal part $\overline{\mathbf{D}}'^{-1}(\mathrm{i}\omega)$ is expected to be the dominant term and its elements can be expressed as:

$$\left\{\overline{\mathbf{D}}'^{-1}(\mathrm{i}\omega)\right\}_{jj} = \frac{1}{-\omega^2\left(1+M'_{\mu_{jj}}\right)+2\mathrm{i}\omega\omega_j\zeta_j+\omega_j^2} \quad [8.57]$$

In the above equation, we defined the modal damping factors as:

$$\left\{\overline{\mathbf{C}}'\right\}_{jj} = 2\omega_j\zeta_j \quad [8.58]$$

Substituting the approximate expression of $\mathbf{D}'^{-1}(\mathrm{i}\omega)$ from equation [8.56] into the expression of the transfer function matrix in equation [8.54], we have:

$$\mathbf{H}(\mathrm{i}\omega) = \left[\mathbf{X}\mathbf{D}'^{-1}(\mathrm{i}\omega)\mathbf{X}^T\right] \approx \overline{\mathbf{H}}'(\mathrm{i}\omega) - \Delta\mathbf{H}'(\mathrm{i}\omega) \quad [8.59]$$

where

$$\overline{\mathbf{H}}'(\mathrm{i}\omega) = \mathbf{X}\overline{\mathbf{D}}'(\mathrm{i}\omega)\mathbf{X}^T = \sum_{k=1}^{n}\frac{\mathbf{x}_k\mathbf{x}_k^T}{-\omega^2\left(1+M'_{\mu_{kk}}\right)+2\mathrm{i}\omega\omega_k\zeta_k+\omega_k^2} \quad [8.60]$$

and

$$\Delta\mathbf{H}'(\mathrm{i}\omega) = \mathbf{X}\overline{\mathbf{D}}'^{-1}(\mathrm{i}\omega)\Delta\mathbf{D}'(\mathrm{i}\omega)\overline{\mathbf{D}}'^{-1}(\mathrm{i}\omega)\mathbf{X}^T \quad [8.61]$$

Considering that the matrix $\Delta\mathbf{D}'(\mathrm{i}\omega)$ has only off-diagonal terms, expanding the matrix multiplications a general term of the previous matrix can be expressed as:

$$\Delta H'_{ij}(\mathrm{i}\omega) = \sum_{l=1}^{n}\sum_{k\neq l}^{n} \times \frac{x_{il}\Delta D'_{lk}(\mathrm{i}\omega)x_{jk}}{\left(-\omega^2(1+M'_{\mu_{ll}})+2\mathrm{i}\omega\omega_l\zeta_l+\omega_l^2\right)\left(-\omega^2(1+M'_{\mu_{kk}})+2\mathrm{i}\omega\omega_k\zeta_k+\omega_k^2\right)} \quad [8.62]$$

Equation [8.59], therefore, completely defines the transfer function of the damped non-local system in terms of the classical normal modes. This can be useful in practice as all the quantities arise in this expression can be obtained from a conventional finite element software. We only need the non-local part of the mass matrix as derived in section 8.2. Some notable features of the expression of the approximate transfer function matrix in equation [8.59] are:

– for lightly damped systems, from equation [8.60] observe that the transfer function will have peaks around the non-local natural frequencies derived in the previous section. This justifies the consistency of the approximation used in this chapter;

– the decomposition in equation [8.52] indicates that error in the transfer function depends on two components. They include the off-diagonal part of the modal non-local mass matrix $\Delta\mathbf{M}'_\mu$ and the off-diagonal part of the modal damping matrix $\Delta\mathbf{C}'$. While the error in the damping term is present for non-proportionally damped local systems, the error due to the non-local modal mass matrix is unique to the non-local system;

– for a proportionally damped system $\Delta\mathbf{C}' = \mathbf{O}$. For this case, error in the transfer function only depends on $\Delta\mathbf{M}'_\mu$;

– in general, error in the transfer function is expected to be higher for higher frequencies as both $\Delta\mathbf{C}'$ and $\Delta\mathbf{M}'_\mu$ are weighted by frequency ω.

The expressions of the non-local natural frequencies [8.44], non-local normal modes [8.48] and non-local transfer function matrix [8.59] allow us to understand the dynamic characteristic of a non-local system in a qualitative and quantitative manner in the light of equivalent local systems. Next, we illustrate these new expressions by numerical examples of nanoscale structures.

8.5. Numerical examples

8.5.1. *Axial vibration of a single-walled carbon nanotube*

A single-walled carbon nanotube (SWCNT) is considered to examine the accuracy of the non-local finite element formulation and approximate expressions of the natural frequencies, normal modes and transfer functions. A zigzag (7, 0) SWCNT with Young's modulus $E = 6.85$ TPa, $L = 25$ nm, density $\rho = 9.517 \times 10^3$ kg/m^3 and thickness $t = 0.08$ nm is taken from [MUR 11b]. The system considered here is shown in Figure 8.4.

For a carbon nanotube with chirality (n_i, m_i), the diameter can be given by:

$$d_i = \frac{r}{\pi}\sqrt{n_i^2 + m_i^2 + n_i m_i} \qquad [8.63]$$

where $r = 0.246$ nm. The diameter of the SWCNT shown in Figure 8.4 is 0.55 nm. A constant modal damping factor of 1% for all the modes is assumed. By comparing with MD simulation results [CHO 11a, CHO 10], it was observed that $e_0a = 1$ nm is the optimal value of the non-local parameter. In this study, however, we consider a range of values of e_0a within 0–2 nm to understand its role in the accuracy of the dynamic characteristics of the system.

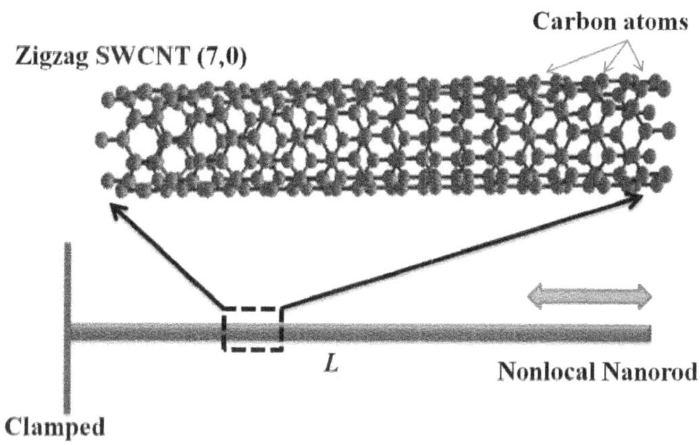

Figure 8.4. *Axial vibration of a zigzag (7, 0) single-walled carbon nanotube (SWCNT) with clamped-free boundary condition. For a color version of the figure, see www.iste.co.uk/murmu/non-local.zip*

We consider clamped-free boundary condition for the SWCNT. Undamped non-local natural frequencies can be obtained [AYD 09] as:

$$\lambda_j = \sqrt{\frac{EA}{m}} \frac{\sigma_j}{\sqrt{1+\sigma_j^2(e_0a)^2}}, \quad \text{where} \quad \sigma_j = \frac{(2j-1)\pi}{2L}, \quad j = 1, 2, \cdots \qquad [8.64]$$

EA is the axial rigidity and m is the mass per unit length of the SWCNT. For the finite element analysis, the SWCNT is divided into 200 elements. The dimension of each of the system matrices becomes 200×200, that is $n = 200$. The global mass matrices $\mathbf{M}_0$ and $\mathbf{M}_\mu$ are obtained by assembling the element mass matrix given by [8.4]. For this case, it turns out (see element stiffness matrix in [8.34]) that the non-local part of the mass matrix is actually proportional to the stiffness matrix, that is $\mathbf{M}_{\mu} \propto \mathbf{K}$. Therefore, the condition for the existence of classical normal modes for the undamped system given by equation [8.34] is exactly satisfied in this case. This in turn implies that the error in the approximate expressions in section 8.3.2 should be zero as $M_{\mu_{kl}} = 0, \forall k \neq l$. We give numerical results to demonstrate that the theory for the existence of classical normal modes for non-local system derived in section 8.3.1 and the approximate expressions derived in section 8.3.2 are consistent.

In Figure 8.5, the natural frequencies obtained using the analytical expression [8.64] are compared with direct finite element simulation results. The frequency values are normalized with respect to the first local natural frequency ω_1.

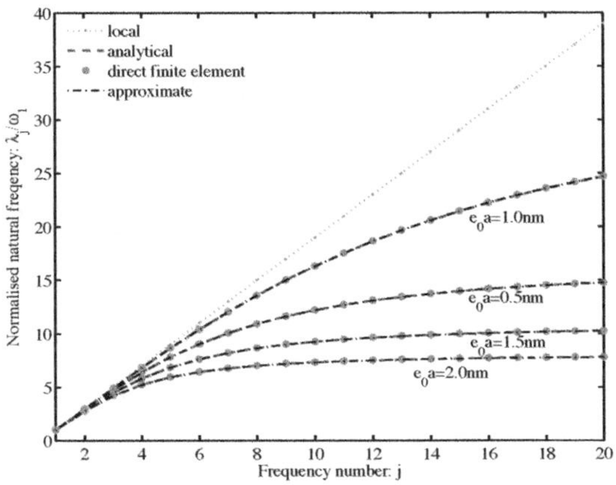

Figure 8.5. *The variation of first 20 undamped natural frequencies for the axial vibration of SWCNT. Four representative values of e_0a (in nm) are considered. For a color version of the figure, see www.iste.co.uk/murmu/non-local.zip*

The first 20 non-local natural frequencies are shown in Figure 8.5, where four values of e_0a, namely 0.5, 1.0, 1.5 and 2.0 nm, have been used. In the same figure, natural frequencies obtained using the direct finite element method and the results obtained using the approximate expression [8.44] are also shown. It can be observed that the values obtained using three different approaches coincide for this problem. Natural frequencies corresponding to the underlying local system are shown in Figure 8.5. Local frequencies are qualitatively different from non-local frequencies as they increase linearly with the number of modes. Non-local frequencies on the other hand approach to a constant value with increasing modes. This upper bound is known as the asymptotic frequency [ADH 13] and is given by $\lambda_{\max} = \frac{1}{(e_0 a)}\sqrt{\frac{EA}{m}}$. It is worth noting that the approximate expression of the natural frequency given by equation [8.44] is able to capture the asymptotic frequency for the axial vibration of SWCNT. Therefore, equation [8.44] can be used to understand both quantitative and qualitative behavior of the natural frequencies of a non-local system.

In Figure 8.6, mode shapes corresponding to modes 2, 5, 6 and 9 are shown for four values of the non-local parameter.

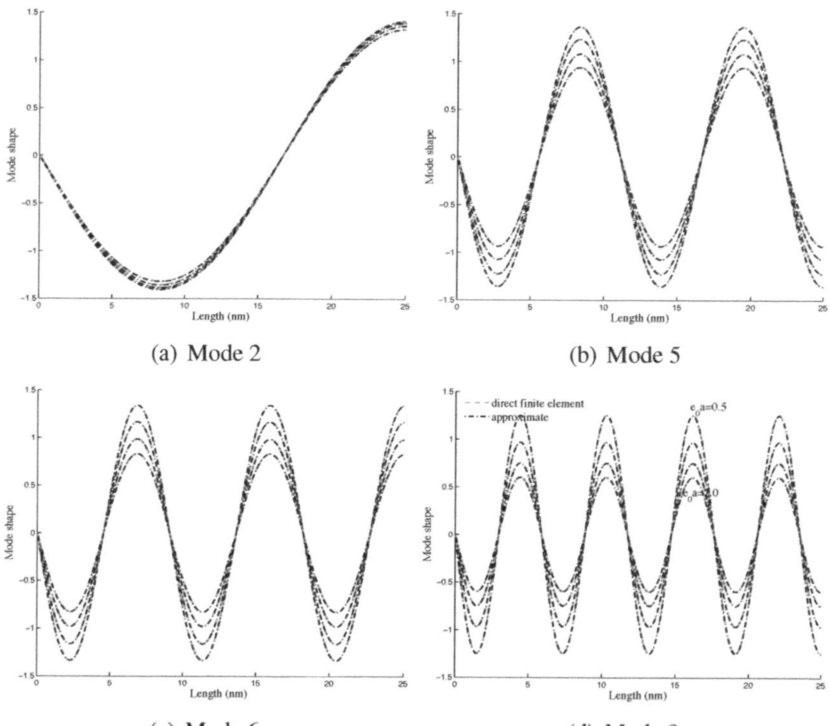

(a) Mode 2 (b) Mode 5

(c) Mode 6 (d) Mode 9

Figure 8.6. *Four selected mode shapes for the axial vibration of SWCNT. Exact finite element results are compared with the approximate analysis based on local eigensolutions. In each subplot, four different values of e_0a, namely 0.5, 1.0, 1.5 and 2.0 nm, have been used (see subplot d). For a color version of the figure, see www.iste.co.uk/murmu/non-local.zip*

These mode numbers are selected for illustration only. The results obtained from the direct finite element are compared with the approximate expression given by equation [8.48]. The mode shapes obtained by both approaches agree with each other well.

Finally, in Figure 8.7, the frequency response function of the tip of the SWCNT is shown for the four representative values of the non-local parameter.

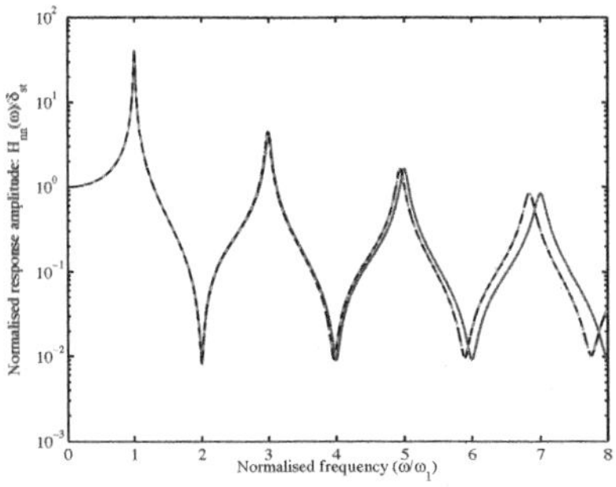

(a) $e_0a = 0.5$nm

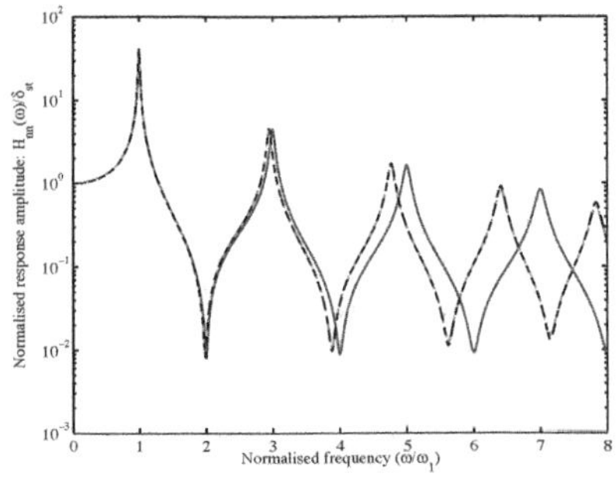

(b) $e_0a = 1.0$nm

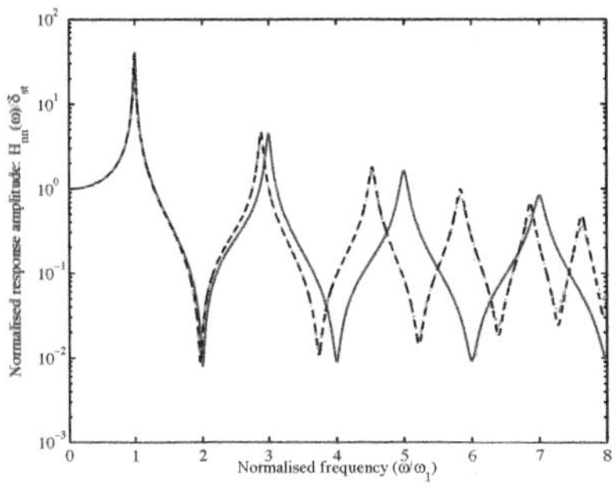

(c) $e_0a = 1.5$nm

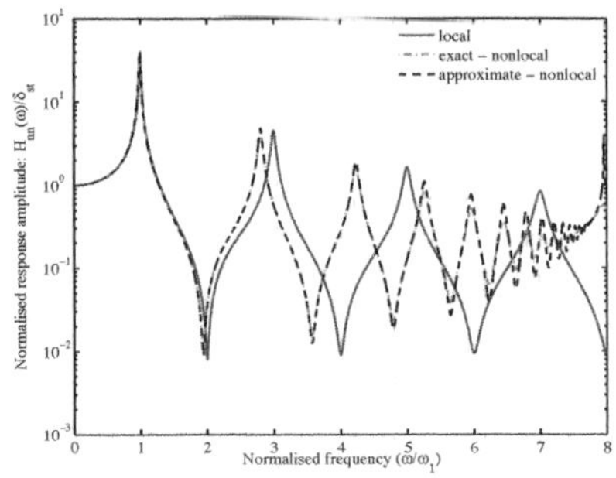

(d) $e_0a = 2.0$nm

Figure 8.7. *Amplitude of the normalized frequency response of the SWCNT at the tip for different values of* e_0a. *Exact finite element results are compared with the approximate analysis based on local eigensolutions. For a color version of the figure, see www.iste.co.uk/murmu/non-local.zip*

On the x-axis, excitation frequency normalized with respect to the first local frequency is considered. The frequency response is normalized by the static response δ_{st} (response when the excitation frequency is zero). The frequency response function of the underlying local model is also plotted to show the difference between the local and non-local responses. For the non-local system, the frequency response is obtained by the direct finite element method and the approximation derived in section 8.4. As a proportional damping model is assumed, the off-diagonal part of the modal damping matrix is a null matrix. For this case, the approximate solution matches exactly to the results obtained from the direct finite element method.

8.5.2. *Bending vibration of a double-walled carbon nanotube*

A double-walled carbon nanotube (DWCNT) is considered to examine the bending vibration characteristics. An armchair (5, 5), (8, 8) DWCNT with Young's modulus $E = 1.0$ TPa, $L = 30$ nm, density $\rho = 2.3 \times 10^3$ kg/m^3 and thickness $t = 0.35$ nm is considered as in [MUR 12b]. The inner and the outer diameters of the DWCNT are, respectively, 0.68 and 1.1 nm. The system considered here is shown in Figure 8.8.

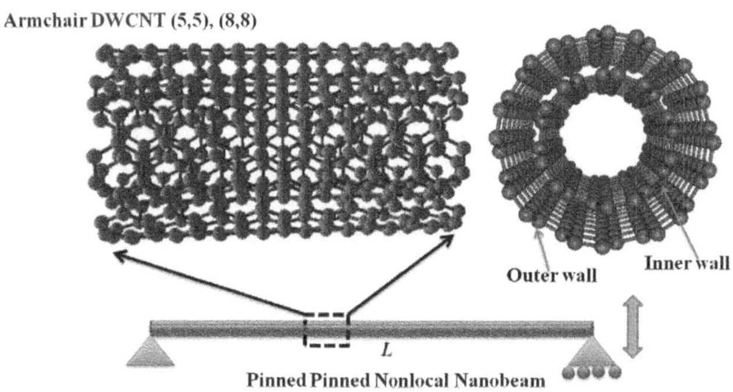

Figure 8.8. *Bending vibration of an armchair (5, 5), (8, 8) double-walled carbon nanotube (DWCNT) with pinned-pinned boundary condition. For a color version of the figure, see www.iste.co.uk/murmu/non-local.zip*

We consider pinned-pinned boundary conditions for the DWCNT. Undamped non-local natural frequencies can be obtained [AYD 09] as:

$$\lambda_j = \sqrt{\frac{EI}{m}} \frac{\beta_j^2}{\sqrt{1 + \beta_j^2 (e_0 a)^2}} \quad \text{where} \quad \beta_j = j\pi/L, \quad j = 1, 2, \cdots \quad [8.65]$$

EI is the bending rigidity and m is the mass per unit length of the DWCNT. For the finite element analysis, the DWCNT is divided into 100 elements. The dimension of each of the system matrices becomes 200×200, that is $n = 200$. The global mass matrices $\mathbf{M}_0$ and $\mathbf{M}_\mu$ are obtained by assembling the element mass matrix given by [8.9]. Unlike the case of the axial vibration of rods, the non-local part of the mass matrix is not proportional to the stiffness matrix. Therefore, the condition for the existence of classical normal modes for the undamped system given by equation [8.34] is not satisfied for this case. This numerical study, therefore, quantifies the accuracy of the approximate expression proposed in this chapter.

The natural frequencies obtained using the analytical expression [8.65] are compared with the direct finite element simulation in Figure 8.9. The frequency values are normalized with respect to the first local natural frequency.

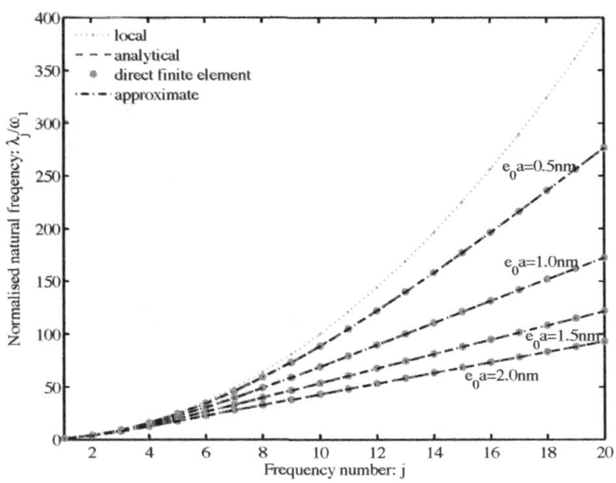

Figure 8.9. *The variation of first 20 undamped natural frequencies for the bending vibration of DWCNT. Four representative values of e_0a (in nm) are considered. For a color version of the figure, see www.iste.co.uk/murmu/non-local.zip*

The first 20 non-local natural frequencies are shown for four distinct values of e_0a, namely 0.5, 1.0, 1.5 and 2.0 nm. In the same figure, natural frequencies obtained using the direct finite element method and the results obtained using the approximate expression [8.44] are also shown. It can be observed that the values obtained using three different approaches almost coincide for this problem. Natural frequencies corresponding to the underlying local system are shown in Figure 8.9. Local frequencies are qualitatively different from non-local frequencies as they increase quadratically with the number of modes. Non-local frequencies on the other hand increase linearly with the number of modes. The approximate expression of the natural frequency given by equation [8.44] is able to capture this crucial qualitative difference.

In Figure 8.10, mode shapes corresponding to modes 2, 5, 6 and 9 are shown for four values of the non-local parameter.

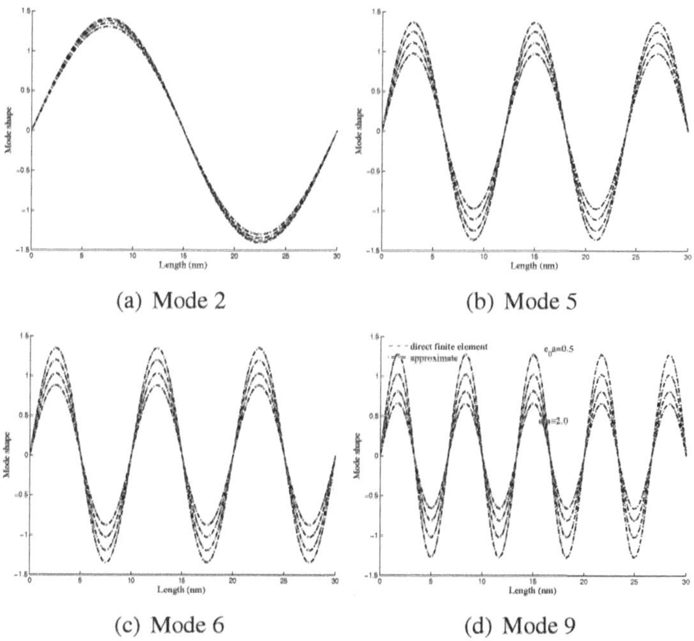

(a) Mode 2 (b) Mode 5

(c) Mode 6 (d) Mode 9

Figure 8.10. *Four selected mode shapes for the bending vibration of DWCNT. Exact finite element results are compared with the approximate analysis based on local eigensolutions. In each subplot, four different values of* e_0a*, namely 0.5, 1.0, 1.5 and 2.0 nm have been used (see subplot d). For a color version of the figure, see www.iste.co.uk/murmu/non-local.zip*

These mode numbers are selected for illustration only. The results obtained from the direct finite elements are compared with the approximate expression given by equation [8.48]. The mode shapes obtained by both approaches agree with each other.

In Figure 8.11, the amplitude of the frequency response function $H_{ij}(\omega)$ for $i = 6, j = 8$ is shown for the four representative values of the non-local parameter.

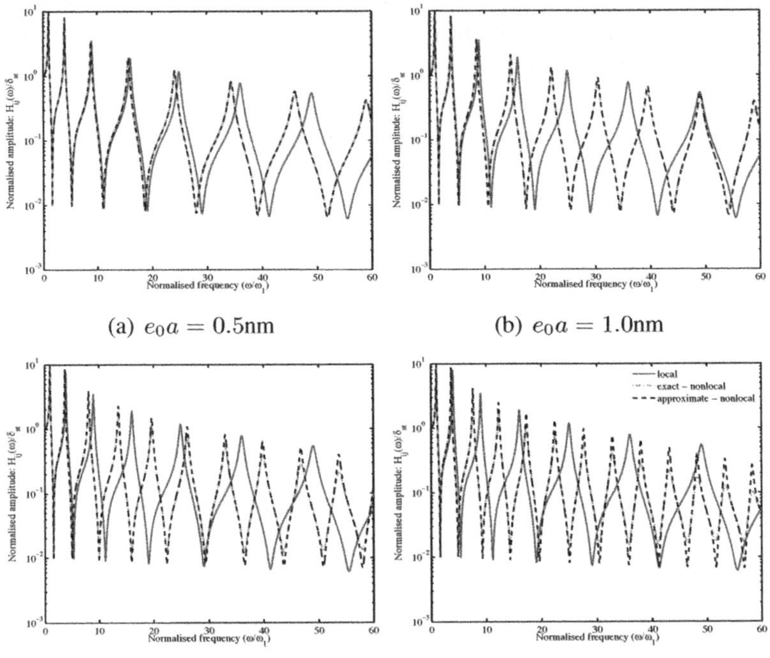

(a) $e_0a = 0.5$nm (b) $e_0a = 1.0$nm

(c) $e_0a = 1.5$nm (d) $e_0a = 2.0$nm

Figure 8.11. *Amplitude of the normalized frequency response of the DWCNT $H_{ij}(\omega)$ for $i = 6, j = 8$ for different values of e_0a. Exact finite element results are compared with the approximate analysis based on local eigensolutions. For a color version of the figure, see www.iste.co.uk/murmu/non-local.zip*

On the x-axis, excitation frequency normalized with respect to the first local frequency is considered. The frequency response is normalized by the static response d_{st}. The frequency response function of the underlying local model is also plotted to show the difference between the local and non-local response. For the non-local system, the frequency response is obtained by the direct finite element method and the approximation derived in section 8.4. As proportional damping model is assumed, the off-diagonal part of the modal damping matrix is a null matrix. For this case, the approximate solution matches closely to the results obtained from the direct finite element method. The dynamic response of the non-local system becomes very different from the corresponding local system for higher frequency values and higher values of the non-local parameter e_0a. The proposed approximate expression of the transfer function given in equation [8.59] can be used to understand this significant different behavior for the bending vibration of DWCNT.

8.5.3. *Transverse vibration of a single-layer graphene sheet*

A rectangular SLGS is considered to examine the transverse vibration characteristics of nanoplates. The graphene sheet is of dimension $L = 20$ nm, $W = 15$ nm and Young's modulus $E = 1.0$ TPa, density $\rho = 2.25\times 10^3$ kg/m^3, Poisson's ratio $\nu = 0.3$ and thickness $h = 0.34$ nm is considered as in [CHO 11b]. The system considered here is shown in Figure 8.12.

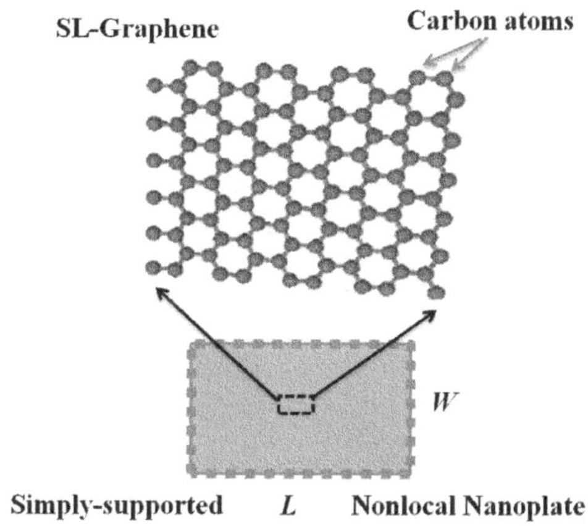

Figure 8.12. *Transverse vibration of a rectangular ($L = 20$ nm, $W = 15$ nm) single-layer graphene sheet (SLGS) with simply-supported boundary condition along the four edges. For a color version of the figure, see www.iste.co.uk/murmu/non-local.zip*

We consider simply-supported boundary condition along the four edges for the SLGS. Undamped non-local natural frequencies can be obtained [KIA 11, KAR 10] as:

$$\lambda_{ij} = \sqrt{\frac{D}{m}} \frac{\beta_{ij}^2}{\sqrt{1 + \beta_{ij}^2 (e_0a)^2}} \quad \text{where} \quad \beta_{ij} = \sqrt{(i\pi/L)^2 + (j\pi/W)^2}, \quad i, j = 1, 2, \cdots \qquad [8.66]$$

D is the bending rigidity and m is the mass per unit area of the SLGS. For the finite element analysis, the DWCNT is divided into 20×15 elements. The dimension of each of the system matrices becomes 868×868, that is $n = 868$. The global mass matrices $\mathbf{M}_0$ and $\mathbf{M}_\mu$ are obtained by assembling the element mass matrix given by [8.18]. Like the case of the bending vibration of nanobeams, the non-local part of the mass matrix is not proportional to the stiffness matrix. Therefore, the condition for the existence of classical normal modes for the undamped system given by equation [8.34] is not satisfied for this case. Among the three types of systems considered here, only the nanorod satisfies the condition of existence of classical normal modes.

In Figure 8.13, the natural frequencies obtained using the analytical expression [8.66] are compared with the direct finite element simulation. The frequency values are normalized with respect to the first local natural frequency.

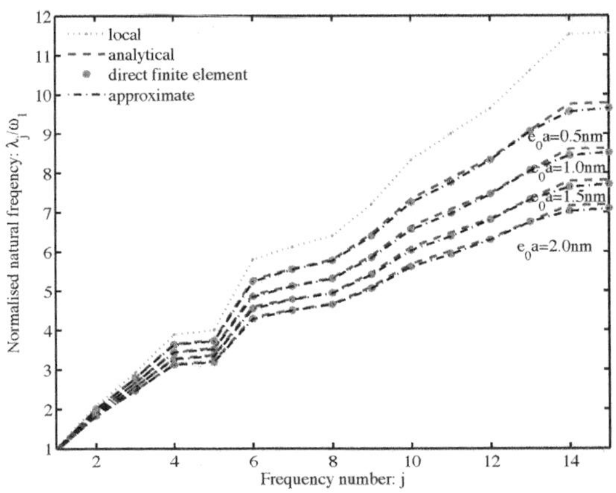

Figure 8.13. *The variation of first 15 undamped natural frequencies for the transverse vibration of SLGS. Four representative values of e_0a (in nm) are considered. For a color version of the figure, see www.iste.co.uk/murmu/non-local.zip*

The first 15 non-local natural frequencies are shown for four distinct values of e_0a, namely 0.5, 1.0, 1.5 and 2.0 nm. In the same figure, natural frequencies obtained using the direct finite element method and the results obtained using the approximate expression [8.44] are also shown. It can be observed that the values obtained using three different approaches are very close. Natural frequencies corresponding to the underlying local system are shown in Figure 8.13. Local frequencies diverge significantly from the non-local frequencies for higher frequency indices. The approximate expression of the natural frequency given by equation [8.44] is able to capture this quantitative difference very well.

In Figure 8.14, mode shapes corresponding to modes 2, 4, 5 and 6 are shown when the non-local parameter $e_0a = 2$ nm. We have selected the highest value of e_0a as this leads to maximum inaccuracy of the proposed approximate expressions. Results obtained from the direct finite element and the approximate expression given by equation [8.48] are shown in these graphs.

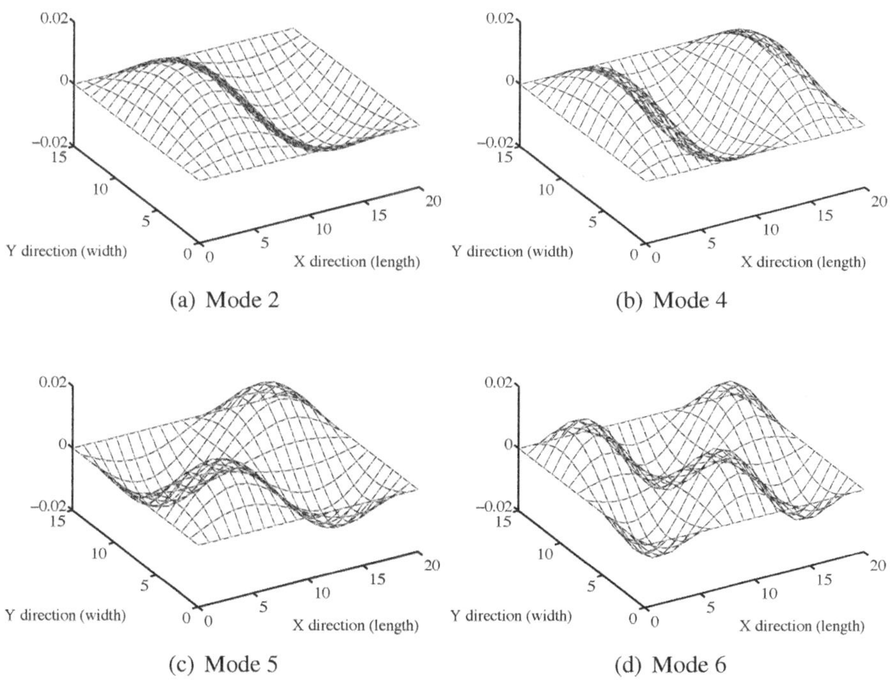

Figure 8.14. *Four selected mode shapes for the transverse vibration of SLGS for $e_0a = 2$ nm. Exact finite element results (solid line) are compared with the approximate analysis based on local eigensolutions (dashed line). For a color version of the figure, see www.iste.co.uk/murmu/non-local.zip*

These mode numbers are selected for illustration only. Results obtained from the direct finite element and the approximate expression given by equation [8.48] are shown in these plots. The mode shapes obtained by both approaches agree to each other well.

Finally, in Figure 8.15, the amplitude of the frequency response function $H_{ij}(\omega)$ for $i = 475, j = 342$ is shown for the four representative values of the non-local parameter.

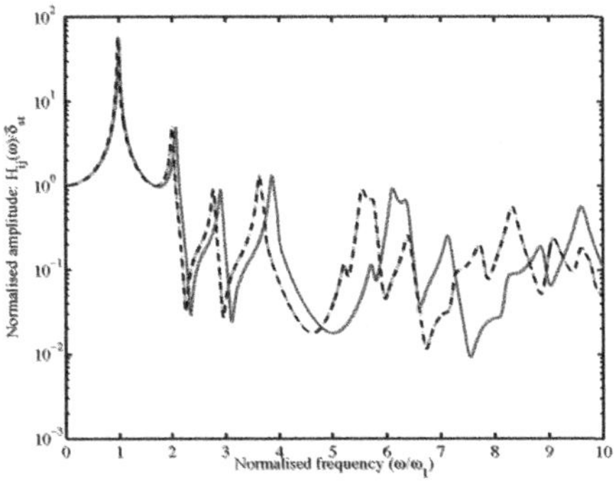

(a) $e_0a = 0.5$nm

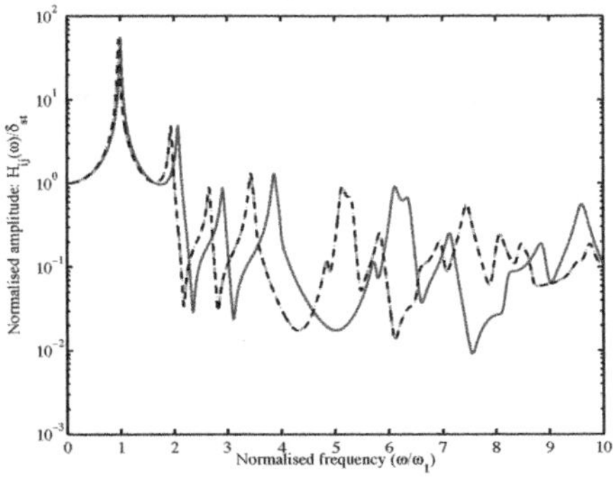

(b) $e_0a = 1.0$nm

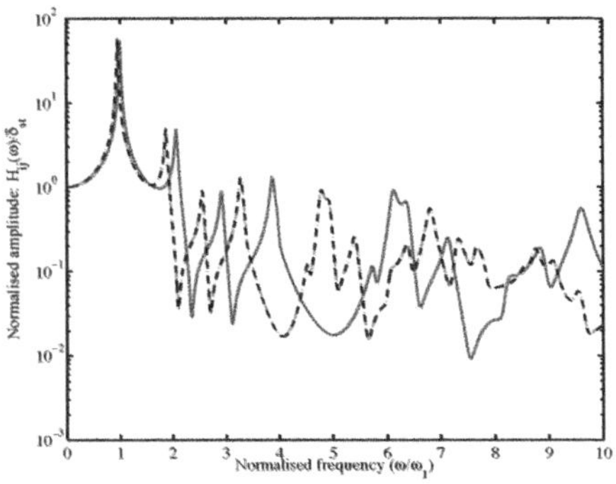

(c) $e_0a = 1.5$nm

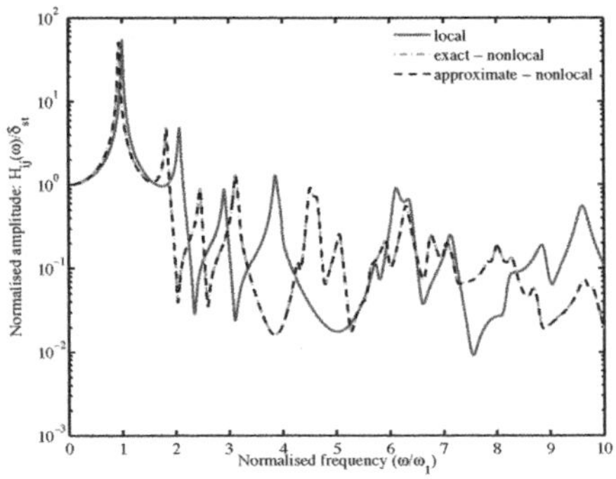

(d) $e_0a = 2.0$nm

Figure 8.15. *Amplitude of the normalized frequency response $H_{ij}(\omega)$ for $i = 475, j = 342$ of the SLGS for different values of e_0a. Exact finite element results are compared with the approximate analysis based on local eigensolutions. For a color version of the figure, see www.iste.co.uk/murmu/non-local.zip*

On the x-axis, excitation frequency normalized with respect to the first local frequency is considered. The frequency response is normalized by the static response d_{st} (that is the response when the excitation frequency is zero rad/s). The frequency response function of the underlying local model is also plotted to show the difference between the local and non-local response. For the non-local system, the frequency response is obtained by the direct finite element method and the approximation derived in section 8.4. As proportional damping model is assumed, the off-diagonal part of the modal damping matrix is a null matrix. For this case, the approximate solution matches exactly to the results obtained from the direct finite element method. The dynamic response of the non-local system becomes very different from the corresponding local system for higher frequency values and higher values of the non-local parameter e_0a. The proposed approximate expression of the transfer function given in equation [8.59] can be used to understand this significant different behavior.

8.6. Summary

Non-local elasticity is a promising theory for the modeling of nanoscale dynamical systems such as carbon nantotubes and graphene sheets. A finite element approach is proposed for dynamic analysis of general non-local structures. Explicit closed-form expressions of element mass and stiffness matrices of nanorods, nanobeams and nanoplates have been derived. The mass matrix can be decomposed into two parts, namely the classical local mass matrix $\mathbf{M}_0$ and a non-local part denoted by $\mathbf{M}_\mu$. The non-local part of the mass matrix is scale-dependent and vanishes for systems with large length scales. New analytical approaches have been developed to understand the dynamic behavior of general discrete non-local systems. Approximate expressions for non-local natural frequencies, mode shapes and frequency response functions have been derived. Major theoretical contributions made in this chapter include the following results:

– an undamped non-local system will have classical normal modes provided that the non-local part of the mass matrix satisfies the condition $\mathbf{K}\mathbf{M}_0^{-1}\mathbf{M}_\mu = \mathbf{M}_\mu\mathbf{M}_0^{-1}\mathbf{K}$ where $\mathbf{K}$ is the stiffness matrix;

– a viscously damped non-local system with damping matrix $\mathbf{C}$ will have classical normal modes provided that $\mathbf{C}\mathbf{M}_0^{-1}\mathbf{K} = \mathbf{K}\mathbf{M}_0^{-1}\mathbf{C}$ and $\mathbf{C}\mathbf{M}_0^{-1}\mathbf{M}_\mu = \mathbf{M}_\mu\mathbf{M}_0^{-1}\mathbf{C}$ in addition to the previous condition;

– natural frequency of a general non-local system can be expressed as $\lambda_j \approx \frac{\omega_j}{\sqrt{1+M'_{\mu_{jj}}}}, \forall j = 1, 2, \cdots$ where ω_j are the corresponding local frequencies and $M'_{\mu_{jj}}$ are the elements of non-local part of the mass matrix in the modal coordinate;

– every non-local normal mode can be expressed as a sum of two principal components as $\mathbf{u}_j \approx \mathbf{x}_j + (\sum_{k \neq j}^{n} \frac{\lambda_j^2}{(\lambda_k^2 - \lambda_j^2)} \frac{M'_{\mu_{kj}}}{(1+M'_{\mu_{kk}})} \mathbf{x}_k), \forall j = 1, 2, \cdots$. One of them is parallel to the corresponding local mode $\mathbf{x}_j$ and the other is orthogonal to it.

The theoretical results obtained in this chapter are applied to three representative problems, namely (i) axial vibration of an SWCNT, (ii) bending vibration of a DWCNT and (iii) transverse vibration of an SLGS. These three systems are modeled by non-local rods, beams and plates, respectively. Among these three systems, only the non-local rod model satisfies the condition of existence of classical normal modes. For the other two systems, it was observed that the proposed approximate expressions of non-local natural frequencies, mode shapes and frequency response functions provide acceptable accuracy. The results obtained in this chapter give physical insights into the dynamic behavior of discrete non-local systems which can be understood in the light of well-known dynamic behavior of the underlying local systems.

The finite element method developed in this chapter is based on static shape functions. In the next chapter, dynamic shape functions are employed.

9

Dynamic Finite Element Analysis of Non-local Rods: Axial Vibration

The formulation in the previous chapter was based on static shape functions. Here, we extend the formulation to include dynamic shape functions.

This chapter gives an overview of dynamic finite element methods applied to non-local nanostructures. We consider a simple example of a non-local rod. Free and forced axial vibration of damped non-local rods are presented. Two types of non-local damping models, namely, strain-rate-dependent viscous damping and velocity-dependent viscous damping, are considered. A frequency-dependent dynamic finite element method (FEM) is developed to obtain the forced vibration response. Frequency-adaptive complex-valued shape functions are proposed to obtain the dynamic stiffness matrix in closed form. The stiffness and mass matrices of non-local rods are also obtained using the conventional FEM. Results from the dynamic FEM and conventional FEM are compared. Using an asymptotic analysis, it is shown that, unlike its local counterpart, a non-local rod has a maximum cutoff frequency. A closed-form exact expression for this maximum frequency as a function of the non-local parameter has been presented for undamped and damped systems. The frequency response function obtained using the proposed dynamic FEM shows an extremely high modal density near the maximum frequency. This leads to clustering of resonance peaks which is not easily obtainable using classical finite element analysis.

9.1. Introduction

Research on size-dependent structural theories for the accurate design and analysis of microstructures and nanostructures is growing rapidly [AKG 11, KAH 10, KON 09, CHH 11]. This is because, although molecular dynamic (MD) simulation is justified for the analysis of nanostructures [CHO 11b, CHO 10a] (such as nanorods, nanobeam, nanoplates, nanoshells and nanocones), the approach is computationally exorbitant for nanostructures with large numbers of atoms. This calls for the use of conventional continuum mechanics [BER 09] and finite elements in analysis of nanostructures. However, although the classical continuum modeling approach is considered scale-free, it fails to account for the effects arising from the small scale where "size-effects" are prominent.

Nanoscale experiments demonstrate that the mechanical properties of nanodimensional materials are much influenced by size effects or scale effects [ZHA 09, BIE 06]. Size effects are related to atoms and molecules that constitute the materials. Furthermore, atomistic simulations have also reported size effects on the magnitudes of resonance frequency and buckling loads of nanoscale objects such as nanotubes and graphene [CHO 11a, SCA 08]. The application of classical continuum approaches is thus questionable in the analysis of nanostructures such as nanorods, nanobeams and nanoplates. Examples of nanorods and nanobeams include carbon and boron nanotubes, while examples of nanoplates can be graphene sheets or gold nanoplates.

One widely promising size-dependent continuum theory is the non-local elasticity theory pioneered in [ERI 83] which brings in the scale effects and underlying physics within the formulation. The theory is already elaborated in Chapter 1. Non-local elasticity theory contains information related to the forces between atoms, and the internal length scale in structural, thermal and mechanical analyses. In the non-local elasticity theory, the small-scale effects are captured by assuming that the stress at a point is a function of the strains at all points in the domain. Non-local theory considers long-range interatomic interaction and yields results dependent on the size of a body [ERI 83]. Some drawbacks of classical continuum theory can be efficiently avoided and the size-dependent phenomena can be reasonably explained by non-local elasticity. Recent literature shows that the theory of non-local elasticity is being increasingly used for reliable and fast analysis of nanostructures. Studies include non-local analysis of nanostructures, e.g. nanobeams [PIS 03, ZHA 10, LU 07], nanoplates [KIA 11], carbon nanotubes (CNTs)

[KAR 10], graphene [SAM 11], microtubules [CIV 11] and nanorings [WAN 08].

Recently, due to greater interest in nanotechnology, various one-dimensional nanostructures have been realized. They include nanodots, nanorods, nanowires, nanobelts, nanotubes, nanobridges, nanonails, nanowalls and nanohelixes. Among all the one-dimensional nanostructures, nanotubes, nanorods and nanowires are the most widely studied (see Chapter 1). This is because of the easy material formation and device applications. One important one-dimensional nanostructure is nanorods. Nanorods [WEN 03] are one-dimensional objects ranging from 1 to 3,000 nm in length. They can be grown from various methods, including (i) vapor phase synthesis [RED 07b], (ii) metal-organic chemical vapor deposition [PAR 02] and (iii) hydrothermal synthesis [CHO 04]. Nanorods have found an application in a variety of nanodevices, including ultraviolet photodetectors, nanosensors, transistors, diodes and light emitting diode (LED) arrays.

Axial vibration experiments can be used for the determination of Young's modulus of CNTs. Generally, the flexural modes occur at low frequencies. However, vibrating nanobeams (CNTs) may also have longitudinal modes at relatively high frequencies and can be of very practical significance in high-operating frequencies. Nanorods, when used as electromechanical resonators, can be externally excited and exhibit axial vibrations. Furthermore, for a moving nanoparticle inside a single-walled carbon nanotube (SWCNT), the SWCNT generally vibrates both in the transverse and longitudinal directions. The longitudinal vibration is generated because of the friction existing between the outer surface of the moving nanoparticle and the inner surface of the SWCNT. It is also reported [SAP 06] that transport measurements on suspended SWCNTs show the signatures of phonon-assisted tunneling, influenced by longitudinal vibration (stretching) modes. Chowdhury *et al.* [CHO 10b] have reported sliding axial modes for multiwalled carbon nanotubes (MWCNTs). Tong *et al.* [TON 09] have considered axial buckling of MWCNTs with heterogeneous boundary conditions.

Only limited research work on non-local elasticity has been devoted to the axial vibration of nanorods. Aydogdu [AYD 09] developed a non-local elastic rod model and applied it to investigate the small-scale effect on the axial vibration of clamped–clamped and clamped–free nanorods. Filiz and Aydogdu [FIL 10] applied the axial vibration of non-local rod theory to CNT heterojunction systems. Narendra and Gopalkrishnan [NAR 10] have studied the wave propagation of non-local nanorods. Recently, Murmu and Adhikari [MUR 10] have studied the axial vibration analysis of a double-nanorod-system. In this chapter, we will be referring to a nanorod as a non-local rod, so as to distinguish it from a local rod.

Several computational techniques have been used for solving the non-local governing differential equations. These techniques include Navier's method [PRA 09], differential quadrature method (DQM) [MAL 11] and the Galerkin technique [FIR 11]. Recently, attempts have been made to develop a finite element method (FEM) based on non-local elasticity. The upgraded FEM in contrast to other aforementioned methods can effectively handle more complex geometry, material properties as well as boundary and/or loading conditions. Pisano *et al.* [PIS 09] reported a finite element procedure for non-local integral elasticity. Recently, some motivating work on a finite element approach based on non-local elasticity was reported [PHA 10]. The majority of the reported works consider free vibration studies where the effect of non-locality on the eigensolutions has been studied. However, forced vibration response analysis of non-local systems has received very little attention.

Based on the above discussion, in this chapter we develop the dynamic FEM based on non-local elasticity with the aim of considering dynamic response analysis. The dynamic FEM belongs to the general class of spectral methods for linear dynamical systems [DOY 89]. This approach, or approaches very similar to this, is known by various names such as the dynamic stiffness method [PAZ 80, BAN 85, BAN 89, BAN 92b, BAN 92a, FER 93a, FER 93b, BAN 95, MAN 98, BAN 97, ADH 00], spectral FEM [DOY 89, GOP 07] and dynamic FEM [HAS 99, HAS 00]. Some of the key features of the method are:

– the mass distribution of the element is treated in an exact manner in deriving the element dynamic stiffness matrix;

– the dynamic stiffness matrix of one-dimensional structural elements, taking into account the effects of flexure, torsion, axial and shear deformation, and damping, is exactly determinable, which, in turn, enables the exact vibration analysis by an inversion of the global dynamic stiffness matrix;

– the method does not employ eigenfunction expansions and, consequently, a major step of the traditional finite element analysis, namely, the determination of natural frequencies and mode shapes, is eliminated which automatically avoids the errors due to series truncation;

– since modal expansion is not employed, *ad hoc* assumptions concerning the damping matrix being proportional to the mass and/or stiffness are not necessary;

– the method is essentially a frequency-domain approach suitable for steady-state harmonic or stationary random excitation problems;

– the static stiffness matrix and the consistent mass matrix appear as the first two terms in the Taylor expansion of the dynamic stiffness matrix in the frequency parameter.

So far, the dynamic FEM has been applied to classical local systems only. In this chapter, we generalize this approach to non-local systems. One of the novel features of the analysis proposed here is the employment of frequency-dependent complex non-local shape functions for damped systems. This in turn enables us to obtain the element stiffness matrix using the usual weak form of the FEM.

The chapter is organized as follows. In section 9.2, we introduce the equation of motion of axial vibration of undamped and damped rods. Natural frequencies and their asymptotic behaviors for both cases are discussed for different boundary conditions. The conventional and the dynamic FEM are developed in section 9.3. Closed-form expressions are derived for the mass and stiffness matrices. In section 9.4, the proposed methodology is applied to an armchair SWCNT for illustration. Theoretical results, including the asymptotic behaviors of the natural frequencies, are numerically illustrated. Finally, in section 9.5 some conclusions are drawn based on the results obtained in this chapter.

9.2. Axial vibration of damped non-local rods

9.2.1. *Equation of motion*

The equation of motion of axial vibration for a damped non-local rod can be expressed as:

$$\begin{aligned} &EA\frac{\partial^2 U(x,t)}{\partial x^2} + \widehat{c}_1\left(1-(e_0a)_1^2\frac{\partial^2}{\partial x^2}\right)\frac{\partial^3 U(x,t)}{\partial x^2 \partial t} \\ &= \widehat{c}_2\left(1-(e_0a)_2^2\frac{\partial^2}{\partial x^2}\right)\frac{\partial U(x,t)}{\partial t} \\ &\qquad + \left(1-(e_0a)^2\frac{\partial^2}{\partial x^2}\right)\left\{m\frac{\partial^2 U(x,t)}{\partial t^2} + F(x,t)\right\} \qquad [9.1] \end{aligned}$$

This is an extension of the equation of motion of an undamped non-local rod for axial vibration [MUR 10, RED 07a, AYD 09]. Here, EA is the axial rigidity, m is the mass per unit length, e_0a is the non-local parameter [ERI 83], $U(x,t)$ is the axial displacement, $F(x,t)$ is the applied force, x is the spatial variable and t is the time. The constant $\widehat{c}_1$ is the strain-rate-dependent viscous damping coefficient and $\widehat{c}_2$ is the velocity-dependent viscous damping coefficient. The parameters $(e_0a)_1$ and $(e_0a)_2$ are non-local parameters related to the two damping terms, respectively. For simplicity, we have not taken into account any non-local effect related to the damping. Although this can be mathematically incorporated in the analysis, the determination of these non-local parameters is beyond the scope of this work and therefore only local interaction for the damping is adopted. Thus, in the following analysis, we consider $(e_0a)_1 = (e_0a)_2 = 0$. Assuming harmonic response as:

$$U(x,t) = u(x)\exp\left[\mathrm{i}\omega t\right] \qquad [9.2]$$

and considering free vibration, from equation [9.3] we have:

$$\left(1 + \mathrm{i}\omega\frac{\widehat{c}_1}{EA} - \frac{m\omega^2}{EA}(e_0a)^2\right)\frac{d^2u}{dx^2} + \left(\frac{m\omega^2}{EA} - \mathrm{i}\omega\frac{\widehat{c}_2}{EA}\right)u(x) = 0 \qquad [9.3]$$

Following the damping convention in dynamic analysis [MEI 97], we consider stiffness and mass proportional damping. Therefore, we express the damping constants as:

$$\widehat{c}_1 = \zeta_1(EA) \quad \text{and} \quad \widehat{c}_2 = \zeta_2(m) \qquad [9.4]$$

where ζ_1 and ζ_2 are the stiffness and mass proportional damping factors. Substituting these, from equation [9.3] we have:

$$\frac{d^2u}{dx^2} + \alpha^2 u = 0 \qquad [9.5]$$

where:

$$\alpha^2 = \frac{\left(\omega^2 - \mathrm{i}\zeta_2\omega\right)/c^2}{\left(1 + \mathrm{i}\omega\zeta_1 - (e_0a)^2\omega^2/c^2\right)} \qquad [9.6]$$

with

$$c^2 = \frac{EA}{m} \qquad [9.7]$$

It can be noticed that α^2 is a complex function of the frequency parameter ω. In the special case of undamped systems when damping coefficients ζ_1 and ζ_1 go to zero, α^2 in equation [9.6] reduces to $\frac{\Omega^2}{1-(e_0a)^2\Omega^2}$, which is a real function of ω.

9.2.2. *Analysis of damped natural frequencies*

Natural frequencies of undamped non-local rods have been discussed in the literature [AYD 09]. Natural frequencies of damped systems receive little attention. The damped natural frequency depends on the boundary conditions. We denote a parameter σ_k as:

$$\sigma_k = \frac{k\pi}{L}, \quad \text{for clamped–clamped boundary conditions} \qquad [9.8]$$

$$\text{and} \quad \sigma_k = \frac{(2k-1)\pi}{2L}, \quad \text{for clamped–free boundary conditions} \qquad [9.9]$$

Following the conventional approach [MEI 97], the natural frequencies can be obtained from:

$$\alpha = \sigma_k \qquad [9.10]$$

Taking the square of this equation and denoting the natural frequencies as ω_k we have:

$$\left(\omega_k^2 - \mathrm{i}\zeta_2\omega_k\right) = \sigma_k^2 c^2 \left(1 + \mathrm{i}\omega_k\zeta_1 - (e_0a)^2\omega_k^2/c^2\right) \qquad [9.11]$$

Rearranging, we obtain:

$$\omega_k^2 \left(1 + \sigma_k^2(e_0a)^2\right) - \mathrm{i}\omega_k \left(\zeta_2 + \zeta_1\sigma_k^2c^2\right) - \sigma_k^2c^2 = 0 \qquad [9.12]$$

This is a very generic equation and many special cases can be obtained from this as follows:

– *undamped local systems:* this case can be obtained by substituting $\zeta_1 = \zeta_2 = 0$ and $e_0a = 0$. From equation [9.12], we therefore obtain:

$$\omega_k = \sigma_k c \qquad [9.13]$$

which is the classical expression [MEI 97].

– *undamped non-local systems:* this case can be obtained by substituting $\zeta_1 = \zeta_2 = 0$. Solving equation [9.12], we therefore obtain:

$$\omega_k = \frac{\sigma_k c}{\sqrt{1 + \sigma_k^2 (e_0 a)^2}} \tag{9.14}$$

which is obtained in [AYD 09].

– *damped local systems:* this case can be obtained by substituting $\zeta_1 = \zeta_2 = 0$. Solving equation [9.12], we obtain:

$$\omega_k = \mathrm{i}\left(\zeta_2 + \zeta_1 \sigma_k^2 c^2\right)/2 \pm \sigma_k c \sqrt{1 - \left(\zeta_1 \sigma_k c + \zeta_2/(\sigma_k c)\right)^2/4} \tag{9.15}$$

Therefore, the decay rate is $\left(\zeta_2 + \zeta_1 \sigma_k^2 c^2\right)/2$ and damped oscillation frequency is $\sigma_k c \sqrt{1 - \left(\zeta_1 \sigma_k c + \zeta_2/(\sigma_k c)\right)^2/4}$. We observe that damping effectively reduces the oscillation frequency.

For the general case of a non-local damped system, the damped frequency can be obtained by solving equation [9.12] as:

$$\omega_k = \frac{\mathrm{i}\left(\zeta_2 + \zeta_1 \sigma_k^2 c^2\right)}{2\left(1 + \sigma_k^2 (e_0 a)^2\right)} \pm \frac{\sigma_k c}{\sqrt{1 + \sigma_k^2 (e_0 a)^2}} \sqrt{1 - \frac{\left(\zeta_1 \sigma_k c + \zeta_2/(\sigma_k c)\right)^2}{4\left(1 + \sigma_k^2 (e_0 a)^2\right)}} \tag{9.16}$$

Therefore, the decay rate is given by $\frac{\left(\zeta_2 + \zeta_1 \sigma_k^2 c^2\right)}{2\left(1 + \sigma_k^2 (e_0 a)^2\right)}$ and the damped oscillation frequency is given by:

$$\omega_{d_k} = \frac{\sigma_k c}{\sqrt{1 + \sigma_k^2 (e_0 a)^2}} \sqrt{1 - \frac{\left(\zeta_1 \sigma_k c + \zeta_2/(\sigma_k c)\right)^2}{4\left(1 + \sigma_k^2 (e_0 a)^2\right)}} \tag{9.17}$$

It can be observed that the non-local damped system has the lowest natural frequencies. Note that the expressions derived here are general in terms of the boundary conditions.

9.2.3. *Asymptotic analysis of natural frequencies*

In this chapter, we are interested in dynamic response analysis of damped non-local rods. As a result, the behavior of natural frequencies across a wide frequency range is of interest. An asymptotic analysis is conducted here to understand the frequency behavior in the high-frequency limit. We first consider the undamped natural frequency given by equation [9.12]. To obtain asymptotic values, we rewrite the frequency equation in [9.12] and take the mathematical limit $k \to \infty$ to obtain:

$$\lim_{k\to\infty} \omega_k = \lim_{k\to\infty} \frac{c}{\sqrt{\frac{1}{\sigma_k^2} + (e_0 a)^2}} = \frac{c}{(e_0 a)} = \frac{1}{(e_0 a)}\sqrt{\frac{EA}{m}} \tag{9.18}$$

This is obtained by noting the fact that for $k \to \infty$, for both sets of boundary conditions we have $\sigma_k \to \infty$. The result in equation [9.18] shows that there exists an "upper limit" of frequency in non-local systems. This upper limit of frequency is an inherent property of a non-local system. It is a function of material properties only and is independent of the boundary conditions and the length of the rod. The smaller the value of $e_0 a$, the larger this upper limit becomes. Eventually, for a local system when $e_0 a = 0$, the upper limit becomes infinite, which is well known.

Now, we turn our attention to the oscillation frequency of the damped system. Rewriting the expressions for the oscillation frequency from equation [9.17] and taking the limit as $k \to \infty$, we obtain:

$$\begin{aligned}\lim_{k\to\infty} \omega_{d_k} &= \lim_{k\to\infty} \frac{c}{\sqrt{\frac{1}{\sigma_k^2} + (e_0 a)^2}} \sqrt{1 - \frac{\left(\zeta_1 c + \frac{1}{\sigma_k^2}(\zeta_2/c)\right)^2}{4\left(\frac{1}{\sigma_k^2} + (e_0 a)^2\right)}} \\ &= \frac{c}{(e_0 a)}\sqrt{1 - \left(\frac{\zeta_1 c}{2 e_0 a}\right)^2}\end{aligned} \tag{9.19}$$

Therefore, the upper frequency limit for the damped systems is lower than that of the undamped system. It is interesting to note that it is independent of the mass proportional damping ζ_2. Only the stiffness proportional damping has an effect on the upper frequency limit. Equation [9.19] can also be used to obtain an asymptotic critical damping factor. For vibration to continue, the term within the square root in equation [9.19] must be greater than zero. Therefore, the asymptotic critical damping factor for non-local rods can be obtained as:

$$(\zeta_1)_{\text{crit}} = \frac{2e_0a}{c} \quad [9.20]$$

In practical terms, this implies that the value of ζ_1 should be less than this value for high-frequency vibration. Again, observe that like the upper frequency limit, the asymptotic critical damping factor is a function of material properties only and is independent of the boundary conditions and the length of the rod.

The spacing between the natural frequencies is important for dynamic response analysis as the shape of the frequency response function depends on the spacing. Because k is an index, the derivative $\frac{d\omega_k}{dk}$ is not meaningful as k is an integer. However, in the limit $k \to \infty$, we can obtain mathematically $\frac{d\omega_k}{dk}$ and it would mean the rate of change of frequencies with respect to the counting measure. This in turn is directly related to the frequency spacing. For the local rod, it is well known that frequencies are uniformly spaced. This can be seen by differentiating ω_k in equation [9.13] as:

$$\lim_{k\to\infty} \frac{d\omega_k}{dk} = c\frac{d\sigma_k}{dk}, \quad \text{where} \quad \frac{d\sigma_k}{dk} = \frac{\pi}{L} \quad [9.21]$$

for both sets of boundary conditions. For non-local rods, from equation [9.14] we have:

$$\lim_{k\to\infty} \frac{d\omega_k}{dk} = \lim_{k\to\infty} \frac{d}{dk}\left(\frac{c}{\sqrt{\frac{1}{\sigma_k^2} + (e_0a)^2}}\right) \quad [9.22]$$

$$= \lim_{k\to\infty} \frac{\pi}{L}\frac{c}{(\frac{1}{\sigma_k^2} + (e_0a)^2)^{3/2}\sigma_k^2} = \lim_{k\to\infty} \frac{\pi}{L}\frac{c}{(e_0a)^3}\frac{1}{\sigma_k^2} = 0$$

The limit in the preceding equation goes to zero because $\sigma_k \to \infty$ for $k \to \infty$. This shows that unlike local systems, for large values of k, the undamped natural frequencies of non-local rods will tend to cluster together. A similar conclusion can be drawn by considering the damped natural frequencies also. Next, we consider classical and dynamic FEMs for dynamic response calculations.

9.3. Dynamic finite element matrix

9.3.1. *Classical finite element of non-local rods*

We first consider standard finite element analysis of the non-local rod. Recently, Phadikar and Pradhan [PHA 10] proposed a variational-formulation-based finite element approach for nanobeams and nanoplates. Let us consider an element of length L with axial stiffness EA and mass per unit length m. An element for the damped axially vibrating rod is shown in Figure 9.1.

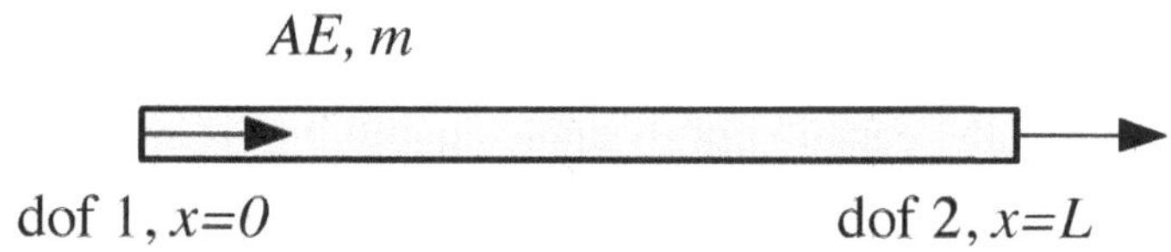

Figure 9.1. *An element for the axially vibrating rod with damping. The axial rigidity AE and mass per unit length m are assumed to be uniform along the length. The element has two degrees of freedom and the displacement field within the element is complex and frequency-dependent*

This element has two degrees of freedom and there are two shape functions, $N_1(x)$ and $N_2(x)$. The shape function matrix for the axial deformation can be given by [PET 98]:

$$\mathbf{N}(x) = [N_1(x), N_2(x)]^T = [1 - x/L, x/L]^T \qquad [9.23]$$

Using this, the stiffness matrix can be obtained using the conventional variational formulation as:

$$\mathbf{K}_e = EA \int_0^L \frac{d\mathbf{N}(x)}{dx} \frac{d\mathbf{N}^T(x)}{dx} \mathrm{d}x = \frac{EA}{L} \begin{bmatrix} 1 & -1 \\ -1 & 1 \end{bmatrix} \qquad [9.24]$$

The mass matrix for the non-local element can be obtained as:

$$\begin{aligned}\mathbf{M}_e &= m\int_0^L \mathbf{N}(x)\mathbf{N}^T(x)\mathrm{d}x + m(e_0a)^2\int_0^L \frac{d\mathbf{N}(x)}{dx}\frac{d\mathbf{N}^T(x)}{dx}\mathrm{d}x \\ &= \frac{mL}{6}\begin{bmatrix}2 & 1\\ 1 & 2\end{bmatrix} + mL(e_0a/L)^2\begin{bmatrix}1 & -1\\ -1 & 1\end{bmatrix} \\ &= mL\begin{bmatrix}1/3+(e_0a/L)^2 & 1/6-(e_0a/L)^2\\ 1/6-(e_0a/L)^2 & 1/3+(e_0a/L)^2\end{bmatrix}\end{aligned} \quad [9.25]$$

For the special case when the rod is local, the mass matrix derived above reduces to the classical mass matrix as $e_0a = 0$.

9.3.2. *Dynamic finite element for damped non-local rod*

The first step for the derivation of the dynamic element matrix is the generation of dynamic shape functions. The dynamic shape functions are obtained such that the equation of dynamic equilibrium is satisfied exactly at all points within the element. Similarly to the classical FEM, assume that the frequency-dependent displacement within an element is interpolated from the nodal displacements as:

$$u_e(x,\omega) = \mathbf{N}^T(x,\omega)\hat{\mathbf{u}}_e(\omega) \quad [9.26]$$

Here, $\hat{\mathbf{u}}_e(\omega) \in \mathbb{C}^n$ is the nodal displacement vector, $\mathbf{N}(x,\omega) \in \mathbb{C}^n$ is the vector of frequency-dependent shape functions and $n = 2$ is the number of the nodal degrees-of-freedom. Suppose that $s_j(x,\omega) \in \mathbb{C}, j = 1,2$ are the basis functions which exactly satisfy equation [9.9]. It can be shown that the shape function vector can be expressed as:

$$\mathbf{N}(x,\omega) = \mathbf{\Gamma}(\omega)\mathbf{s}(x,\omega) \quad [9.27]$$

where the vector $\mathbf{s}(x,\omega) = \{s_j(x,\omega)\}^T, \forall j = 1,2$ and the complex matrix $\mathbf{\Gamma}(\omega) \in \mathbb{C}^{2\times 2}$ depends on the boundary conditions.

In order to obtain $\mathbf{s}(x,\omega)$, first assume that:

$$u(x) = \bar{u}\exp[kx] \quad [9.28]$$

where k is the wave number. Substituting this in equation [9.5], we have:

$$k^2 + \alpha^2 = 0 \quad \text{or} \quad k = \pm \mathrm{i}\alpha \qquad [9.29]$$

In view of the solutions in equation [9.29], the complex displacement field within the element can be expressed by a linear combination of the basis functions $e^{-\mathrm{i}\alpha x}$ and $e^{\mathrm{i}\alpha x}$ so that in our notations $\mathbf{s}(x,\omega) = \left\{e^{-\mathrm{i}\alpha x}, e^{\mathrm{i}\alpha x}\right\}^T$. Therefore, it is more convenient to express $\mathbf{s}(x,\omega)$ in terms of trigonometric functions. Considering $e^{\pm \mathrm{i}\alpha x} = \cos(\alpha x) \pm \mathrm{i}\sin(\alpha x)$, the vector $\mathbf{s}(x,\omega)$ can be alternatively expressed as:

$$\mathbf{s}(x,\omega) = \begin{Bmatrix} \sin(\alpha x) \\ \cos(\alpha x) \end{Bmatrix} \in \mathbb{C}^2 \qquad [9.30]$$

Considering unit axial displacement boundary condition as $u_e(x = 0, \omega) = 1$ and $u_e(x = L,\omega) = 1$, after some elementary algebra, the shape function vector can be expressed in the form of equation [9.27] as:

$$\mathbf{N}(x,\omega) = \mathbf{\Gamma}(\omega)\mathbf{s}(x,\omega), \quad \text{where} \quad \mathbf{\Gamma}(\omega) = \begin{bmatrix} -\cot(\alpha L) & 1 \\ \operatorname{cosec}(\alpha L) & 0 \end{bmatrix} \in \mathbb{C}^{2\times 2} \qquad [9.31]$$

Simplifying this, we obtain the dynamic shape functions as:

$$\mathbf{N}(x,\omega) = \begin{bmatrix} -\cot(\alpha L)\sin(\alpha x) + \cos(\alpha x) \\ \operatorname{cosec}(\alpha L)\sin(\alpha x) \end{bmatrix} \qquad [9.32]$$

Taking the limit as ω goes to 0 (this is the static case), it can be shown that the shape function matrix in equation [9.32] reduces to the classical shape function matrix given by equation [9.23]. Therefore, the shape functions given by equation [9.32] can be viewed as the generalization of the non-local dynamical case.

The stiffness and mass matrices can be obtained similarly to the static finite element case discussed before. Note that for this case, all the matrices become complex and frequency-dependent. It is more convenient to define the dynamic stiffness matrix as:

$$\mathbf{D}_e(\omega) = \mathbf{K}_e(\omega) - \omega^2\mathbf{M}_e(\omega) \qquad [9.33]$$

so that the equation of dynamic equilibrium is:

$$\mathbf{D}_e(\omega)\widehat{\mathbf{u}}_e(\omega) = \widehat{\mathbf{f}}(\omega) \qquad [9.34]$$

In equation [9.33], the frequency-dependent stiffness and mass matrices can be obtained as:

$$\mathbf{K}_e(\omega) = EA\int_0^L \frac{d\mathbf{N}(x,\omega)}{dx}\frac{d\mathbf{N}^T(x,\omega)}{dx}\mathrm{d}x \quad \text{and} \quad \mathbf{M}_e(\omega) = m\int_0^L \mathbf{N}(x,\omega)\mathbf{N}^T(x,\omega)\mathrm{d}x \qquad [9.35]$$

After some algebraic simplifications [MAN 98, ADH 11], it can be shown that the dynamic stiffness matrix is given by the following closed-form expression:

$$\mathbf{D}_e(\omega) = EA\alpha\begin{bmatrix}\cot(\alpha L) & -\mathrm{cosec}(\alpha L)\\ \mathrm{cosec}(\alpha L) & \cot(\alpha L)\end{bmatrix} \qquad [9.36]$$

This is in general a 2×2 matrix with complex entries. The frequency response of the system at the nodal point can be obtained by simply solving equation [9.34] for all frequency values. The calculation only involves inverting a 2×2 complex matrix and the results are exact with only one element for any frequency value. This is a significant advantage of the proposed dynamic finite element approach compared to the conventional finite element approach discussed in section 9.3.1.

So far, we did not explicitly consider any forces within the element. A distributed body force can be considered following the usual finite element approach [PET 98] and replacing the static shape functions by the dynamic shape functions [9.32]. Suppose that $p_e(x,\omega), x \in [0, L]$ is the frequency-depended distributed body force. The element nodal forcing vector can be obtained as:

$$\mathbf{f}_e(\omega) = \int_0^L p_e(x,\omega)\mathbf{N}(x,\omega)\mathrm{d}x \qquad [9.37]$$

As an example, if a point harmonic force of magnitude p is applied at length $b < L$, then $p_e(x,\omega) = p\delta(x - b)$ where $\delta(\bullet)$ is the Dirac delta function. The element nodal force vector becomes:

$$\mathbf{f}_e(\omega) = p \int_0^L \delta(x - b)\mathbf{N}(x,\omega)\mathrm{d}x \qquad [9.38]$$

$$= p \begin{Bmatrix} -\cot(\alpha L)\sin(\alpha b) + \cos(\alpha b) \\ \operatorname{cosec}(\alpha L)\sin(\alpha b) \end{Bmatrix}$$

Next, we illustrate the formulation derived in this section using an example.

9.4. Numerical results and discussions

We consider an SWCNT to illustrate the theory developed in this chapter. An armchair (5, 5) SWCNT with Young's modulus $E = 6.85$ TPa, $L = 25$ nm, density $\rho = 9.517 \times 10^3$ kg/m^3 and thickness t = 0.08 nm is considered as in [MUR 11]. We consider only mass proportional damping, such that the damping factor $\zeta_2 = 0.05$ and $\zeta_1 = 0$. By comparing with MD simulation results [CHO 11a, CHO 10a], it was observed that $e_0a = 1$ nm is the optimal value of the non-local parameter. In this study, however, we consider a range of values of e_0a within 0–2 nm to understand its role in the dynamic response. Although the role of the non-local parameter on the natural frequencies has been investigated, its effect on the dynamic response is relatively unknown. It is assumed that the SWCNT is fixed at one end and we are interested in the frequency response at the free end due to harmonic excitation. Using the dynamic finite element approach, only one "finite element" is necessary as the equation of motion is solved exactly. We consider natural frequencies and dynamic responses of the CNT due to a harmonic force at the free edge.

First, we look into the nature of the novel non-local dynamic shape functions employed in this study. In Figure 9.2, the amplitudes of the two dynamic shape functions as a function of frequency for $e_0a = 0.5$ nm are shown.

For convenience, the shape functions are plotted against normalized frequency:

$$\widehat{\omega} = \omega/\omega_{1l} \qquad [9.39]$$

and normalized length coordinate x/L. Here, ω_{1l} is the first natural frequency of the local rod [MEI 97], given by:

$$\omega_{1l} = \frac{\pi}{2L}\sqrt{\frac{EA}{m}} \qquad [9.40]$$

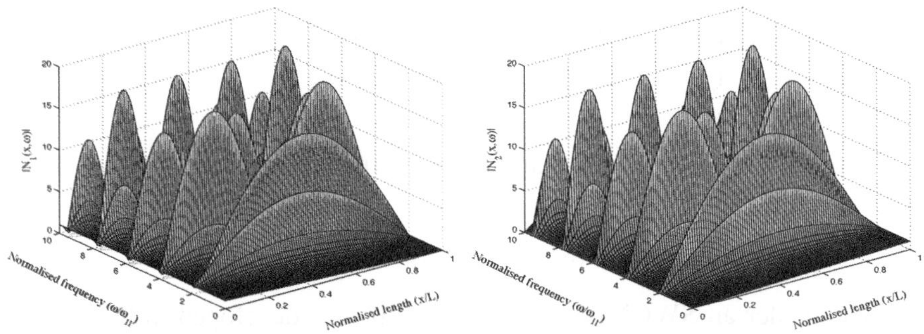

(a) Shape function $N_1(x,\omega)$ (b) Shape function $N_2(x,\omega)$

Figure 9.2. *Amplitude of the dynamic shape functions as a function of frequency for* $e_0a = 0.5$ *nm. For a color version of the figure, see www.iste.co.uk/murmu/non-local.zip*

Similar plots for $e_0a = 2.0$ nm are shown in Figure 9.3 to examine the influence of the non-local parameter on the dynamic shape functions.

The plots of the shape functions show the following interesting features:

– at zero frequency (this is for the static case), the shape functions reduce to the classical linear functions given by equation [9.23]. It can be observed that $N_1(0,0) = 1, N_2(L,0) = 0$ and $N_2(0,0) = 0, N_2(L,0) = 1$;

– for an increasing frequency, the shape functions become nonlinear in x and adapt themselves according to the vibration modes. We can observe multiple modes in the higher frequency range. This nonlinearity in the shape functions is the key for obtaining the exact dynamic response using the proposed approach;

– Figures 9.2 and 9.3 also show the role of the non-local parameter. In Figure 9.3, we can observe more number of modes in the high-frequency range. This is due to the fact that natural frequency of the non-local rod reduced with the increase in the value of the non-local parameter.

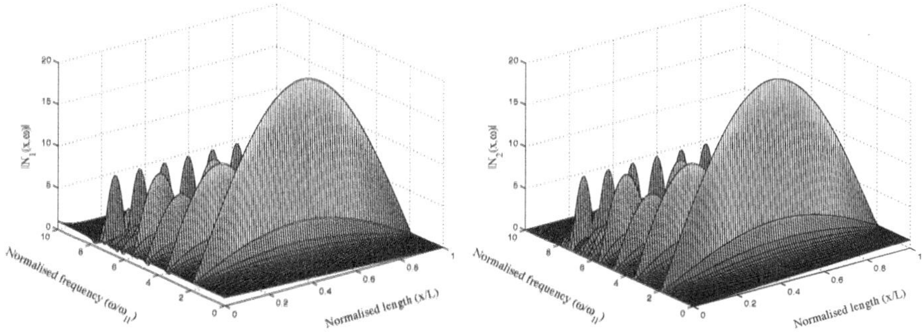

(a) Shape function $N_1(x,\omega)$ (b) Shape function $N_2(x,\omega)$

Figure 9.3. *Amplitude of the dynamic shape functions as a function of frequency for* $e_0a = 2.0$ *nm. For a color version of the figure, see www.iste.co.uk/murmu/non-local.zip*

The shape functions used for the proposed dynamic finite element analysis do not provide the natural frequencies directly. By considering the undamped system (that is by substituting $\zeta_2 = 0$ and $\zeta_1 = 0$ in equation [9.6]), setting the denominator of any of the shape functions to zero and solving the resulting equation, we can obtain the natural frequencies. Following this, from equation [9.32] we have $\sin(\alpha L) = 0$. This in turn will give an expression identical to the frequency equation in [9.1]. In Figure 9.4, the first 10 normalized natural frequencies and the normalized displacement amplitude of the dynamic response at the tip of the SWCNT are shown.

The normalized displacement amplitude is defined by:

$$\delta(\omega) = \frac{\widehat{u}_2(\omega)}{u_{\text{static}}} \tag{9.41}$$

where u_{static} is the static response at the free edge given by $u_{\text{static}} = FL/EA$. Assuming that the amplitude of the harmonic excitation at the free edge is F, the dynamic response can be obtained using the equation of dynamic equilibrium [9.34] as:

$$\widehat{u}_2(\omega) = \frac{F}{EA\alpha\cot(\alpha L)} = \frac{F\tan(\alpha L)}{EA\alpha} \tag{9.42}$$

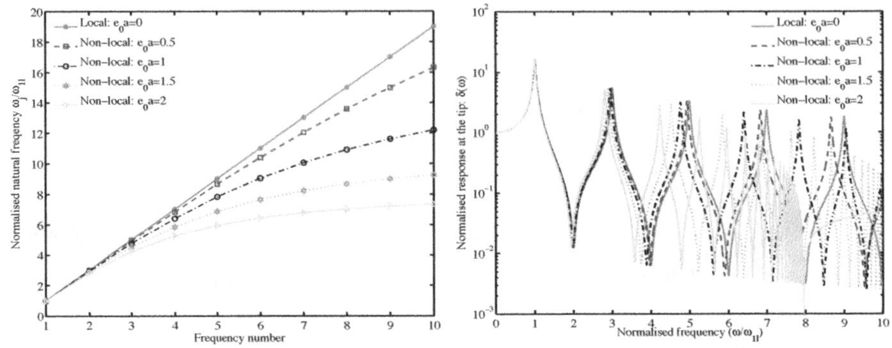

(a) Normalized undamped natural frequency (ω_j/ω_{1l})

(b) Normalized dynamic frequency response amplitude

Figure 9.4. *The variation of undamped natural frequency and dynamic response for different values of e_0a. The local case along with four representative values of e_0a (in nm) is considered. For a color version of the figure, see www.iste.co.uk/murmu/non-local.zip*

Therefore, the normalized displacement amplitude in equation [9.43] is given by:

$$\delta(\omega) = \frac{\widehat{u}_2(\omega)}{u_{\text{static}}} = \left(\frac{F\tan(\alpha L)}{EA\alpha}\right)/(FL/EA) = \frac{\tan(\alpha L)}{\alpha L} \qquad [9.43]$$

The frequency axis of the response amplitude in Figure 9.4(b) is normalized similarly to the plots of the shape functions given earlier. The frequency plot in Figure 9.4(a) clearly shows that the natural frequencies decrease with an increasing value of the non-local parameter e_0a. One interesting feature arising for larger values of e_0a is that the frequency curve effectively becomes "flat". This implies that the natural frequencies reach a terminal value as shown by the asymptotic analysis in section 9.2.3. Using equation [9.18], for large values of k, the normalized natural frequency plotted in Figure 9.4(a) would approach:

$$\frac{\omega_k}{\omega_{1l}} \approx \frac{2/\pi}{(e_0a/L)} \qquad [9.44]$$

Therefore, for $e_0a = 2$ nm, we have $\omega_{k\max} \leq 7.957$. Clearly, the smaller the value of e_0a, the larger this upper limit becomes. The consequence of this upper limit can be seen in the frequency response amplitude plot in Figure 9.4(b). For higher values of e_0a, increasingly resonance peaks are clustered within a frequency band. Indeed, in equation [9.23], we have proved that, asymptotically, the spacing between the natural frequencies goes to zero. This implies that higher natural frequencies of a non-local system are very closely spaced. In Figure 9.4(b), this fact can be observed in the frequency band $7 \lesssim \widehat{\omega} \lesssim 8$ for the case when $e_0a = 2$ nm. The same behavior is expected for other values of e_0a in the higher frequency ranges. It is worth pointing out that the frequency response curve for the case of $e_0a = 2.0$ nm is invalid after $\widehat{\omega} > 8$ as it is beyond the maximum frequency limit. From Figure 9.4(b), it can also be seen that the resonance peak shifts to the left for increasing values of e_0a. This shift corresponds to the reduction in the natural frequencies as shown in Figure 9.4(a).

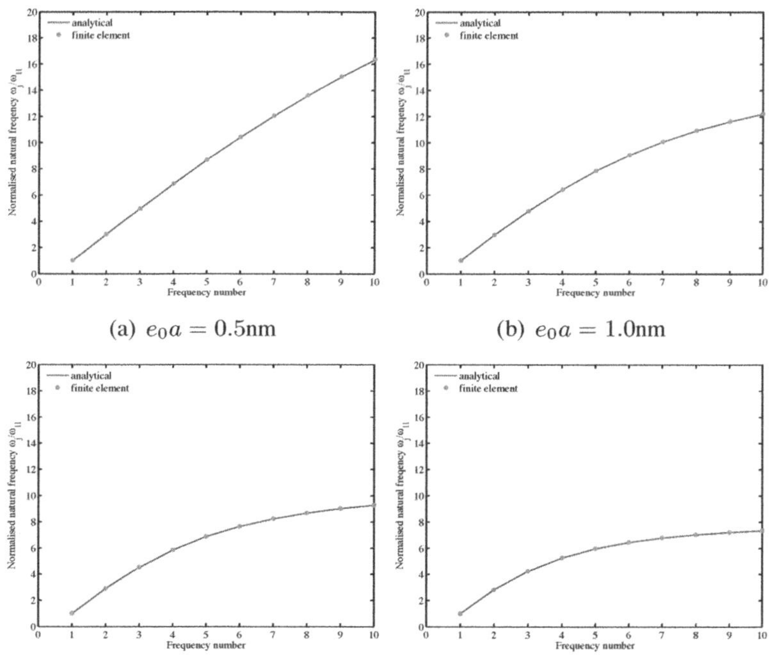

(a) $e_0a = 0.5$nm (b) $e_0a = 1.0$nm

(c) $e_0a = 1.5$nm (d) $e_0a = 2.0$nm

Figure 9.5. *Normalized natural frequency (ω_j/ω_{1l}) at the tip for different values of e_0a. Analytical results are compared with the finite element (with 100 elements) results. For a color version of the figure, see www.iste.co.uk/murmu/non-local.zip*

Now, we compare the results from the dynamic finite element and conventional FEMs. The natural frequencies can be obtained using the non-local FEM developed in section 9.3.1. By assembling the element stiffness and mass matrices given by equation [9.24] and [9.25] and solving the resulting matrix eigenvalue problem $\mathbf{K}\phi_j = \omega_j^2\mathbf{M}\phi_j, j = 1, 2, \cdots$, we can obtain both the eigenvalues and eigenvectors (denoted by ϕ_j here). For the numerical calculation, we used 100 elements. This in turn results in global mass and stiffness matrices of dimension 200×200. In Figure 9.5, the natural frequencies obtained from the non-local FEM are compared with the analytical expression in equation [9.14] for four values of e_0a within the range of 0.5–2.0 nm.

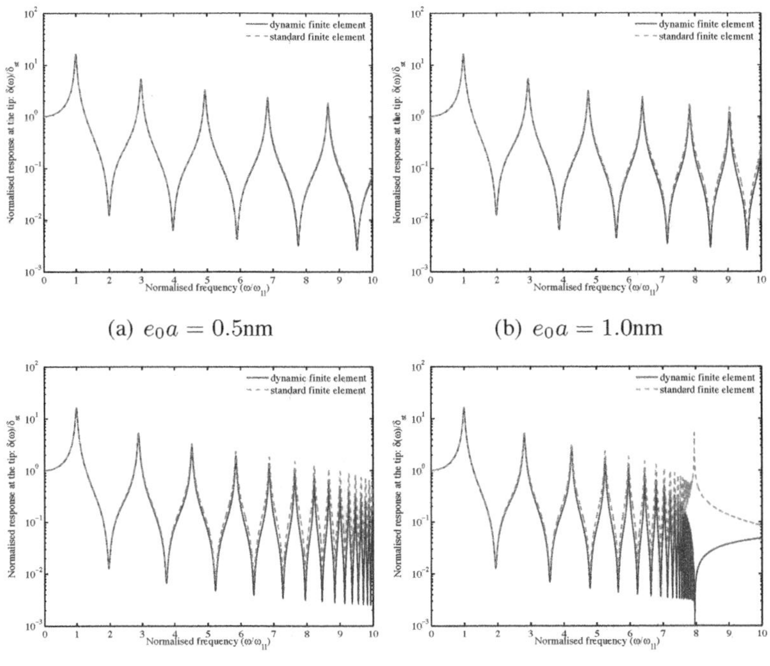

(a) $e_0a = 0.5$nm (b) $e_0a = 1.0$nm

(c) $e_0a = 1.5$nm (d) $e_0a = 2.0$nm

Figure 9.6. *Amplitude of the normalized dynamic frequency response at the tip for different values of e_0a. Dynamic finite element results (with one element) are compared with the conventional finite element results (with 100 elements). For a color version of the figure, see www.iste.co.uk/murmu/non-local.zip*

An excellent agreement was found for the first 10 natural frequencies. However, the results become quite different for the dynamic response as shown in Figure 9.6. In the numerical calculations, 10^5 points are used in the frequency axis. The frequency response functions from the standard finite element were obtained using the classical modal series method [MEI 97]. For small values of e_0a, the results from the dynamic finite element and standard finite element method agree well, as seen in Figure 9.6(a) and 9.6(b). The discrepancies between the methods increase for higher values of e_0a as seen in Figure 9.6(c) and 9.6(d). Note that the results from the dynamic finite element approach are exact as they do not suffer from error arising due to finite element discretization. For higher values of e_0a, increasing numbers of natural frequencies lie within a given frequency range. As a result, a very fine mesh is necessary to capture the high number of modes. If the given frequency is close the maximum cutoff frequency derived in [9.18], then a very high number of finite elements will be necessary (theoretically, infinitely many and there exist an infinite number of frequencies upto the cutoff frequency). In such a situation effectively, the conventional finite element analysis breaks down, as seen in Figure 9.6(d) in the range $7 \leq \widehat{\omega} < 8$. The proposed dynamic finite element is effective in these situations as it does not suffer from discretization errors as in the conventional FEM.

9.5. Summary

In this chapter, a novel dynamic finite element approach for axial vibration of damped non-local rods is proposed. Strain rate-dependent viscous damping and velocity-dependent viscous damping are considered. Damped and undamped natural frequencies for general boundary conditions are derived. An asymptotic analysis is used to understand the behavior of the frequencies and their spacings in the high-frequency limit. Frequency-dependent complex-valued shape functions are used to obtain the dynamic stiffness matrix in closed form. The dynamic response in the frequency domain can be obtained by inverting the dynamic stiffness matrix. The stiffness and mass matrices of the non-local rod were also obtained using the conventional FEM. In the special case when the non-local parameter becomes zero, the expression of the mass matrix reduces to the classical case. The proposed method is numerically applied to the axial vibration of a (5,5) CNT. Some of the key contributions from this chapter are:

– unlike local rods, non-local rods have an upper cutoff natural frequency. Using an asymptotic analysis, it was shown that for an undamped rod, the natural frequency $(\omega_{k_{\max}}) \to \frac{1}{(e_0 a)}\sqrt{\frac{EA}{m}}$. This maximum frequency does not depend on the boundary conditions or the length of the rod;

– near to the maximum frequency, the spacing between the natural frequencies becomes very small. This in turn leads to clustering of the resonance peaks near the maximum frequency;

– for the oscillation frequency of damped systems, the upper cutoff frequency is given by $(\omega_{k_{\max}}) \to \frac{c}{(e_0 a)}\sqrt{1-\left(\frac{\zeta_1 c}{2e_0 a}\right)^2}$ where $c = \sqrt{EA/m}$ and ζ_1 is the stiffness proportional damping factor arising from the strain rate-dependent viscous damping constant. The velocity-dependent viscous damping has no affect on the maximum frequency of the damped rod;

– the asymptotic critical damping factor for non-local rods is given by $(\zeta_1)_{\text{crit}} = 2e_0 a\sqrt{\frac{m}{EA}}$.

The natural frequencies and the dynamic response obtained using the conventional finite element approach were compared with the results obtained using the dynamic finite method. Good agreement between the two methods was found for small values of the non-local parameter. For larger values of the non-local parameter, the conventional finite element approach is unable to capture the dynamics due to very high modal density near to the maximum frequency. In this case, the proposed dynamic finite element approach provides a simple and robust alternative.

This chapter, and all the chapters before it, developed basic theoretical foundations for nanostructures. In the next chapter, we show an application in nanosensors.

10

Non-local Nanosensor Based on Vibrating Graphene Sheets

Single-layer graphene sheets (SLGSs) as a nanoscale label-free mass sensor are discussed in this chapter. A mathematical framework according to non-local elasticity is considered. The non-local elasticity incorporates the small-scale effects of non-locality in the analysis. Rectangular graphene resonators are assumed to be in cantilevered configuration. Closed-form non-local equations are derived for the frequency shift due to the added mass based on four types of different mass loadings. From the potential and kinetic energy of the mass-loaded graphene sheets, generalized non-dimensional calibration constants are proposed for an explicit relationship among the added mass, non-local parameter and frequency shift. These equations, based on non-local elasticity, are used for sensing the added mass (e.g. adenosine biofragment). Molecular mechanics simulation is used to validate the new non-local sensor equations. The optimal values of span of non-local parameters are used and compared with the molecular mechanics simulation results. The non-local approach generally predicts the frequency shift accurately compared to the local approach in most cases. Numerical results show the importance of considering the distributed nature of the added mass while using the non-local theory. The performance of the sensor is governed on the spatial distribution of the attached mass on the graphene sheet. Our discussion of the numerical results illustrates that the sensitivity of graphene sensors is in the order of Gigahertz/zeptograms. Furthermore, the superiority of considering non-local effects in the sensor analysis for most of the cases is illustrated.

10.1. Introduction

In today's world, there exist various kinds of threats originating from different households and industrial applications. Many of these threats are in the form of gases, chemicals and biofragments. In this context, effective detection systems or sensors are required. Sensing of biological objects is also useful for drug-delivery. The best sensors are the ones which are able to detect small numbers of molecules/atoms of the gas or chemical which needs to be detected [PUM 10, HUA 08], with least amount of change in measured quantity. Among the various types of sensors, solid state sensors have received significant attention. These sensors include solid electrolyte sensors [SHA 10], catalytic sensors [ANG 08], biosensors [MOH 08] and semi-conducting oxide gas sensors [ARS 09, QAZ 08, YU 08, MOR 08].

A new class of sensors was developed following the discovery of fullerene and carbon nanotubes (CNT). Electronic sensors made of fullerene are very sensitive to any adsorbed molecule. Electron transport through CNT is influenced by the functionalization of side walls and, therefore, by controlling the defect sites, we can enhance the sensitivity of the sensor. However, the recent discovery of graphene [KOS 09, HIR 09] has opened a completely new area that promises ultra-sensitive and ultra-fast electronic sensors due to low electrical noise material. Graphene is made up of a single layer of carbon atoms packed into a two-dimensional (2D) honeycomb lattice. It has attracted tremendous attention in both its 2D and one-dimensional (1D) forms, the latter being obtained by patterning the layer into a strip or ribbon. Scanning probe microscopy of graphene ribbons [BIE 09] revealed bright stripes along its edges, suggesting a large density of states at the edge, near Fermi level. The electronic properties of graphene [CHO 11b] defined by their quasi-1D electronic confinement and the shape of the ribbon ends [ROS 08] indicate remarkable applications in graphene-based devices [SAN 09]. Graphene has similarities with many properties of CNT [KOS 09]. However, due to their planner structure, some of the properties seem to be easier to manipulate than CNTs. Being one atom thick, graphene comes in direct contact with substrate, thus interface state should play an important role in sensing. Building and designing such nanosensors that are able to make measurements of external deposited agents with ultra-high resolution [JIA 10, LI 08] is one of the main goals in the field of nanomechanics and is the main topic of this chapter.

Resonance-based sensors [CHO 09, LI 04, ADH 10, JOS 11] offer significant potential of achieving the high-fidelity requirement of many sensing applications. The principle of mass detection using resonators is based on the fact that the resonant frequency is sensitive to the resonator mass. The resonator mass includes the self-mass of the resonator and the attached mass. The change of the attached mass on the resonator causes a shift to the resonant frequency. The key issue of mass detection is in quantifying the change in the resonant frequency due to the added mass. In this chapter, we derive the calibration constants necessary for using SLGS as a nanomechanical resonator in nanosized mass sensors. In particular, we develop a new analytical approach for graphene sensors using non-local elasticity theory (Figure 10.1). Graphene is modeled as non-local thin plate. Natural vibrations of SLGS with biofragment is discussed in section 10.3. Four types of mass loadings are considered and closed-form equations have been derived for the frequency shift due to the added mass. Sensor equations and sensitivity calculations are discussed in section 10.4. A molecular mechanics approach based on the universal force field (UFF) model [RAP 92, CHO 11c] is used in section 10.5 to validate the new results derived in this chapter. The results obtained using the analytical approach are discussed for three cases of mass distributions.

10.2. Free vibration of graphene sheets

The vibration of single and multiple layer graphene sheets has been investigated by several authors, using either continuum mechanics approaches [KIT 05], equivalent lattice structures using atomistic-continuum models representing the C–C bonds [HAS 09], or molecular dynamics approaches combined with continuum mechanics for thickness identification [GUP 10]. The out-of-plane deformation of SLGS has been considered using the continuum mechanics models [KIT 05, DUA 09], together with continuum and truss-like structural assemblies [ODE 02, CHO 10b, TSE 05, SCA 08, SCA 10].

Among the number of previous works, we consider the thin elastic plate model [KIT 05] for vibration analysis of SLGSs due to its simplicity. The equation of motion of the transverse-free vibration of a non-local thin elastic plate [TIM 40, SOE 04] of dimension $c \times b$ (Figure 10.1b) can be expressed as:

$$D\nabla^4 u + m\left(1 - (e_0 a)^2 \nabla^2\right)\left\{\frac{\partial^2 u}{\partial t^2}\right\}, \quad [10.1]$$

$$0 \leq x \leq c;\ 0 \leq y \leq b.$$

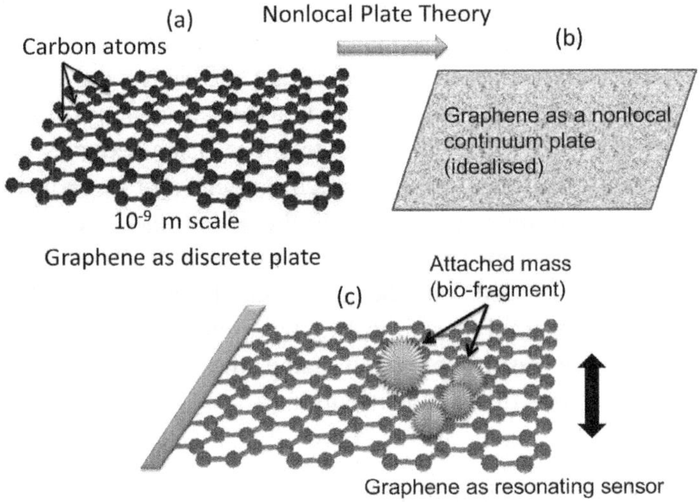

Figure 10.1. *a) Schematic diagram of single-layer graphene sheets, b) non-local continuum plate as a model for graphene sheets and c) resonating graphene sheets sensors with attached biofragment molecules such as adenosine. For a color version of the figure, see www.iste.co.uk/murmu/non-local.zip*

Here, $u \equiv u(x, y, t)$ is the transverse deflection, $\nabla^2 = \left(\frac{\partial^2}{\partial x^2} + \frac{\partial^2}{\partial x^2}\right)$ is the differential operator, x, y are the coordinates, t is the time, ρ is the mass density per area and bending rigidity is defined by:

$$D = \frac{Eh^3}{12(1 - \nu^2)} \quad [10.2]$$

E is the Young's modulus, h is the thickness and ν is the Poisson ratio. We consider rectangular graphene sheets with cantilevered (clamped at one edge) boundary condition. For some practical device, a graphene layer would be supported on some kind of substrate. This would alter the bending rigidity in equation [10.2], if the proposed theory were to be used. We consider rectangular graphene sheets with cantilevered (clamped at one edge) boundary conditions (Figure 10.1c). For SLGSs, the thickness is generally assumed as one atomic thick (though it can be scattered), normally at 0.34 nm. The term e_0a is the dimensional non-local parameter or scale coefficient. Term e_0a is a parameter associated with the material, where a is the intrinsic characteristic length such as lattice parameter, distance between C–C atoms and granular length. Here, e_0 is a constant for calibrating the model with experimental results and other validated models. The parameter e_0 is estimated such that the relations of the non-local elasticity model could deliver satisfactory approximation to the atomic dispersion curves of the plane waves with those obtained from the atomistic lattice dynamics. The non-local parameter is not unique and depends on various conditions. Researchers are currently working to find an optimum value of non-local parameter for nanotubes and graphene sheets. For CNTs and graphene sheets, the e_0a is assumed in the range of 0–2.0 nm. One important way of obtaining the values of non-local parameter is by evaluating from the molecular dynamics simulation.

10.2.1. *Vibration of SLGS without attached mass*

We first calculate the natural frequency of vibration of SLGS without any attached masses in order to obtain the frequency shift. Cantilever boundary condition is considered in this study. We are primarily interested in the first vibration mode of the system. Following Blevins [BLE 84], the first natural frequency (in rad/s) of a rectangular plate of dimension $a \times b$ can be expressed as:

$$\omega_0^2 = \left(\frac{\pi^4 D}{c^4 \rho}\right) \frac{0.0313}{0.2268} \qquad [10.3]$$

In the above result, and in the subsequent analysis, it is assumed that the graphene sheets are in a perfect planar configuration. However, spontaneous ripples and wrinkling have been identified by Meyer *et al.* [MEY 07] on free-standing graphene sheets through vacuum between metal struts. Edge stress-induced warping and instability has been investigated using analytical [SHE 08], first principles study [HUA 09] and numerical methods [GIL 10]. Since the SLGS considered here is clamped at one edge, the substrate at the clamping edge may contain atomistic-scale defects resulting in wrinkles along that edge. Inaccuracies arising due to the wrinkling of SLGS have been neglected in this chapter. The vibration mode-shape for the first mode of vibration of the planar SLGS is given by:

$$w(x,y) = 1 - \cos\left(\pi x/2c\right) \qquad [10.4]$$

The natural frequency of the system can be alternatively obtained using the energy principle. Assuming the harmonic motion, the kinetic energy of the non-local vibrating plate can be expressed by:

$$T = \omega^2 \int_A w^2(x,y)\rho \mathrm{d}A + \omega^2 \int_A (e_0 a)^2 \left\{\nabla w(x,y)\right\}^2 \rho \mathrm{d}A \qquad [10.5]$$

Here, ω denotes the frequency of oscillation and A denotes the area of the plate. Using the expression of $w(x,y)$ in equation [10.4], we have:

$$T = \frac{1}{2}\omega^2 (cb\rho)\left(\frac{3\pi - 8}{2\pi} + \frac{\mu^2\pi^2}{8}\right) \qquad [10.6]$$

where the non-dimensional length-scale parameter:

$$\mu = \frac{e_0 a}{c} \qquad [10.7]$$

The potential energy can be obtained as:

$$\begin{aligned} U = \frac{D}{2}\int_A &\left\{\left(\frac{\partial^2 w}{\partial x^2} + \frac{\partial^2 w}{\partial y^2}\right)^2 \right. \\ &\left. -2(1-\nu)\left[\frac{\partial^2 w}{\partial x^2}\frac{\partial^2 w}{\partial y^2} - \left(\frac{\partial^2 w}{\partial x\,\partial y}\right)^2\right]\right\} \mathrm{d}A \end{aligned} \qquad [10.8]$$

Using the expression of $w(x,y)$ in [10.4], we have:

$$U = \frac{D}{2}\rho \int_0^c \int_0^b \left(\frac{\partial^2 w}{\partial x^2}\right)^2 \mathrm{d}x\,\mathrm{d}y = \frac{1}{2}\frac{\pi^4 D}{c^3} b(1/32) \qquad [10.9]$$

Considering the energy balance, which is $T_{\max} = U_{\max}$, from equation [10.7] and [10.9] the resonance frequency can be obtained as:

$$\omega_0^2 = \left(\frac{\pi^4 D}{c^4 \rho}\right) \frac{1/32}{(3\pi - 8)/2\pi + \mu^2\pi^2/8} \qquad [10.10]$$

This matches well with the numerical value reported in equation [10.3]. Next, we calculate the kinetic energy due to additional attached mass and use the energy principle to obtain the modified resonance frequency.

10.3. Natural vibration of SLGS with biofragment

Several predominantly 1D systems such as CNTs [BOI 09, PEN 06] and boron nitride nanotubes [CHO 11a] have been used as nanomechanical resonators for mass sensing. Due to the 2D nature of the graphene sheet, the SLGS resonator offers more flexibility in terms of attaching different types of molecules at different spatial locations. By exploiting the spatial spread of a graphene sheet, SLGS sensors can be designed such that it can effectively replace an array of cantilever sensors based on nanotubes only. A 2D surface offers huge opportunities for attaching molecules to the graphene sheet.

We consider four possible arrangements by which biomolecules can be attached with the graphene sheet. The four cases considered here are:

– case (a): masses at the cantilever tip in a line;

– case (b): masses in a line along the width;

– case (c): masses in a line along the length;

– case (d): masses in a line with an arbitrary angle.

Representations of these four cases are shown in Figure 10.2.

It is certainly possible to have mass distributions other than the four considered here. The method developed in this chapter is general and can be applied to more general cases if the geometry is known. The reason behind the selection of the four cases in Figure 10.2 is to develop an understanding on the behavior of the proposed sensor for typical mass distributions.

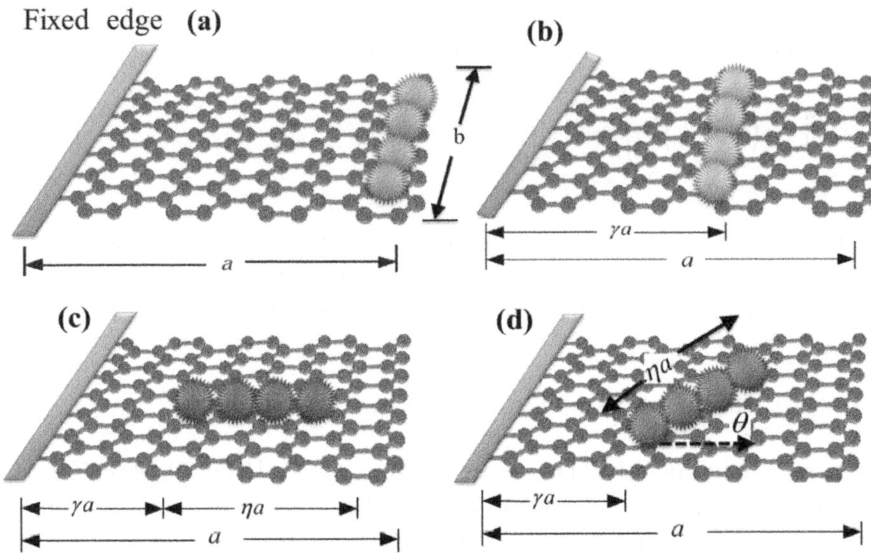

Figure 10.2. *Configurations of attached masses in a cantilevered SLGS resonator: a) masses at the cantilever tip in a line b) masses in a line along the width, c) masses in a line along the length and d) masses in a line with an arbitrary angle. For a color version of the figure, see www.iste.co.uk/murmu/non-local.zip*

From the practical point of view, for label-free sensors it is generally not possible to *a priori* know exactly the spatial location of the attached molecules [BOI 09]. High-resolution imaging techniques can be used to capture the spatial location of the attached molecules. Alternatively, many nanomechanical biosensors use specific coatings on the resonators to make them sensitive to particular biomolecules. In this approach, the sensor will not be label-free but the spatial location of the attachment region will be *a priori* known. The attachment region can even be optimally designed taking the shape and size of the molecules to be detected into account. The analysis presented below is valid for both of these approaches. The analytical results to be derived precisely quantify the effect of spatial location of the attached molecules (not just their mass) on the performance of the sensor.

10.3.1. *Attached masses are at the cantilever tip*

We first consider the case when the attached masses are at the cantilever tip in a line as shown in Figure 10.2(a). Assuming that the total attached mass is M, the combined kinetic energy of the SLGS and the attached mass can be obtained as:

$$\begin{aligned} T_a &= \frac{1}{2}\omega^2 \left\{ \int_A \left[w^2(x,y) + (e_0 a)^2 \left\{ \nabla w(x,y) \right\}^2 \right] \rho \mathrm{d}A \right. \\ &\qquad \left. + M w^2(x,y)|_{x=c} \right\} \\ &= \frac{1}{2}\omega^2 \left\{ M_g \left(\frac{3\pi - 8}{2\pi} + \frac{\mu^2 \pi^2}{8} \right) + M \right\} \end{aligned} \qquad [10.11]$$

where:

$$M_g = \rho c b \qquad [10.12]$$

is the mass of the graphene sheet. Considering the energy balance, which is $T_{\max} = U_{\max}$, from equations [10.11] and [10.9] the resonance frequency can be obtained as:

$$\begin{aligned} \omega_a^2 &= \frac{\frac{1}{2} \frac{\pi^4 D}{c^3} b (1/32)}{\frac{1}{2} \left\{ M_g \left(\frac{3\pi - 8}{2\pi} + \frac{\mu^2 \pi^2}{8} \right) + M \right\}} \\ &= \left(\frac{\pi^4 D}{c^4 \rho} \right) \frac{1/32}{(3\pi - 8)/2\pi + \mu^2 \pi^2 / 8 + M/M_g} \end{aligned} \qquad [10.13]$$

This equation shows how the added mass M reduces the resonance frequency.

10.3.2. *Attached masses arranged in a line along the width*

Next, we consider the case where the attached masses are arranged in a line along the width as shown in Figure 10.2(b). It is assumed that the masses are at a distance of γc, $\gamma \leq 1$, from the fixed edge of the graphene sheet. The kinetic energy of the system can be obtained as:

$$\begin{aligned} T_b &= \frac{1}{2}\omega^2 \left\{ \int_A \left[w^2(x,y) + (e_0 a)^2 \left\{ \nabla w(x,y) \right\}^2 \right] \rho \mathrm{d}A \right. \\ &\qquad \left. + M w^2(x,y)|_{x=\gamma c} \right\} \\ &= \frac{1}{2}\omega^2 \left\{ M_g \left(\frac{3\pi - 8}{2\pi} + \frac{\mu^2 \pi^2}{8} \right) + M \alpha_b \right\} \end{aligned} \qquad [10.14]$$

where the factor:

$$\alpha_b = (1 - \cos(\pi\gamma/2))^2 \qquad [10.15]$$

10.3.3. *Attached masses arranged in a line along the length*

For the case shown in Figure 10.2(c), we assume that the length of the attached mass is ηc and its density along the length is uniform. We also consider that the mass is placed at a distance of γc from the fixed edge. Since the mass always remains within the graphene sheet, both $\gamma \leq 1$, $\eta \leq 1$ and additionally $\gamma + \eta \leq 1$. The kinetic energy of the system can be obtained as:

$$\begin{aligned} T_c &= \frac{1}{2}\omega^2 \left\{ \int_A \left[w^2(x,y) + (e_0 a)^2 \left\{ \nabla w(x,y) \right\}^2 \right] \rho \mathrm{d}A \right. \\ &\quad \left. + \int_{\gamma a}^{(\gamma+\eta)c} \frac{M}{\eta a} w^2(x,y) \mathrm{d}x \right\} \qquad [10.16] \\ &= \frac{1}{2}\omega^2 \left\{ M_g \left(\frac{3\pi - 8}{2\pi} + \frac{\mu^2 \pi^2}{8} \right) + M\alpha_c \right\} \end{aligned}$$

where, by calculating the above integral, we have:

$$\begin{aligned} \alpha_c = \{&3\pi\eta + [\sin((\gamma+\eta)\pi) - \sin(\gamma\pi)] \\ &-8[\sin((\gamma+\eta)\pi/2) - \sin(\gamma\pi/2)]\}/2\pi\eta \end{aligned} \qquad [10.17]$$

In the above expression, we use $\frac{M}{\eta c}$ as the mass per unit length of the attached object.

10.3.4. *Attached masses arranged with arbitrary angle*

The most general case of the attached object is shown in Figure 10.2(d). Depending on the value of θ, it reduces to the two previous cases. The values of γ and η should be such that the mass remains within the graphene sheet. This implies that $\gamma + \eta\cos(\theta) \leq 1$ and $\eta c \leq b\csc(\theta)$. Considering the mass per unit length along the x-axis as $\frac{M}{\eta c\cos(\theta)}$, the kinetic energy of the system can be obtained as:

$$T_d = \frac{1}{2}\omega^2 \left\{ \int_A \left[w^2(x,y) + (e_0 a)^2 \{\nabla w(x,y)\}^2 \right] \rho \mathrm{d}A \right.$$
$$\left. + \int_{\gamma a}^{(\gamma+\eta\cos(\theta))c} \frac{M}{\eta a \cos(\theta)} w^2(x,y) \mathrm{d}x \right\} \quad [10.18]$$
$$= \frac{1}{2}\omega^2 \left\{ M_g \left(\frac{3\pi - 8}{2\pi} + \frac{\mu^2\pi^2}{8} \right) + M\alpha_c \right\}$$

where:

$$\alpha_d = \{3\pi\eta\cos(\theta) + [\sin((\gamma+\eta\cos(\theta))\pi) - \sin(\gamma\pi)]$$
$$-8[\sin((\gamma+\eta\cos(\theta))\pi/2) - \sin(\gamma\pi/2)]\}\, 2\pi\eta\cos(\theta) \quad [10.19]$$

Taking the limit $\theta \to \pi/2$ and $\theta \to 0$, equation [10.14] reduces to equation [10.14] and [10.16], respectively.

The resonance frequency corresponding to cases (b), (c) and (d) can be obtained using the energy principle used for case (a). Considering the energy balance, the resonance frequency can be expressed in a general form as:

$$\omega_{b,c,d}^2 = \frac{\frac{1}{2}\frac{\pi^4 D}{c^3} b(1/32)}{\frac{1}{2}\left\{ cb\rho \left(\frac{3\pi-8}{2\pi} + \frac{\mu^2\pi^2}{8} \right) + \alpha_{b,c,d} M \right\}}$$
$$= \left(\frac{\pi^4 D}{c^4 \rho} \right) \frac{1/32}{(3\pi - 8)/2\pi + \mu^2\pi^2/8 + \beta\alpha_{b,c,d}} \quad [10.20]$$

Here, the ratio of the added mass:

$$\beta = \frac{M}{M_g} \quad [10.21]$$

and $\alpha_{b,c,d}$ are the factors which depend on the mass distribution as defined before.

10.4. Sensor equations and sensitivity analysis

For notational convenience, we express the natural frequency of the mass-loaded graphene sheet as:

$$\omega_n^2 = \left(\frac{\pi^4 D}{c^4 \rho} \right) \frac{1/32}{(3\pi - 8)/2\pi + \mu^2\pi^2/8 + \beta\alpha_n} \quad [10.22]$$

where α_n stands for different values of α, which in turn depend on the distribution of the attached mass.

Taking the ratio with the resonance frequency of the graphene sheet without any mass in [10.10], we have:

$$\frac{\omega_n}{\omega_0} = \frac{1}{\sqrt{1 + c_n\beta}} \qquad [10.23]$$

Here, the calibration constant c_n is given by:

$$c_n = \frac{8\pi\alpha_n}{4(3\pi - 8) + \mu^2\pi^3} \qquad [10.24]$$

Considering $\omega = 2\pi f$, the frequency shift in Hz due to the added mass can be obtained as:

$$\Delta f = f_0 - f_n = 2\pi(\omega_0 - \omega_n) \qquad [10.25]$$

The relative frequency shift [CHO 09, ADH 10] can be obtained from [10.25] as:

$$\frac{\Delta f}{f_0} = 1 - \frac{1}{\sqrt{1 + c_n\beta}} \qquad [10.26]$$

Using this expression, the relative-added mass of the biofragment can be obtained from the frequency shift as:

$$\beta = \frac{1}{c_n\left(1 - \frac{\Delta f}{f_0}\right)^2} - \frac{1}{c_n} \qquad [10.27]$$

The normalized sensitivity of the graphene-based sensor can be obtained by the differentiation of equation [10.26] as:

$$\frac{\partial\left(\frac{\Delta f}{f_0}\right)}{\partial\beta} = \frac{c_n}{2(1 + c_n\beta)^{3/2}} \qquad [10.28]$$

The dimensional sensitivity (in Hz/gm) can be obtained from equation [10.28] as:

$$\frac{\partial\left(\Delta f\right)}{\partial M}=\frac{f_0}{M_g}\frac{c_n}{2}\left(1+c_n\beta\right)^{-3/2} \quad [10.29]$$

The absolute maximum sensitivity of a graphene-based sensor, therefore, is given by:

$$\left.\frac{\partial\left(\Delta f\right)}{\partial M}\right|_{\max}=\frac{f_0}{M_g}\frac{c_n}{2} \quad [10.30]$$

This implies that for a given graphene sheet, a higher value of c_n will result in a sensor with higher sensitivity. The four calibration constants given are plotted for different values of the normalized length η in Figure 10.3 and non-local parameter e_0a.

It can be seen that, among the four cases considered here, case (a), that is when the masses are at the cantilever tip, is the most sensitive case since the value of the calibration constant is highest for this case. Therefore, for graphene-based sensors in the cantilever configuration, it is desirable to place the mass at the free edge.

10.5. Analysis of numerical results

In this section, we validate the sensor equations derived in section 10.4 based on the frequency shift. We use armchair SLGS of length 4.12 nm and width 2.21 nm. The mass of the SLGS is 7.57 zeptogram (1 zg = 10^{-21} g) and in the cantilever configuration its first natural frequency is 23.96 GHz. The added biofragment is considered to be adenosine, which is a nucleoside composed of a molecule of adenine attached to a ribose sugar molecule. Adenosine plays an important role in biochemical processes, such as energy transfer as adenosine triphosphate (ATP) and adenosine diphosphate (ADP) as well as in signal transduction as cyclic adenosine monophosphate. It is also an inhibitory neurotransmitter, believed to play a role in promoting sleep and suppressing arousal, with levels increasing with each hour an organism is awake. Here, we investigate the possibility of detecting adenosine using a cantilevered SLGS. The mass of each adenosine molecule is 0.44 zg, which is about 6% of the mass of graphene.

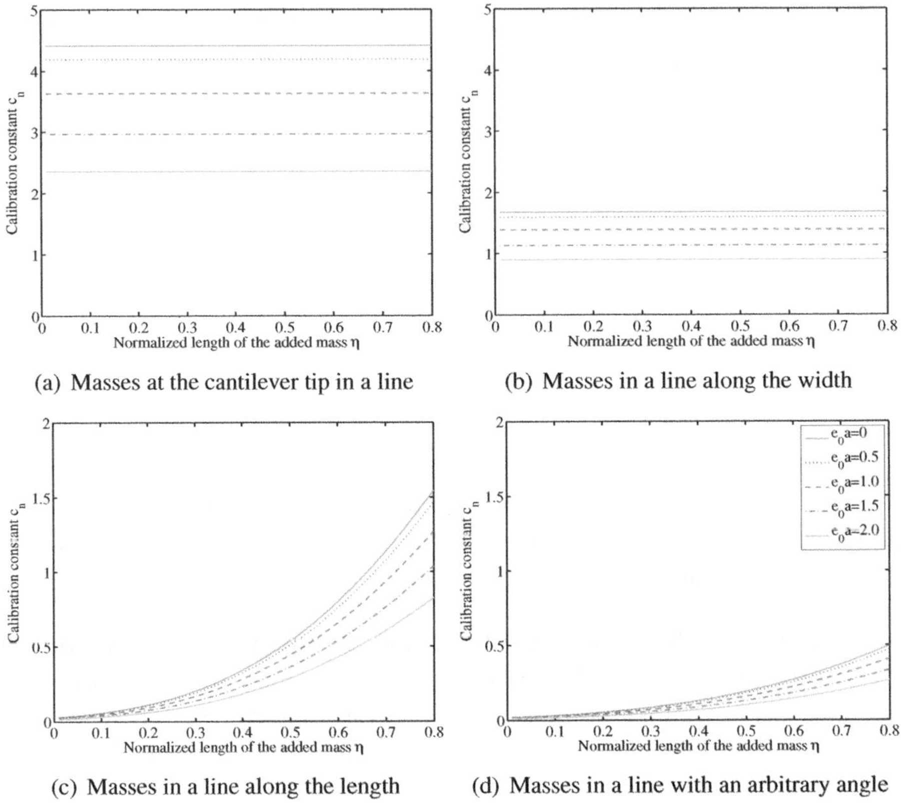

(a) Masses at the cantilever tip in a line

(b) Masses in a line along the width

(c) Masses in a line along the length

(d) Masses in a line with an arbitrary angle

Figure 10.3. *Calibration constants corresponding to four cases. Case a) is independent of* γ *and* η*. For case b), the value of* $\gamma = 0.75$ *is considered. For case c), the value of* $\gamma = 0.25$ *is considered and the normalized length* η *is varied between 0 and 0.75. For case d), the value of angle* $\theta = \pi/4$ *is considered and rest of the values are kept same as case (c). The length of the SLGS is assumed to be 5 nm for the purpose of calculating* $\mu = e_0a/L$*. For a color version of the figure, see www.iste.co.uk/murmu/non-local.zip*

The added mass and the corresponding frequency-shift are calculated from the molecular mechanics approach. The molecular mechanics approach used here [CHO 11d] is more accurate compared to the atomistic finite element method [SCA 08, SCA 10]. We used the Gaussian software [FRI] in conjunction with the UFF model developed by Rappe *et al.* [RAP 92]. The calculation of frequency and their validation for CNTs were given in [CHO 10a]. A detailed study on transverse vibration of SLGSs using the methodology adopted is illustrated in [CHO 11d]. Here, we calculate the natural frequency of the bare graphene sheet and graphene sheet with attached biofragments. From these two sets of frequencies, the frequency shifts are calculated. These frequency shifts in turn are considered as experimental observations and used in the sensor equation [10.27].

Figure 10.4 shows the normalized-free vibration response at the tip of the graphene sheet due to unit initial displacement obtained from molecular mechanics simulation. Responses of the graphene sheet with and without added mass are shown. Here, $T_0 = 2\pi/\Omega_0$ is the time period of oscillation without any added mass. A damping factor of 10% is considered for this plot. The shaded area represents the motion of all the mass loading cases considered. Added mass results the system to move slowly.

The value of the mass predicted by this equation is then compared with the known values used in the molecular mechanics simulations. The analytical approach is verified with the exact approach in Figure 10.5 for case (a) when the added molecules are at the cantilever tip in a line.

It can be seen that the results from the energy-based analytical approach match the exact results obtained from the molecular mechanics reasonably well. The maximum sensitivity obtained from equation [10.30] is calculated as 6.9761 GHz/zg. When the added mass is very high (when mass ratio is more than 0.3), the proposed analytical approach becomes less accurate. This is expected as for such high value of the added mass, the assumed deformation shape in equation [10.4] is not strictly applicable as it was derived from SLGS without any added mass. As a result, the energy expressions and consequently the frequency estimates become inaccurate. This analysis showed that the proposed expressions are accurate up to added mass weighing 35% of the SLGS mass. Beyond this, the proposed approach starts to lose accuracy. This is, however, not a severe limitation as 35% of the SLGS mass can be adequate for practical applications. If higher mass needs to be identified, we can simply use a larger SLGS within the sensor device. We, therefore, conclude that the mass of SLGS should be approximately more than 3 times the mass to be detected in order to reliably use the proposed analytical expressions.

Figure 10.6 shows the identified masses from the frequency-shift for case (b), that is when the added adenosine molecules are arranged in a line along the width.

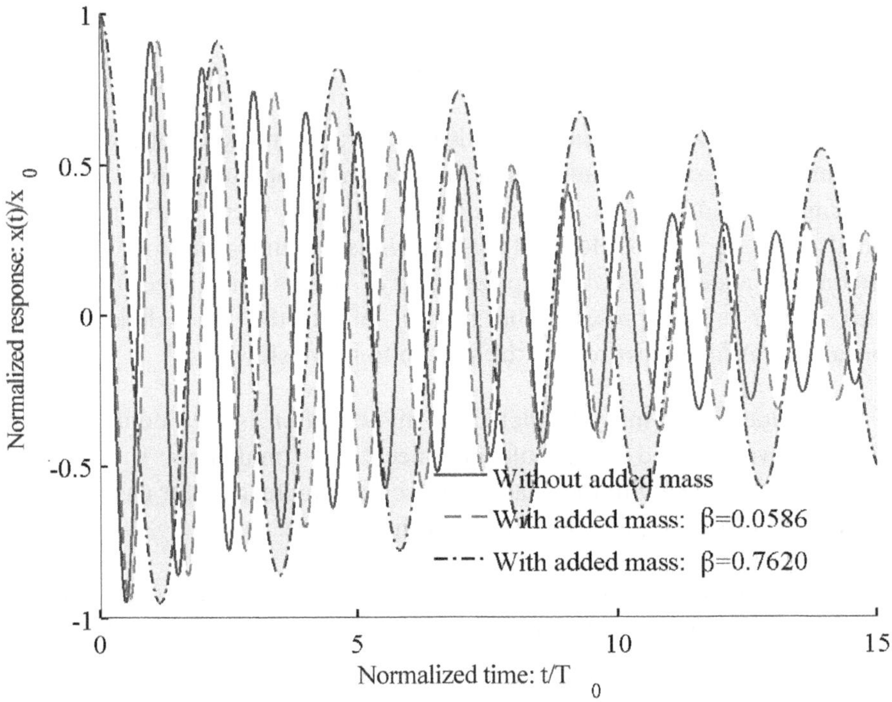

Figure 10.4. *Free vibration response at the tip of the graphene sheet due to the unit initial displacement obtained from molecular mechanics simulation. Here, T_0 is the time period of oscillation without any added mass. The shaded area represents the motion of all the mass loading cases considered for case (a). For a color version of the figure, see www.iste.co.uk/murmu/non-local.zip*

For the analytical calculation, we use $\gamma = 0.85$. The maximum sensitivity obtained from equation [10.30] is calculated as 4.0992 GHz/zg. It can be seen that, except a few cases, the results from the energy-based analytical approach match the exact results obtained from the molecular mechanics reasonably well. Compared to the previous case, it can also be seen that for a given value of added mass, the relative frequency shift is less for this case. This implies that the SLGS-based sensor for this mass distribution is less sensitive compared to the case when the mass was placed at the edge.

Identified masses corresponding to case (c) are shown in Figure 10.7.

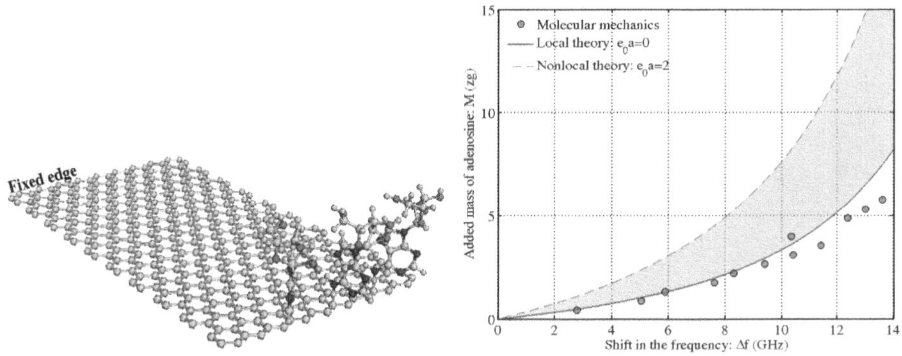

(a) SLGS with adenosine molecules at the cantilever tip in a line

(b) Identified mass from the frequency shift

Figure 10.5. *Identified attached masses from the frequency-shift of a cantilevered SLGS resonator for case (a). The SLGS mass is 7.57 zg and the mass of each adenosine molecule is 0.44 zg. The proposed approach is validated using data from the molecular mechanics simulations. Up to 12 adenosine molecules are attached to the graphene sheet. For a color version of the figure, see www.iste.co.uk/murmu/non-local.zip*

This is the most general case when the added masses are arranged in a line with an arbitrary angle. Case (c) is not shown as it is a special case of case (c). For the results shown in Figure 10.7, we consider $\gamma = 0.25$ and $\eta = 0.6$. The maximum sensitivity obtained from equation [10.30] is calculated as 1.7401 GHz/zg. The proposed approach generally captures the trend of the added mass but the accuracy is lower compared to the last two cases. This can be explained from the fact that the sensitivity is the lowest for this case. Identified masses corresponding to case (d) is shown in Figure 10.8.

This is the most general case when the added masses are arranged in line with an arbitrary angle. Case (c) is not shown as it is a special case of case (d). For the results shown in Figure 10.8, we consider $\gamma = 0.25$, $\eta = 0.7$ and $\theta = \pi/6$. The maximum sensitivity obtained from equation [10.30] is calculated as 1.7401 GHz/zg. The proposed approach generally captures the trend of the added mass with that of molecular mechanics approach. Here, we can see that the non-local sensor equations cover most of the added mass compared to local sensor equations.

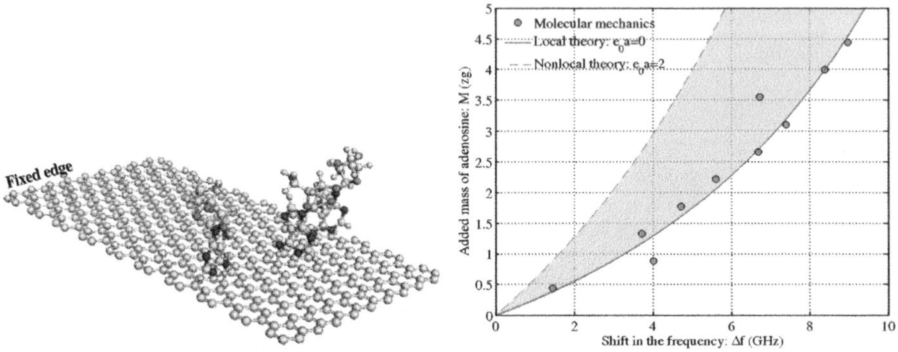

(a) SLGS with adenosine molecules in a line along the width

(b) Identified mass from the frequency shift, $\gamma = 0.85$

Figure 10.6. *Identified attached masses from the frequency-shift of a cantilevered SLGS resonator for case b). The proposed approach is validated using data from the molecular mechanics simulations. Up to 10 adenosine molecules are attached to the graphene sheet. For a color version of the figure, see www.iste.co.uk/murmu/nonlocal.zip*

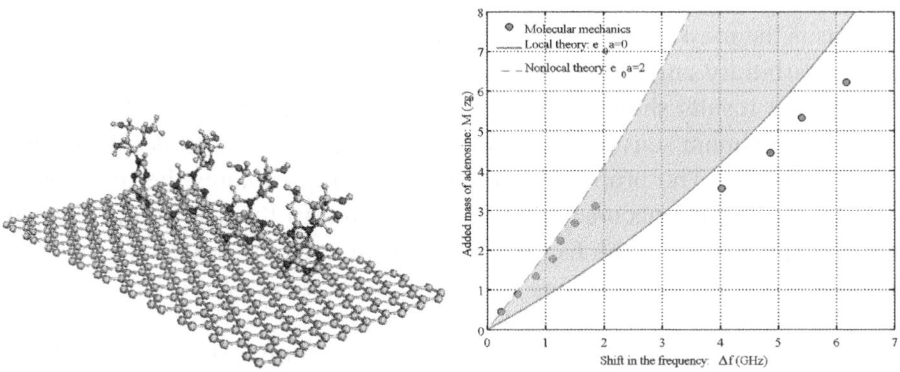

(a) SLGS with adenosine molecules in a line along the length

(b) Identified mass from the frequency shift, $\gamma = 0.25$ and $\eta = 0.6$

Figure 10.7. *Identified attached masses from the frequency-shift of a cantilevered SLGS resonator for case (c). The proposed approach is validated using data from the molecular mechanics simulations. Up to 11 adenosine molecules are attached to the graphene sheet. For a color version of the figure, see www.iste.co.uk/murmu/nonlocal.zip*

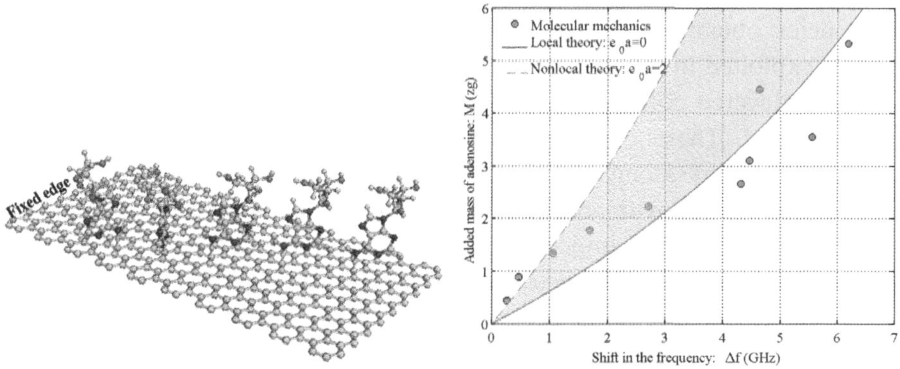

(a) SLGS with adenosine molecules in a line with an arbitrary angle

(b) Identified mass from the frequency shift, $\gamma = 0.25$, $\eta = 0.7$ and $\theta = \pi/6$

Figure 10.8. *Identified attached masses from the frequency-shift of a cantilevered SLGS resonator for case (d). The proposed approach is validated using data from the molecular mechanics simulations. Up to 10 adenosine molecules are attached to the graphene sheet. For a color version of the figure, see www.iste.co.uk/murmu/non-local.zip*

An analytical method for calculating the frequency-shift in a graphene-based resonator is shown using non-local elasticity. Validation is carried out with molecular mechanics simulation results. Up to 60% of mass loading is considered in the numerical examples involving four possible loading patterns. In general, the proposed non-local theory is accurate up to 35% of mass loading. The trend in resonating graphene sensors is to increase the sensitivity by making the resonators smaller. The theoretical model is reasonable, however, it cannot be denied that many technological problems still, such as manufacturability, readout and functionalization, are present. The work presented here could act as an input to understand such graphene sensors once they become realizable. This is a case where the theory might be ahead of the fabrication capabilities, but it will be a valuable model to have for evaluating future cantilever graphene sensors.

10.6. Summary

In this chapter, we theoretically investigate the application of non-local elasticity for the possibility of using SLGS as a nanoscale label-free mass sensor. Non-local theory is an appropriate theory as it considers small-scale effects at nanoscale such as in graphene. The shift in the resonance frequencies obtained via non-local sensor equations due to the additional mass is exploited in the proposed sensor. The SLGS resonator is assumed to be in cantilevered configuration. Four physically realistic mass distributions are addressed. These comprise masses at the (i) cantilever tip in a line, (ii) masses in a line along the width, (iii) masses in a line along the length and (iv) masses in a line with an arbitrary angle. Since the non-local parameter in non-local elasticity is an important parameter and the exact value is unknown, we choose a span for the value of parameter. This comprises analysis without non-local effects too. It is observed that the performance of the sensor depends on the spatial distribution of the attached mass on the graphene sheet with and without non-local effects. Explicit closed-form non-local analytical expressions have been derived to detect the added mass from the frequency-shift. Sensitivities of the proposed sensor for different mass distributions are compared. A molecular mechanics-based approach is used to validate the analytical sensor equations. We used the UFF force field model, wherein the force field parameters are estimated using the general rules based on the element, its hybridization and connectivity. Acceptable agreements between the proposed approach and the molecular mechanics simulations have been observed.

Numerical results indicate that the new equations derived in this chapter are acceptable when the added mass is up to one-third of the mass of the SLGS cantilever. Our analysis shows that by placing the adenosine at the edge of the graphene sheet results in the most sensitive sensor. This observation is related to the modal profile of the graphene, i.e. that a mass at the tip is moving up and down more vigorously than a mass at a different position. Further research will include the dynamics of the subgrade which is essential for immobilizing the biomolecules on to the graphene sheets. Further, we see the superiority of considering non-local effects in the analysis as it matches with the molecular mechanics approach for most of the cases. Significant work is also necessary to physically realize a graphene-based mass sensing resonator where the analytical non-local expressions developed in this chapter would be utilized.

11

Introduction to Molecular Dynamics for Small-scale Structures

In this chapter, we present a review of molecular dynamics (MD) for use with small-scale structures. The simulation results can be compared with the prediction of mechanical behavior of carbon nanotubes (CNTs) and graphene sheets by upgraded continuum theory such as non-local elasticity theory.

11.1. Background

Molecular dynamics (MD) is a highly versatile simulation method for generating the trajectories of multiple, interacting particles, through numerical integration of Newton's laws of motion. Developed independently by Alder and Wainwright [ALD 59] and Rahman [RAH 64], it has since been used in many diverse fields within chemistry, physics, biology and engineering. Individual molecules, fluids or solids can be modeled, as can phase transitions, and system properties can be extracted such as temperature, pressure, heat capacity and thermal conductivity [HAI 92]. Compared to *ab initio* quantum mechanical simulations, MD provides a coarser representation of reality, but this allows thousands or even millions of atoms to be simulated in a reasonable timeframe. Provided due care is taken in the choice of potential energy function, boundary conditions, time step, relaxation period and other model parameters, very useful predictions concerning material response can be made. In the present chapter, we will focus on the use of MD for modeling small-scale solid structures. The treatment will be necessarily brief and for further details, readers are referred to the many excellent texts on the subject such as those of Haile [HAI 92], Leach [LEA 01], Hinchliffe [HIN 03], Frenkel and Smit [FRE 02] and Rapaport [RAP 04]. In an effort to differentiate the current work from the many previous works on this topic, the descriptions will be aimed at an engineering audience, new to the topic, and the focus will be on modeling of nanostructures such as CNTs and graphene. As such, comprehensiveness and rigor will be sacrificed in favor of simplified explanations and extra detail on quite specific issues, where it is felt that an insight can be gained by doing so.

11.2. Overview of the molecular dynamics simulation method

The MD simulation method is quite straightforward in principle, the only complex part being the calculation of the forces between the atoms or molecules making up the system at an instant in time. This is done by taking the gradient of a potential energy function, generally referred to as the "potential", which describes the potential energy of the system as a function of the positions of the atoms/molecules.

To illustrate things, we will assume that our task is to model a CNT (or CNTs) or graphene sheet(s) under some kind of mechanical loading. The entire MD process for this problem could be coded with a few thousand lines of code, but common practice would be to use a public domain or commercial code for the "solver" (which has many different potentials available, and is optimized for parallel processing), and perhaps a custom "pre-processor" for creating the initial relaxed geometry of the system (to be fed into the solver). A custom preprocessor is particularly useful for putting in things such as vacancy defects, dislocation defects (e.g. so-called Stone–Wales defects) and interwall bonds. We will assume that this is our methodology and focus on the requirements for such a preprocessor, which will serve to illustrate the overall MD method.

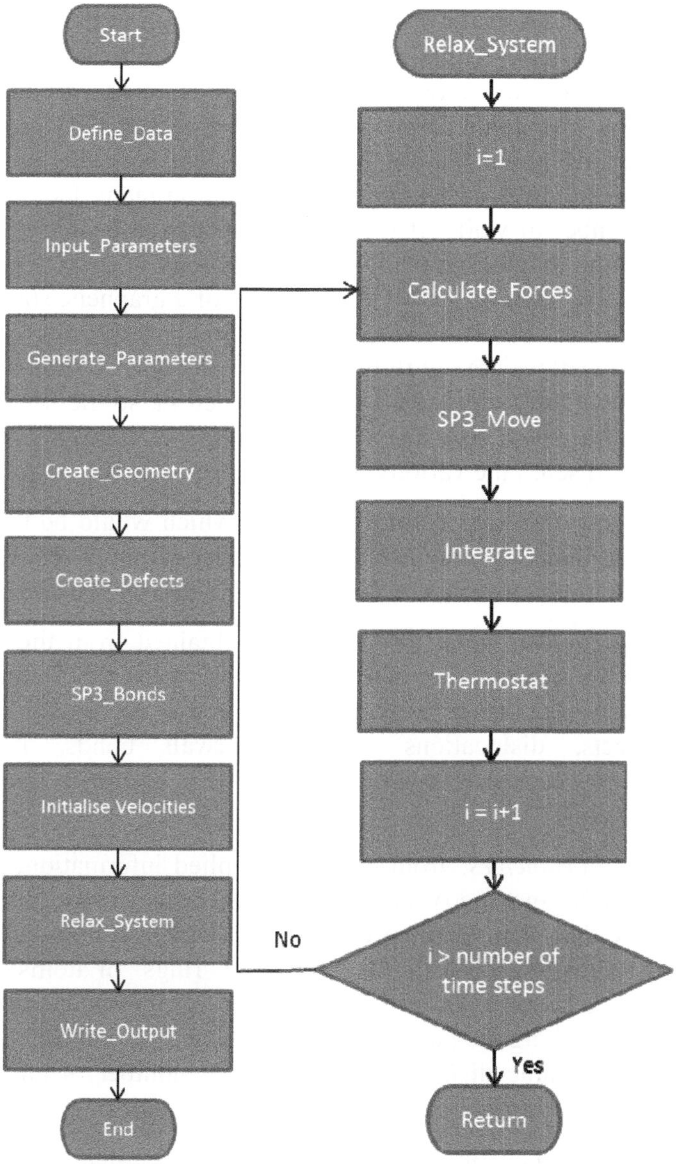

Figure 11.1. *Flowchart of MD "preprocessor" for modeling CNT(s) or graphene sheets(s). Preprocessor creates the geometry, adds defects or interwall bonds if desired, and relaxes the system to equilibrium at the desired temperature. a) Top-level chart, b) chart for Relax_System step*

A high-level flowchart for the preprocessor is given in Figure 11.1(a). An important point to note is that defining an equilibrated or "relaxed" CNT geometry is not just a matter of assigning predefined carbon (C) atom coordinates in a hexagonal arrangement, with constant C-C bond lengths. This is the starting point, but the final equilibrium positions of the atoms, and the corresponding bond lengths, will depend on many factors including wall radius, number of walls (if we include a potential handling long-range interactions), tube length, temperature and locations of defects or interwall bonds (if any). The ends of a CNT or the edges of a graphene sheet will also have some localized curvature when in equilibrium due to the fact that the atoms at these locations are not fully surrounded by other C atoms. Thus, our preprocessor must have a potential function coded in, so the forces between the atoms can be calculated, and the equilibrated or relaxed configuration found. We now discuss the various steps in Figure 11.1(a):

– *Define_Data:* this step defines constants which would be the same for all simulations, such as nominal C-C bond length in a CNT, interwall spacing, C atom mass, parameters in the potential energy function and so on.

– *Input_Parameters:* here, information is obtained from the user about this particular simulation, e.g. number of CNT walls, CNT chirality (e.g. armchair or zig-zag), inner wall radius, approximate length, number of vacancy defects, dislocations and/or interwall bonds, the desired temperature, time step size, number of time steps to use in the relaxation process and so on.

– *Generate_Parameters:* from the user-supplied information, parameters related to the CNT geometry are calculated. These include a more exact length estimate than that supplied by the user (based on nominal C-C bond lengths), the radius of each wall, the number of "rings" of atoms in the axial direction (a ring being a group of atoms having approximately the same length coordinate along the nanotube; for graphene, this will be just a line of atoms having the same distance from one end), number of atoms in each ring, wall and overall in the CNT, number of bonds and so on.

– *Create_Geometry:* for each wall, the *initial* coordinates of each atom are now calculated and stored, and a neighbor list is developed, i.e. initially, each atom will be assigned a list of its three nearest neighbors, all of which will be in the same wall. The coordinates of 1-2 rings of atoms at either end are also stored at this point to facilitate the application of boundary conditions and loads later on.

– *Create_Defects:* this step generates vacancies and/or Stone–Wales dislocations. We could write the code to place the defects at specific locations, or at random locations at the defect density indicated by the user. If locations are generated randomly, locations near defects already created should be rejected (unless it is desired to allow multi-atom vacancies). The nearest neighbor lists then have to be updated for the new defective structure.

– *SP3_Bonds_Identify:* a feature of multi-wall CNTs is the ease with which walls in multi-wall CNTs can slide over each other. This is due to relatively weak van der Waals forces between atoms in adjacent walls, which are separated by quite a large gap (~3.4Å). This could be advantageous in some applications, but from a structural point of view (e.g. for use of CNTs in a structural composite material) it is a significant disadvantage as it precludes load transfer between walls, which may render most of the walls structurally ineffective [BYR 10, BYR 09]. Load transfer can be greatly improved if even a small number of covalent bonds exist between walls. These can sometimes occur naturally, but they can also be deliberately induced *via* electron or ion irradiation [OBR 13a, OBR 13b, MCC 14], which either pushes C atoms in adjacent walls into closer proximity causing a direct bond to occur between them, or knocks C atoms out of the CNT lattice into the interwall space, which can then form bonds with both walls, establishing a bridge between them. In the case of direct bonds, each C atom in the bond becomes fully conjugated (i.e. bonded to four other C atoms) and the hybridization changes from sp^2 to sp^3. In our sample preprocessor, we include a way to form direct sp^3 bonds between walls. The *SP3_Bonds_Identify* step begins the process by identifying and logging the atoms involved in sp^3 bonds. We define the "inter-wall bond density" λ_i between any pair of walls as the total number of interwall bonds between walls i and i+1 divided by the total number of atoms in walls i and i+1. The user can specify an overall density in the *Input_Parameters* step above, or if desired, a different density could be specified between each pair of walls. Assuming that an overall CNT interwall bond density is supplied, the *SP3_Bonds_Identify* step picks a random shell, ring and then atom, i. Before moving on, it then checks if i, or any neighbor of i is already a member of an sp^3 bond, in which case it goes back and picks a different random atom. It can also be coded to reject atoms near the ends of the tube if it is desired to avoid sp^3 bonds in these regions. Having found a valid "seed" for an sp^3 bond, it then loops over all atoms in the next outer-most shell to find the closest atom to i, which we will call j. It then also checks if j is already a member of an sp^3 bond (e.g. with the *next* outer-most wall), in which case it starts over with a new *i* atom. Once a valid pair of atoms, *i* and *j*, is found, atom *j* is now added as a *fourth* member of atom *i*'s neighbor list (recall that CNT lattice atoms which are not bonded to other walls have only three members in their neighbor list).

– *Initialise_Velocities:* here, we set the initial velocities of the atoms. A typical method is as follows. Calculate the standard deviation of the Maxwell velocity distribution [HAI 92]:

$$v_0 = \sqrt{\frac{k_B T}{m}} \qquad [11.1]$$

where k_B is the Boltzmann's constant, T is the user-defined temperature of the system and m is the mass of a carbon atom. The Maxwell distribution is the expected equilibrium distribution of velocities at temperature *T*, according to the Maxwell–Boltzmann law [HAI 92], and is Gaussian with a mean value of zero. However, there is no need to assign an initial velocity distribution carefully [RAP 04], as the Maxwell distribution should evolve naturally during equilibration. Thus, it is typical to initially assign a uniform velocity distribution [FRE 02]. Using a random number generator, for each atom *i*, we set each of the *x*, *y* and *z* components of its velocity to a random value within the interval $[-v_0, v_0]$. We then determine the actual initial temperature, T_{act} by using the actual initial velocities that we have assigned to get the initial kinetic energy, E and applying [HAI 92]:

$$T_{act} = \frac{2}{3}\frac{E}{nk_B} \qquad [11.2]$$

The next step in the top-level flowchart *Relax_System* performs the system relaxation, and is discussed in more detail below. The last step is *Write_Output* which outputs the coordinates of the atoms in the relaxed system for input to a solver such as the public domain code LAAMPS [PLI 95].

The *Relax_System* step is expanded upon in Figure 11.1(b). We see a loop is executed *n* times where *n* is the user-defined number of time steps.

– *Calculate_Forces:* this step is the heart of the MD simulation. It is the most complex aspect to understand and by far the most time-consuming part of the calculation computationally. The task is to calculate the force on each atom in the system based on the current position of each atom in the system. The force on atom i is obtained as the gradient of the potential energy of the system with respect to a movement of atom i, or:

$$\vec{F}_i = -grad\left(V_{SYS}\right) = -\left[\frac{\partial V_{SYS}}{\partial x_i}\hat{\imath} + \frac{\partial V_{SYS}}{\partial y_i}\hat{\jmath} + \frac{\partial V_{SYS}}{\partial z_i}\hat{k}\right] \qquad [11.3]$$

Engineers may find it useful to consider this equation in the context of a simple spring-mass system such as the one shown in Figure 11.2. Clearly, moving mass m_3, for example, left or right, while keeping all other masses fixed, will increase the potential energy in springs 3 *and* 4 (compressing one and stretching the other), which according to equation [11.3] leads to a force from each of the two springs on m_3 opposing the applied motion. This is as we expect.

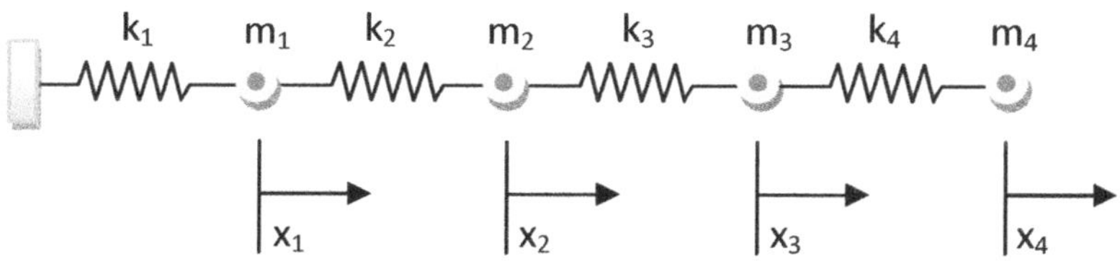

Figure 11.2. *Simple spring-mass system*

A key aspect of this simple system is that the potential energy in each spring depends only on the two masses it is attached to. This type of potential energy function is called a "two-body" or "pair" potential. For this type of potential, a movement of mass i can only affect the potential energy *in the springs to which it is attached.* However, equation [11.3] clearly allows for a more complex situation. It says that if a movement of atom i causes an increase in potential energy *anywhere* in the system, that will result in a resisting force on i. In the simple spring-mass system of Figure 11.2, a movement of m_3 cannot lead to a change in potential energy in spring 1 or 2, so we only need to consider potential energy changes in springs (bonds) between i and its immediate neighbors when using [11.3] to calculate the force on atom *i*. However, this is not how things work on an atomic level. For one thing, the potential energy in the bond between atoms *i* and *j* depends not only on the position of those atoms but also on the positions of other nearby atoms. In particular, bond *angles* affect the potential energy, and angles can only be determined by consideration of the positions of *three* atoms. Thus, pair potentials are not adequate and we need *multi*-body potentials. Figure 11.3 illustrates how the movement of atom *i* in a CNT or graphene sheet affects angles involved in the calculation of the potential energy in the three bonds connected to *i*, i.e. bonds ij, ik_1 and ik_2.

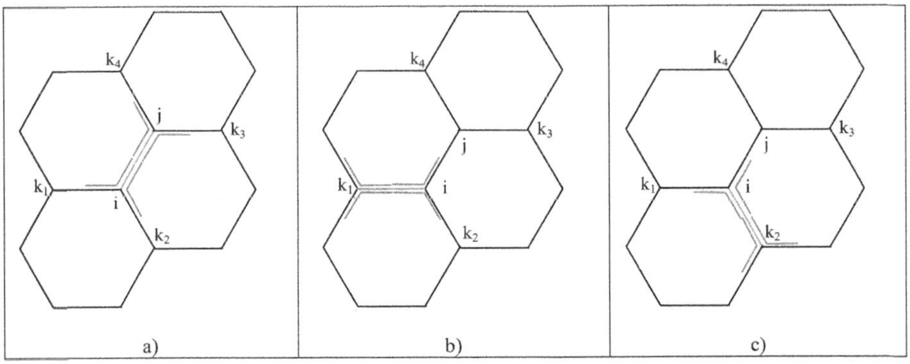

Figure 11.3. *Angles (in red) affected by movement of atom i, involved in calculation of the potential energy in the bonds (in blue) between i and its three nearest neighbors, a) V_{ij}, b) V_{ik_1} and c) V_{ik_2}. In each case, four angles are affected. If either or both of i, j are sp3 bonded to another wall, five or six angles will be affected, respectively. For a color version of the figure, see www.iste.co.uk/murmu/non-local.zip*

However, an implication of using a multi-body potential is that movement of atom *i* can lead to changes in potential energy in bonds not directly involving *i* (which as noted above is not possible with a pair potential). Figure 11.4 illustrates this; here, movement of atom *i* affects one of the angles involved in calculating the potential energy in bonds jk_3 and jk_4. In fact, movement of atom *i* affects the energy in bonds between each of *i*'s neighbors (j, k_1 and k_2) and *their* neighbors. The *x*-component (for example) of the force on atom *i* becomes:

$$F_{i_x} = -\sum_{j \neq i} \frac{\partial V_{ij}}{\partial x_i} - \sum_{j \neq i} \sum_{k(\neq i,j)} \frac{\partial V_{jk}}{\partial x_i} \qquad [11.4]$$

where the first term considers potential energy changes in bonds between *i* and its nearest neighbors, while the second term considers energy changes in bonds between *i*'s neighbors and *their* neighbors. Similar expressions exist for the *y* and *z* components. The second term is not present in pair potentials, so clearly the use of a multi-body potential considerably increases the number of calculations required to determine the force on atom *i*.

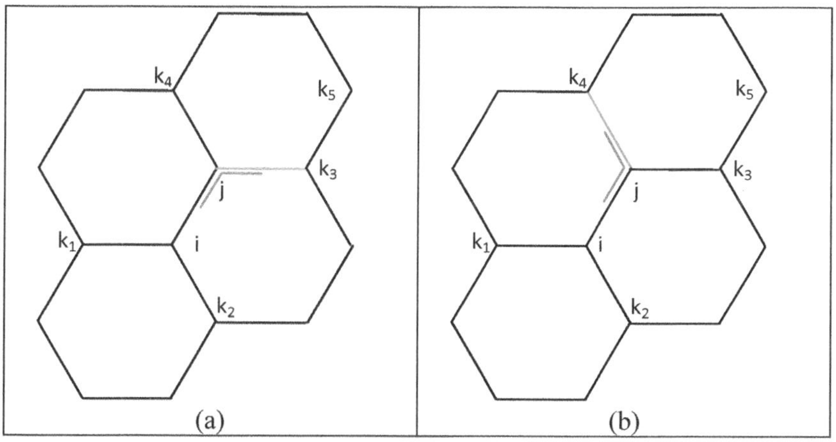

Figure 11.4. *Angles (in red) affected by movement of atom i, involved in calculation of energy in bonds (in blue) between j and two of its neighbors a)* V_{jk_3} *and b)* V_{jk_4}
For a color version of the figure, see www.iste.co.uk/murmu/non-local.zip

The force calculation then is a matter of finding the partial derivatives of a potential energy function (the "potential") which describes the energy in each bond, as a function of the distances between the atoms in the bond, and the aforementioned angles in Figure 11.3 (plus other factors as well in more complex potentials). The potential entirely determines the behavior of the system and is thus the key component in any MD simulation. As such, potentials are under continuous development, and new or improved potentials are published every year. Many potentials incorporate additional sophistication, such as a dependence on whether the bonds are part of a conjugated system (alternating single and double bonds) which requires consideration of non-local effects [BRE 90], or corrections for cases where atoms with different coordinations are bonded together (e.g. imagine atom *i* in Figure 11.3 is also sp^3 bonded to an atom in an adjacent wall; then *i* will have coordination four, while j will have coordination three) [BRE 90]. Potentials are empirical; they usually take a relatively simple mathematical form, involving exponential or power law functions, the parameters of which are tuned to a given type of system (e.g. one involving hydrocarbons) based on experimental and/or quantum mechanical simulation data. Detailed discussion of potentials is beyond the scope of this work, but brief details on one potential originally published by Brenner [BRE 90] and since used in numerous studies on carbon-based systems, in its original or modified forms [BRE 02, PAS 08], will serve to illustrate things.

In Brenner's original potential [BRE 90], the potential energy of the bond between atoms i and j, V_{ij}, is:

$$V_{ij} = V_R(r_{ij}) - \bar{B}_{ij} V_A(r_{ij}) \qquad [11.5]$$

where:

$$V_R(r_{ij}) = f_{ij} \frac{D_e}{S_{ij} - 1} e^{-\sqrt{2S_{ij}}\beta_{ij}(r_{ij} - R_e)} \qquad [11.6]$$

$$V_A(r_{ij}) = f_{ij} \frac{D_e S_{ij}}{S_{ij} - 1} e^{-\sqrt{\frac{2}{S_{ij}}}\beta_{ij}(r_{ij} - R_e)} \qquad [11.7]$$

are, respectively, the repulsive and attractive terms, similar in form to the Morse potential. D_e is the well depth, R_e is the equilibrium distance and β_{ij} and S_{ij} are the fitting parameters. The sum of the repulsive and attractive terms in [11.5] produces a function with high-positive values at very small distances, dropping to a minimum at the equilibrium bond distance and then rising to zero asymptotically at large distances. This results in repulsive forces at distances below the equilibrium bond length and attractive forces at distances above the equilibrium bond length. The so-called "bond-order" term $\bar{B}_{ij}$ in [11.5] accounts for the angles in the bond, as well as the above-mentioned effects of conjugation and bonding between atoms with different coordinations. The remaining parameter in [11.6], [11.7], is the cutoff function:

$$f_{ij} = \begin{cases} 1 & \text{if } r_{ij} < 1.7\,\overset{\circ}{\mathrm{A}} \\ \dfrac{1}{2}\left[1+\cos\left[\dfrac{\pi(r_{ij}-1.7)}{2-1.7}\right]\right] & \text{if } 1.7\,\overset{\circ}{\mathrm{A}} < r_{ij} < 2\,\overset{\circ}{\mathrm{A}} \\ 0 & \text{if } r_{ij} > 2\,\overset{\circ}{\mathrm{A}} \end{cases} \quad [11.8]$$

As mentioned above, the attractive force, [11.5], decays asymptotically to zero at large distances, so that each atom experiences a finite (but in most cases extremely small) force from every other atom in the system. Calculating all the forces in this way would be computationally prohibitive and entirely unnecessary since the force decays rapidly beyond the equilibrium bond distance, so the cutoff function is introduced to truncate the potential to zero beyond a certain difference. Since the calculation of interatomic forces is by far the most computationally expensive part of an MD calculation, this cutoff is generally set so that only the nearest neighbors in the structure are included within the cutoff distance. A value of 2Å is typical, and since the nominal equilibrium bond distance in CNTs is ~1.42Å we see that only the nearest neighbors would be included within this distance. Note that bond-forming and bond-breaking are possible with this potential as atoms enter and exit the cutoff distance. However, we need to be careful when using the potential for this purpose, as the cutoff function can introduce unrealistic forces [BEL 02].

The cutoff function means that we consider atom i in Figure 11.3 to be directly bonded only to atoms within the cutoff distance from i, so that the first sum in equation [11.4] is only over the three bonds between i and its nearest neighbors, i.e. there is no direct bond between i and k_3, for example. Were we to extend the cutoff distance even to include just second-nearest neighbors (of which there would be six in a simple sp^2-bonded CNT wall like in Figure 11.3), the computational cost would be enormous since we recall that all neighbors are included in the second term of equation [11.4]. As it turns out, the results would not even be better, since the potential is not "tuned" to include second-nearest neighbor effects. Recent efforts to improve bond-forming and breaking behavior of this potential [PAS 08] have involved increasing the range of the potential with the addition of a "screening" function to incorporate the screening effect on the attraction between two atoms when a third atom comes within a certain distance of them.

– *SP3_Move:* returning now to the flowchart in Figure 11.1(b), we see the next step is *SP3_Move*. In this step, atoms identified in *SP3_Bond_Identify* (one in one wall, one in an adjacent wall) are moved a very small distance closer together. For these atoms only, the *Integrate* step (below) is skipped. *SP3_Move* is called for each time step until the two atoms move within the bonding distance of the potential, and an sp^3 bond is then formed.

– *Integrate:* in this step, for each atom, the force on the atom found in *Calculate_Forces* is divided by the mass of a C atom to get the acceleration of the atom over the next time step. This is then integrated numerically to get the velocity and displacement of the atom over the next time step. Numerous numerical integration methods exist, but methods generally used for MD simulations are selected because they are "reversible" and highly efficient (due to the computationally intensive nature of the method). Reversibility, in principle means the simulation could be run to its end, and then run backward to its beginning, and the result would be the original configuration of the system (in practice, numerical round-off errors would prevent this occurring). Not all numerical integration methods are reversible. An excellent discussion of integration methods for MD is provided by Haile [HAI 92]. As noted above, the atoms being artificially pushed together in *SP3_Move* to form an sp^3 bond are not treated in *Integrate* as we do not want them to be pulled back into their original lattice position while we are pushing them together.

– *Thermostat:* the final step is to apply a thermostat. Different types exist but a simple approach is to find the kinetic energy of the system from the new atomic velocities just found for this time step. The actual system temperature, T_{act}, is then found from equation [11.2]. The temperature is then adjusted to the user-specified temperature T by simply scaling all the atomic velocities by $\sqrt{T/T_{act}}$. These scaled velocities become the input velocities for the next time step. By this measure, energy is added and subtracted to the system at each time step until the system equilibrates or relaxes to its equilibrium configuration at the desired temperature. Note, the velocities will continue to oscillate (thermal fluctuations) but the energy should settle to an approximately constant value once the system is relaxed. This completes the description of the preprocessor in Figure 11.1.

To perform the actual MD simulation, the relaxed configuration of the system is then input to a solver such as LAAMPS [PLI 95] together with boundary conditions and loads. The same potential should be used for the solver as was used in the preprocessor as different potentials will result in slightly different relaxed configurations. Postprocessing involves visualizing results (for which many public domain codes are available), and plotting quantities such as displacement, force and so on.

11.3. Acknowledgement

Prof. McCarthy wishes to thank Prof. William A. Curtin of École Polytechnique Fédérale de Lausanne for introducing the author to the MD method during a sabbatical and Brown University, and Prof. Zhenhai Xia of University of North Texas who wrote the MD pre-processor which the description in Chapter 11 is based upon.

Bibliography

[ADH 00] ADHIKARI S., MANOHAR C. S., "Transient dynamics of stochastically parametered beams", ASCE *Journal of Engineering Mechanics*, vol. 126, no. 11, pp. 1131–1140, November 2000.

[ADH 06] ADHIKARI S., "Damping modelling using generalized proportional damping", *Journal of Sound and Vibration*, vol. 293, nos. 1–2, pp. 156–170, May 2006.

[ADH 10] ADHIKARI S., CHOWDHURY R., "The calibration of carbon nanotube based bionano sensors", *Journal of Applied Physics*, vol. 107, no. 12, pp. 124322:1-8, 2010.

[ADH 11] ADHIKARI S., "Doubly spectral stochastic finite element method (DSSFEM) for structural dynamics", *ASCE Journal of Aerospace Engineering*, vol. 24, no. 3, pp. 264–276, 2011.

[ADH 13] ADHIKARI S., MURMU T., MCCARTHY M., "Dynamic finite element analysis of axially vibrating nonlocal rods", *Finite Elements in Analysis and Design*, vol. 63, no. 1, pp. 42–50, 2013.

[AH 05] AH C.S., YUN Y.J., PARK H.J. *et al.*, "Size-Controlled Synthesis of Machinable Single Crystalline Gold Nanoplates", *Chemistry of Materials*, vol. 17, no. 22, pp. 5558–5561, 2005.

[AJA 01] AJAYAN P.M., ZHOU O.Z., "Applications of carbon nanotubes", *Carbon Nanotubes,* Springer Berlin Heidelberg, pp. 391–425, 2001.

[AKG 11] AKGOZ B., CIVALEK O., "Strain gradient elasticity and modified couple stress models for buckling analysis of axially loaded micro-scaled beams", *International Journal of Engineering Science*, vol. 49, no. 11, pp. 1268–1280, 2011.

[AKG 12] AKGOZ B., CIVALEK O., "Free vibration analysis for single-layered grapheme sheets in an elastic matrix via modified couple stress theory", *Materials & Design*, vol. 42, no. 164, 2012.

[AKS 11] AKSENCER T., AYDOGDU M., "Levy type solution method for vibration and buckling of nanoplates using nonlocal elasticity theory", *Physica E-Low-Dimensional Systems & Nanostructures*, vol. 43, no. 954, 2011.

[ALD 59] ALDER B.J., WAINWRIGHT T.E., "Studies in molecular dynamics. 1. General method", *Journal of Chemical Physics,* vol. 31, no. 2, pp.

[AMA 10] AMARA K., TOUNSI A., MECHAB I. *et al.*, *Applied Mathematical Modelling,* vol. 34, pp. 3933–3942, 2010.

[AND 04] ANDRIANOV I., AWREJCEWICZ J., "Theory of plates and shells: new trends and applications", *International Journal of Nonlinear Sciences and Numerical Simulation,* vol. 5, pp. 23–36, 2004.

[ANG 08] ANG P. K., CHEN W., WEE A. *et al.*, "Solution-gated epitaxial graphene as pH sensor", *Journal of the American Chemical Society*, vol. 130, no. 44, pp. 14392+, 2008.

[ANS 10a] ANSARI R., RAJABIEHFARD R., ARASH B., "Nonlocal finite element model for vibrations of embedded multi-layered graphene sheets", *Computational Materials Science*, vol. 49, no. 831, 2010.

[ANS 10b] ANSARI R., SAHMANI S., ARASH B., "Nonlocal plate model for free vibrations of single-agered graphene", *Physics Letters A,* vol. 375, pp. 53–62, 2010.

[ANS 11a] ANSARI R., ARASH B., ROUHI H., "Vibration characteristics of embedded multi-layered graphene sheets with different boundary conditions via nonlocal elasticity", *Composite Structures*, vol. 93, no. 2419–2429, 2011.

[ANS 11b] ANSARI R., ROUHI H., SAHMANI S., "Calibration of the analytical nonlocal shell model for vibrations of double-walled carbon nanotubes with arbitrary boundary conditions using molecular dynamics", *International Journal of Mechanical Sciences*, vol. 53, no. 9, pp. 786–792, 2011.

[ANS 12a] ANSARI R., SAHMANI S., "Small scale effect on vibrational response of single-walled carbon nanotubes with diffrent boundary conditions based on nonlocal beam models", *Communications in Nonlinear Science and Numerical Simulation,* vol. 17, pp. 1965–1979, 2012.

[ANS 12b] ANSARI R., GHOLAMI R., ROUHI H., "Vibration analysis of single-walled carbon nanotubes using different gradient elasticity theories", *Composites Part B: Engineering*, vol. 43, no. 8, pp. 2985–2989, 2012.

[ARA 12] ARASH B., WANG Q., "A review on the application of nonlocal elastic models in modeling of carbon nanotubes and graphenes", *Computational Materials Science*, vol. 51, no. 303–313, 2012.

[ARA 13] ARANI A.G., ROUDBARI M.A., "Nonlocal piezoelectric surface effect on the vibration of visco-Pasternak coupled boron nitride nanotube system under a moving nanoparticle", *Thin Solid Films*, vol. 542, pp. 232–241, 2013.

[ARS 09] ARSAT R., BREEDON M., SHAFIEI M. *et al.*, "Graphene-like nano-sheets for surface acoustic wave gas sensor applications", *Chemical Physics Letters*, vol. 467, no. 4–6, pp. 344–347, 2009.

[ART 11] ARTAN R., TEPE A., "Nonlocal effects in curved single-walled carbon nanotubes", *Mechanics of Advanced Materials and Structures*, vol. 18, no. 347, 2011.

[AYD 09a] AYDOGDU M., "Axial vibration of the nanorods with the nonlocal continuum rod model", *Physica E-Low-Dimensional Systems & Nanostructures*, vol. 41, no. 5, pp. 861–864, 2009.

[AYD 09b] AYDOGDU M., "A general nonlocal beam theory: its application to nanobeam bending, buckling and vibration", *Physica E: Low-dimensional Systems and Nanostructures*, vol. 41, no. 9, pp. 1651–1655, 2009.

[AYD 11] AYDOGDU M., FILIZ S., "Modeling carbon nanotube-based mass sensors using axial vibration and nonlocal elasticity", *Physica E-Low-Dimensional Systems & Nanostructures*, vol. 43, no. 1229, 2011.

[AYD 12] AYDOGDU M., "Axial vibration analysis of nanorods (carbon nanotubes) embedded in an elastic medium using nonlocal elasticity", *Mechanics Research Communications*, vol. 43, pp. 34–40, 2012.

[BAB 11] BABAEI H., SHAHIDI A., "Small-scale effects on the buckling of quadrilateral nanoplates based on nonlocal elasticity theory using the Galerkin method", *Archive of Applied Mechanics*, vol. 81, pp. 1051–1062, 2011.

[BAN 85] BANERJEE J.R., WILLIAMS F.W., "Exact Bernoulli-Euler dynamic stiffness matrix for a range of tapered beams", *International Journal for Numerical Methods in Engineering*, vol. 21, no. 12, pp. 2289–2302, 1985.

[BAN 89] BANERJEE J.R., "Coupled bending torsional dynamic stiffness matrix for beam elements", *International Journal for Numerical Methods in Engineering*, vol. 28, no. 6, pp. 1283–1298, 1989.

[BAN 92a] BANERJEE J.R., FISHER S.A., "Coupled bending torsional dynamic stiffness matrix for axially loaded beam elements", *International Journal for Numerical Methods in Engineering*, vol. 33, no. 4, pp. 739–751, 1992.

[BAN 92b] BANERJEE J.R.,WILLIAMS F.W., "Coupled bending-torsional dynamic stiffness matrix for Timoshenko beam elements", *Computer and Structures*, vol. 42, no. 3, pp. 301–310, 1992.

[BAN 95] BANERJEE J. R., WILLIAMS F. W., "Free-vibration of composite beams – an exact method using symbolic computation", *Journal of Aircraft*, vol. 32, no. 3, pp. 636–642, 1995.

[BAN 97] BANERJEE J. R., "Dynamic stiffness formulation for structural elements: a general approach", *Computer and Structures*, vol. 63, no. 1, pp. 101–103, 1997.

[BAU 11] BAUER S., PITTROF A., TSUCHIYA H. *et al.*, *Electrochemistry Communications,* vol. 13, pp. 538–541, 2011.

[BEL 02] BELYTSCHKO T., XIAO S.P., SCHATZ G.C. *et al.*, "Atomistic simulations of nanotube fracture", *Physical Review B*, vol. 65, no. 23, 2002.

[BER 09] BERDICHEVSKY V. L., "Theory of elastic plates and shells", *Variational Principles of Continuum Mechanics II: Applications*, Interaction of Mechanics and Mathematics, Springer, 2009.

[BES 11] BESSEGHIER A., TOUNSI A., HOUARI M. *et al.*, "Thermal effect on wave propagation in double-walled carbon nanotubes embedded in a polymer matrix using nonlocal elasticity", *Physica E-Low-Dimensional Systems & Nanostructures,* vol. 43, pp. 1379–1386, 2011.

[BIE 06] BIENER J., HODGE A.M., HAYES J.R. *et al.*, "Size effects on the mechanical behaviour of nanoporous Au", *Nano Letters*, vol. 6, no. 10, pp. 2379–2382, 2006.

[BIE 09] BIEL B., BLASE X., TRIOZON F. *et al.*, "Anomalous doping effects on charge transport in graphene nanoribbons", *Physical Review Letters*, vol. 102, no. 9, 2009.

[BLE 84] BLEVINS R. D., *Formulas for Natural Frequency and Mode Shape*, Krieger Publishing Company, Malabar, FL, 1984.

[BOI 09] BOISEN A., "Mass spec goes nanomechanical", *Nature Nanotechnology*, vol. 4, no. 7, pp. 404–405, 2009.

[BOR 00] BORESI A.P., CHONG K.P., *Elasticity in Engineering Mechanics*, Wiley-IEEE, 2000.

[BRE 02] BRENNER D.W., SHENDEROVA O.A., HARRISON J.A. *et al.*, "A second-generation reactive empirical bond order (REBO) potential energy expression for hydrocarbons", *Journal of Physics-Condensed Matter,* vol. 14, no. 4, pp. 783–802, 2002.

[BRE 90] BRENNER D.W., "Empirical potential for hydrocarbons for use in simulating the chemical vapor-deposition of diamond films", *Physical Review B,* vol. 42, no. 15, pp. 9458–9471, 1990.

[BRO 07] BRODKA A., KOLOCZEK J., BURIAN A., "Application of molecular dynamics simulations for structural studies of carbon nanotubes", *Journal of Nanoscience and Nanotechnology*, vol. 7, pp. 1505–1511, 2007.

[BYR 09] BYRNE E.M., MCCARTHY M.A., XIA Z. *et al.*, "Multiwall nanotubes can be stronger than single wall nanotubes and implications for nanocomposite design", *Physical Review Letters,* vol. 103, no. 4, Article Number 045502, 2009.

[BYR 10] BYRNE E.M., LETERTRE A., MCCARTHY M.A. *et al.*, "Optimizing load transfer in multiwall nanotubes through interwall coupling: theory and simulation", *Acta Materialia,* vol. 58, no. 19, pp. 6324–6333, 2010.

[CAO 06] CAO G., CHEN X., KYSAR J.W., "Thermal vibration and apparent thermal contraction of single-walled carbon nanotubes", *Journal of the Mechanics and Physics of Solids*, vol. 54, no. 6, pp. 1206–1236, 2006.

[CAU 65] CAUGHEY T.K., O'KELLY M.E.J., "Classical normal modes in damped linear dynamic systems", *Transactions of ASME, Journal of Applied Mechanics*, vol. 32, pp. 583–588, September 1965.

[CHA 09] CHANDRASEKER K., MUKHERJEE S., PACI T.J. *et al.*, "An atomistic-continuum Cosserat rod model of carbon nanotubes", *Journal of the Mechanics and Physics of Solids*, vol. 57, no. 6, pp 932–958, 2009.

[CHA 11a] CHANG T.P., LIU M.F., *Composites Part B: Engineering*, 2011.

[CHA 11b] CHANDRA Y., CHOWDHURY R., SCARPA F. *et al.*, "Vibrational characteristics of bilayer graphene", *Thin Solid Films*, vol. 519, p. 6026, 2011.

[CHA 12] CHANG T.P., "Small scale effect on axial vibration of non-uniform and non-homogeneous nanorods", *Computational Materials Science*, vol. 54, pp. 23–27, 2012.

[CHA 12a] CHANG T., "Finite element analysis on vibration of nanorods with heterogeneous property and non-uniform cross section", *Advanced Science Letters*, vol. 13, no. 360, 2012.

[CHA 12b] CHANG T.-P., "Small-scale effect on axial vibration of non-uniform and non-homogeneous nanorods", *Comp. Mater. Sci.*, vol. 54, pp. 23–27, 2012.

[CHH 11] CHHAPADIA P., MOHAMMADI P., SHARMA P., "Curvature-dependent surface energy and implications for nanostructures", *Journal of the Mechanics and Physics of Solids*, vol. 59, no. 10, pp. 2103–2115, 2011.

[CHI 08] CHIU H.Y., HUNG P., POSTMA H.W.C. *et al.*, "Atomic-scale mass sensing using carbon nanotube resonators", *Nano letters*, vol. 8, no. 12, pp. 4342–4346, 2008.

[CHO 04] CHOY J., JANG E., WON J. *et al.*, "Hydrothermal route to ZnO nanocoral reefs and nanofibers", *Applied Physics Letters*, vol. 84, no. 2, pp. 287–289, 2004.

[CHO 09] CHOWDHURY R., ADHIKARI S., MITCHELL J., "Vibrating carbon nanotube based bio-sensors", *Physica E: Low-Dimensional Systems and Nanostructures*, vol. 42, no. 2, pp. 104–109, December 2009.

[CHO 10a] CHOWDHURY R., ADHIKARI S., WANG C.Y. *et al.*, "A molecular mechanics approach for the vibration of single walled carbon nanotubes", *Computational Materials Science*, vol. 48, no. 4, pp. 730–735, 2010.

[CHO 10b] CHOWDHURY R., WANG C.Y., ADHIKARI S. *et al.*, "Sliding oscillations of multiwall carbon nanotubes", *Physica E: Low-dimensional Systems and Nanostructures*, vol. 42, no. 9, pp. 2295–2300, 2010.

[CHO 10c] CHOWDHURY R., WANG C. Y., ADHIKARI S., "Low frequency vibration of multiwall carbon nanotubes with heterogeneous boundaries", *Journal of Physics D: Applied Physics*, vol. 43, no. 8, p. 085405, 2010.

[CHO 11a] CHOWDHURY R., ADHIKARI S., "Boron nitride nanotubes as zeptogram-scale bio-nano sensors: theoretical investigations", *IEEE Transactions on Nanotechnology*, vol. 10, no. 4, pp. 659–667, 2011.

[CHO 11b] CHOWDHURY R., ADHIKARI S., SCARPA F., "Vibrational analysis of ZnO nanotubes: a molecular mechanics approach", *Applied Physics A*, vol. 102, no. 2, pp. 301–308, 2011.

[CHO 11c] CHOWDHURY R., ADHIKARI S., REES P. *et al.*, "Graphene-based biosensor using transport properties", *Physical Review B*, vol. 83, no. 4, p. 045401, 2011.

[CHO 11d] CHOWDHURY R., ADHIKARI S., SCARPA F. *et al.*, "Transverse vibration of single layer graphene sheets", *Journal of Physics D: Applied Physics*, vol. 44, no. 20, pp. 205401:1–11, 2011.

[CHO 95] CHOPRA N.G., LUYKEN R.J., CHERREY K. *et al.*, "Boron nitride nanotubes", *Science,* vol. 269, no. 5226, pp. 966–967, 18 August 1995.

[CIV 11] CIVALEK O., DEMIR C., "Bending analysis of microtubules using nonlocal Euler-Bernoulli beam theory", *Applied Mathematical Modelling*, vol. 35, no. 5, pp. 2053–2067, 2011.

[CUE 04] CUENOT S., FRETIGNY C., CHAMPAGNE S.D. *et al.*, "Surface tension effect on the mechanical properties of nanomaterials measured by atomic force microscopy", *Physical Review B*, vol. 69, no. 165410, 2004.

[DAI 02] DAI H., "Carbon nanotubes: opportunities and challenges", *Surface Science*, Vol. 500, pp. 218–241, 2002.

[DAN 12] DANESH M., FARAJPOUR A., MOHAMMADI M., "Axial vibration analysis of a tapered nanorod based on nonlocal elasticity theory and differential quadrature method", *Mechanics Research Communications*, vol. 39, no. 1, pp. 23–27, 2012.

[DAW 84] DAWE D., *Matrix and Finite Element Displacement Analysis of Structures*, Oxford University Press, Oxford, UK, 1984.

[DEO 09] DEOTARE P.B., MCCUTCHEON M.W., FRANK I.W. *et al.*, *Appld. Phys. Letter*, vol. 95, p. 031102, 2009.

[DER 07] DE ROSA M.A., LIPPIELLO M., *Mech. Res. Comm.*, vol. 34, p. 438, 2007.

[DOY 89] DOYLE J. F., *Wave Propagation in Structures*, Springer Verlag, New York, 1989.

[DUA 07a] DUAN W.H., WANG C.M., "Exact solutions for axisymmetric bending of micro/nanoscale circular plates based on nonlocal plate theory", *Nanotechnology*, vol. 18, no. 385704, 2007.

[DUA 07b] DUAN W.H., WANG C.M., ZHANG Y.Y., "Calibration of nonlocal scaling effect parameter for free vibration of carbon nanotubes by molecular dynamics", *Journal of Applied Physics*, vol. 101, no. 2, pp. 024305–024305, 2007.

[DUA 09] DUAN W.H., WANG C.M., "Nonlinear bending and stretching of a circular graphene sheet under a central point load", *Nanotechnology*, vol. 20, p. 075702, 2009.

[ECE 07] ECE M., AYDOGDU M., "Nonlocal elasticity effect on vibration of in-plane loaded double-walled carbon nano-tubes", *Acta Mechanica,* vol. 190, pp. 185–195, 2007.

[EIC 09] EICHENFIELD M., CAMACHO R., CHAN J. *et al.*, "A picogram- and nanometer-scale photonic crystal opto-mechanical cavity", *Nature*, vol. 459, p. 550, 2009.

[EIS 87] EISENBERGER M., CLASTORNIK J., "Vibrations and buckling of a beam on a variable Winkler elastic foundation", *J. Sound Vib.*, vol. 115, p. 233, 1987.

[ELI 12] ELISHAKOFF I., *Carbon Nanotubes and Nanosensors: Vibration, Buckling and Balistic Impact*, ISTE, London and John Wiley & Sons, New York, 2012.

[ERI 72a] ERINGEN A.C., "Linear theory of nonlocal elasticity and dispersion of plane waves", *International Journal of Engineering Science*, vol. 10, no. 5, pp. 425–435, 1972.

[ERI 72b] ERINGEN A.C., EDELEN D.G.B., "On nonlocal elasticity", *International Journal of Engineering Science*, vol. 10, no. 3, pp. 233–248, 1972.

[ERI 83] ERINGEN A.C., "On differential-equations of nonlocal elasticity and solutions of screw dislocation and surface waves", *Journal of Applied Physics*, vol. 54, no. 9, pp. 4703–4710, 1983.

[FER 93a] FERGUSON N. J., PILKEY W.D., "Literature review of variants of dynamic stiffness method, Part 1: The dynamic element method", *The Shock and Vibration Digest*, vol. 25, no. 2, pp. 3–12, 1993.

[FER 93b] FERGUSON N. J., PILKEY W.D., "Literature review of variants of dynamic stiffness method, Part 2: Frequency-dependent matrix and other", *The Shock and Vibration Digest*, vol. 25, no. 4, pp. 3–10, 1993.

[FIL 10] FILIZ S., AYDOGDU M., "Axial vibration of carbon nanotube heterojunctions using nonlocal elasticity", *Computational Materials Science*, vol. 49, no. 3, pp. 619–627, 2010.

[FIR 11] FIROUZ-ABADI R.D., FOTOUHI M.M., HADDADPOUR H., "Free vibration analysis of nanocones using a nonlocal continuum model", *Physics Letters A*, vol. 375, no. 41, pp. 3593–3598, 2011.

[FRA 02] FRAYSSE J., MINETT A.I., JASCHINSKI O. *et al.*, "Carbon nanotubes acting like actuators", *Carbon*, vol. 40, no. 10, pp. 1735–1739, 2002.

[FRA 10] FRANK W., DEOTARE P.B., MCCUTCHEON M.W. *et al.*, "Programmable photonic crystal nanobeam cavities", *Optics Exp.*, vol. 18, p. 8705, 2010.

[FRE 02] FRENKEL D., SMIT B., *Understanding Molecular Simulation: From Algorithms to Applications,* 2nd ed., Academic Press, San Diego, CA, 2002.

[FRI 13] FRISCH M.J., TRUCKS G.W., SCHLEGEL H.B. *et al.*, Gaussian 09 Revision A.1, 2013.

[GAL 06] GALOPPINI E., ROCHFORD J., CHEN H. *et al.*, "Fast electron transport in metal organic vapor deposition grown dye-sensitized ZnO nanorod solar cells", *The Journal of Physical Chemistry B*, vol. 110, no. 33, pp. 16159–16161, 2006.

[GEI 07] GEIM A.K., NOVOSELOV K.S., "The rise of graphene", *Nature Materials*, vol. 6, no. 3, pp. 183–191, 2007.

[GEI 09] GEIM K., "Graphene: status and prospects", *Science,* vol. 324, p. 1530, 2009.

[GER 97] GÉRADIN M., RIXEN D., *Mechanical Vibrations*, 2nd edition, John Wiley & Sons, New York, NY, 1997.

[GIL 10] GIL A. J., ADHIKARI S., SCARPA F. *et al.*, "The wrinkling of single-layer grapheme sheets", *Journal of Physics: Condensed Matter*, vol. 22, no. 14, p. 145302, 2010.

[GOP 07] GOPALAKRISHNAN S., CHAKRABORTY A., MAHAPATRA D. R., *Spectral Finite Element Method*, Springer Verlag, New York, 2007.

[GOP 13] GOPALAKRISHNAN S., NARENDAR S., *Wave Propagation in Nanostructures*, Springer, 2013.

[GRA 75] GRAFF K.F., *Wave Motion in Elastic Solids*, Dover Publications, New York, 1975.

[GUA 05] GUO J.G., ZHAO Y.P., "The size-dependent elastic properties of nanofilms with surface effects", *Journal of Applied Physics*, vol. 98, no. 074306, 2005.

[GUP 10] GUPTA S.S., BATRA R.C., "Elastic properties and frequencies of free vibrations of single-layer graphene sheets", *Journal of Computational and Theoretical Nanoscience*, vol. 7, no. 10, pp. 2151–2164, 2010.

[GUZ 07] GUZ I.A., RODGER A.A., GUZ A.N. *et al.*, "Developing the mechanical models for nanomaterials", *Composites: Part A*, vol. 38, pp. 1234–1250, 2007.

[HAI 92] HAILE J.M., *Molecular Dynamics Simulation: Elementary Methods*, Wiley-Interscience, John Wiley and Sons Inc., NY, 1992.

[HAN 10] HAN T., HE P., WANG J. *et al.*, "Molecular dynamics simulation of a single graphene sheet under tension", *New Carbon Materials,* vol. 25, pp. 261–266, 2010.

[HAN 11a] HANCOCK Y., "The story of the 2010 Nobel prize: ground-breaking experiments on graphene", *Journal of Physics D-Applied Physics,* vol. 44, 2011.

[HAN 11b] HAN T.W., HE P.F., WANG J. *et al.*, "Molecular dynamics simulation of a single graphene sheet under tension", *Carbon*, vol. 49, no. 1, pp. 353, 2011.

[HAR 03] HARIK V.M., SALAS M.D. (eds), *Trends in Nanoscale Mechanics, Analysis of Nanostructured Materials and Multiscale Modeling*, Kluwer Academic Publishers, 2003.

[HAS 99] HASHEMI S.M., RICHARD M.J., DHATT G., "A new dynamic finite element (DFE) formulation for lateral free vibrations of Euler-Bernoulli spinning beams using trigonometric shape functions", *Journal of Sound and Vibration*, vol. 220, no. 4, pp. 601–624, 1999.

[HAS 00] HASHEMI S. M., RICHARD M. J., "Free vibrational analysis of axially loaded bending-torsion coupled beams: a dynamic finite element", *Computer and Structures*, vol. 77, no. 6, pp. 711–724, 2000.

[HAS 09] HASHEMNIA K., FARID M., VATANKHAH R., "Vibrational analysis of carbon nanotubes and graphene sheets using molecular structural mechanics approach", *Computational Materials Science*, vol. 47, no. 1, pp. 79–85, 2009.

[HE 04] HE L.H., LIM C.W., WU B.S., "A continuum model for size-dependent deformation of elastic films of nano-scale thickness", *International Journal of Solids and Structures*, vol. 41, no. 847–857, 2004.

[HE 05] HE X.Q., KITIPORNCHAI S., LIEW K.M., "Resonance analysis of multi-layered graphene sheets used as nanoscale resonators", *Nanotechnology*, vol. 16, no. 10, p. 2086, 2005.

[HEI 08a] HEIRECHE H., TOUNSI A., BENZAIR A., *Nanotechnology*, "Scale effect on wave propagation of double-walled carbon nanotubes with intial axial loading", vol. 19, p. 185703, 2008.

[HEI 08b] HEIRECHE H., TOUNSI A., BENZAIR A. *et al.*, "Sound wave propagation in single-walled carbon nanotubes with intial axial stress", *J. Appl. Phys.*, vol. 104, p. 014301, 2008.

[HEI 08c] HEIRECHE H., TOUNSI A., BENZAIR A. *et al.*, "Sound wave propagation in single-walled carbon nanotubes using nonlocal elasticity", *Physica E*, vol. 40, p. 2791, 2008.

[HEI 10] HEIRECHE H., TOUNSI A., BENHASSAINI H. *et al.*, "Nonlocal elasticity effect on vibration characteristics of protein microtubules", *Physica E-Low-Dimensional Systems & Nanostructures*, vol. 42, pp. 2375–2379, 2010.

[HIN 03] HINCHLIFFE A., *Molecular Modelling for Beginners*, John Wiley and Sons Ltd., Chichester, England, 2003.

[HIR 09] HIRSCH A., "Unzipping carbon nanotubes: a peeling method for the formation of graphene nanoribbons", *Angewandte Chemie-International Edition*, vol. 48, no. 36, pp. 6594–6596, 2009.

[HSU 11] HSU J.C., LEE H.L., CHANG W.J., "Longitudinal vibration of cracked nanobeams using nonlocal elasticity theory", *Current Applied Physics*, vol. 11, no. 6, pp. 1384–1388, 2011.

[HU 08] HU Y., LIEW K., WANG Q. *et al.*, "Nonlocal shell model for elastic wave propagtion in single- and double-walled carbon nanotubes", *Journal of the Mechanics and Physics of Solids,* vol. 56, pp. 3475–3485, 2008.

[HUA 08] HUANG B., LI Z., LIU Z. *et al.*, "Adsorption of gas molecules on grapheme nanoribbons and its implication for nanoscale molecule sensor", *Journal of Physical Chemistry C*, vol. 112, no. 35, pp. 13442–13446, 2008.

[HUA 09] HUANG B., LIU M., SU N. *et al.*, "Quantum manifestations of graphene edge stress and edge instability: a first-principles study", *Physical Review Letters*, vol. 102, no. 166404, pp. 1–4, 2009.

[HUA 11] HUANG Z., JIANG X., GUO D. *et al.*, "Controllable synthesis and biomedical applications of silver nanomaterials", *Journal of Nanoscience and Nanotechnology*, vol. 11, no. 11, pp. 9395–9408, 2011.

[HUA 12] HUANG Z., "Nonlocal effects of longitudinal vibration in nanorod with internal long-range interactions", *International Journal of Solids and Structures*, vol. 49, no. 15, pp. 2150–2154, 2012.

[IIJ 91] IIJIMA S., "Helical microtubules of graphitic carbon", *Nature*, vol. 354, pp. 56–58, 07 November 1991.

[IIJ 93] IIJIMA S., ICHIHASHI T., "Single-shell carbon nanotubes of 1-nm diameter", *Nature*, pp. 363–603, 1993.

[JIA 10] JIANG L.-C., ZHANG W.-D., "A highly sensitive nonenzymatic glucose sensor based on CuO nanoparticles-modified carbon nanotube electrode", *Biosensors & Bioelectronics*, vol. 25, no. 6, pp. 1402–1407, 2010.

[JOM 11] JOMEHZADEH E., NOORI H., SAIDI A., "The size-dependent vibration analysis of micro-plates based on a modified couple stress theory", *Physica E-Low-Dimensional Systems & Nanostructures*, vol. 43, no. 877, 2011.

[JON 06] JONES R.M., *Buckling of Bar, Plates and Shells*, Bull Ridge Corporation, 2006.

[JOS 11] JOSHI A.Y., SHARMA S.C., HARSHA S.P., "Zeptogram scale mass sensing using single walled carbon nanotube based biosensors", *Sensors and Actuators A-Physical*, vol. 168, no. 2, pp. 275–280, 2011.

[KAH 10] KAHROBAIYAN M.H., ASGHARI M., RAHAEIFARD M. *et al.*, "Investigation of the size-dependent dynamic characteristics of atomic force microscope microcantilevers based on the modified couple stress theory", *International Journal of Engineering Science*, vol. 48, no. 12, pp. 1985–1994, 2010.

[KAR 10] KARAOGLU P., AYDOGDU M., "On the forced vibration of carbon nanotubes via a non-local Euler-Bernoulli beam model", *Proceedings of the Institution of Mechanical Engineers Part C-Journal of Mechanical Engineering Science*, vol. 224, no. C2, pp. 497–503, 2010.

[KAR 14a] KARLIČIĆ D., KOZIĆ P., PAVLOVIĆ R., "Free transverse vibration of nonlocal viscoelastic orthotropic multi-nanoplate system (MNPS) embedded in a viscoelastic medium", *Composite Structures*, vol. 115, pp. 89–99, 2014.

[KAR 14b] KARLIČIĆ D., MURMU T., CAJIĆ M. *et al.*, "Dynamics of multiple viscoelastic carbon nanotube based nanocomposites with axial magnetic field", *Journal of Applied Physics*, vol. 115, no. 23, p. 234303, 2014.

[KAR 14c] KARLIČIĆ D., ADHIKARI S., MURMU T. *et al.*, "Exact closed-form solution for non-local vibration and biaxial buckling of bonded multi-nanoplate system", *Composites Part B: Engineering*, vol. 66, pp. 328–339, 2014.

[KAR 14d] KARLIČIĆ D., JOVANOVIĆ D., KOZIĆ P. *et al.*, "Thermal and magnetic effects on the vibration of a cracked nanobeam embedded in an elastic medium", *Journal of Mechanics of Materials and Structures*, 2014.

[KAR 14e] KARLIČIĆ D., KOZIĆ P., PAVLOVIĆ R., "Flexural vibration and buckling of single-walled carbon nanotubes using different gradient elasticity theories for Reddy and Huu-Tai beam theories", *Journal of Theoretical and Applied Mechanics*, 2014.

[KAR 15] KARLIČIĆ D., CAJIĆ M., MURMU T. *et al.*, "Nonlocal effects on the longitudinal vibration of a complex multi-nanorod system subjected to the transverse magnetic field", *Meccanica*, pp. 1–17, 2015.

[KAS 97] KASUYA A., SASAKI Y., SAITO Y. *et al.*, "Evidence for size-dependent discrete dispersions in single-wall nanotubes", *Phys. Rev. Lett.*, vol. 78, p. 4434, 1997.

[KEL 09] KELLY S.G., SRINIVAS S., "Free vibrations of elastically connected stretched beams", *Journal of Sound and Vibration*, vol. 326, no. 3, pp. 883–893, 2009.

[KHA 10] KHADEMOLHOSSEINI F., RAJAPAKSE R.K., NOJEH A., "Torsional buckling of carbon nanotubes based on nonlocal elasticity shell models", *Comput. Mater. Sci.*, vol. 48, p. 736, 2010.

[KHA 12] KHADEMOLHOSSEINI F., PHANI A.S., NOJEH A. *et al.*, "Nonlocal continuum modeling and molecular dynamics simulation of torsional vibration of carbon nanotubes", *IEEE Transactions on Nanotechnology*, vol. 11, pp. 34–43, 2012.

[KIA 98] KIANG C., ENDO M., AJAYAN P. *et al.*, "Size effects in carbon nanotubes", *Physical Review Letters*, vol. 81, pp. 1869–1872, 1998.

[KIA 11] KIANI K., "Small-scale effect on the vibration of thin nanoplates subjected to a moving nanoparticle via nonlocal continuum theory", *Journal of Sound and Vibrations*, vol. 330, no. 20, pp. 4896–4914, 2011.

[KIA 13] KIANI K., "Longitudinal, transverse, and torsional vibrations and stabilities of axially moving single-walled carbon nanotubes", *Current Applied Physics*, vol. 13, no. 8, pp. 1651–1660, 2013.

[KIT 05] KITIPORNCHAI S., HE X.Q., LIEW K.M., "Continuum model for the vibration of multilayered graphene sheets", *Physical Review B*, vol. 72, no. 7, pp. 075443 1-075443 6, 2005.

[KON 08] KONG S., ZHOU S., NIE Z. *et al.*, "The size-dependent natural frequency of Bernoulli–Euler micro-beams", *International Journal of Engineering Science*, Vol. 46, pp. 427–437, 2008.

[KON 09] KONG S., ZHOU S., NIE Z. *et al.*, "Static and dynamic analysis of micro beams based on strain gradient elasticity theory", *International Journal of Engineering Science*, vol. 47, no. 4, pp. 487–498, 2009.

[KOS 09] KOSYNKIN D.V., HIGGINBOTHAM A.L., SINITSKII A. *et al.*, "Longitudinal unzipping of carbon nanotubes to form graphene nanoribbons", *Nature*, vol. 458, no. 7240, pp. 872-U5, 2009.

[LAO 12] LAO J.Y., WEN J.G., REN Z.F., "Hierarchical ZnO nanostructures", *Nano Lett.*, vol. 2, no. 11, pp. 1287–1291, 2012.

[LEA 01] LEACH A.R., *Molecular Modelling: Principles and Applications,* 2nd ed., Pearson Education Ltd., Harlow, England, 2001.

[LEE 07] LEE W.C., CHO Y.H., "Bio-inspired digital nanoactuators for photon and biomaterial manipulation", *Current Applied Physics*, vol. 7, pp. 139–146, 2007.

[LEE 10] LEE H.L., HSU J.C., CHANG W.J., "Freqwuency shift of carbon-nanotube-based mass sensor using nonlocal elasticity theory", *Nanoscale Res. Lett.*, vol. 5, no. 11, 2010.

[LI 03a] LI C., CHOU T.W., "Single-walled carbon nanotubes as ultrahigh frequency nanomechanical resonators", *Physical review B*, vol. 68, no. 7, p. 073405, 2003.

[LI 03b] LIU Y., CHEN X., "Evaluations of the effective material properties of carbon nanotube-based composites using nanoscale representative volume element", *Mechanics of Materials*, vol. 35, pp. 69–81, 2003.

[LI 04] LI C. Y., CHOU T. W., "Mass detection using carbon nanotube-based nanomechanical resonators", *Applied Physics Letters*, vol. 84, no. 25, pp. 5246–5248, 2004.

[LI 08] LI Y., QIU X., YANG F. *et al.*, "Ultra-high sensitivity of super carbon-nanotube-based mass and strain sensors", *Nanotechnology*, vol. 19, no. 16, 2008.

[LIM 07] LIM C.W., WANG C.M., "Exact variational nonlocal stress modelling with asymptotic higher order strain gradients for nanobeams", *J. Appl. Phys.*, vol. 101, p. 054312, 2007.

[LIM 12] LIM C.W., LI C., YU J.L., "Free torsional vibration of nanotubes based on nonlocal stress theory", *Journal of Sound and Vibration*, vol. 331, no. 12, pp. 2798–2808, 2012.

[LIN 10] LIN Q., ROSENBERG J., CHANG D. *et al.*, "Coherent mixing of mechanical excitations in nano-optimechanical structures", *Nature Photonics*, vol. 4, p. 236, 2010.

[LIU 03] LIU B., ZENG H.C., "Hydrothermal synthesis of ZnO nanorods in the diameter regime of 50 nm", *Journal of the American Chemical Society*, vol. 125, no. 15, pp. 4430–4431, 2003.

[LIU 09] LIU B., ZENG H.C., "Direct growth of enclosed ZnO nanotubes", *Nano Res*, vol. 2, pp. 201–209, 2009.

[LU 97] LU J.P., *Phys. Rev. Lett.*, vol. 79, p. 1297, 1997.

[LU 06] LU P., LEE H.P., LU C. *et al.*, "Dynamic properties of flexural beams using a nonlocal elasticity model", *Journal of Applied Physics*, vol. 99, no. 073510, 2006.

[LU 07a] LU P., LEE H.P., LU C. *et al.*, "Application of nonlocal beam models for carbon nanotubes", *International Journal of Solids and Structures*, vol. 44, no. 16, pp. 5289–5300, 2007.

[LU 07b] LU P., "Dynamic analysis of axially prestressed micro/nanobeam structures based on nonlocal beam theory", *J. Appd. Phys.*, vol. 101, p. 073504, 2007.

[LU 07c] LU P., ZHANG P.Q., LEE H.P. *et al.*, "Non-local elastic plate theories", *Proceedings of the Royal Society A: Mathematical, Physical and Engineering Science*, vol. 463, no. 2088, pp. 3225–3240, 2007.

[MAL 11] MALEKZADEH P., SETOODEH A.R., BENI A.A., "Small scale effect on the free vibration of orthotropic arbitrary straight-sided quadrilateral nanoplates", *Composite Structures*, vol. 93, no. 7, pp. 1631–1639, 2011.

[MAN 98] MANOHAR C.S., ADHIKARI S., "Dynamic stiffness of randomly parametered beams", *Probabilistic Engineering Mechanics*, vol. 13, no. 1, pp. 39–51, January 1998.

[MCC 14] MCCARTHY, M.A., BYRNE E.M., O'BRIEN N.P. *et al.*, "Improved mechanical performance of CNTs and CNT fibres in nanocomposites through inter-wall and inter-tube coupling", Chapter 1 in TSERPES K., SILVESTRE N. (eds) *Multi-scale Modelling and Analysis of Graphene and Carbon nanotube based Composites*, Springer Series in Materials Science, Springer International Publishing, Switzerland, 2014.

[MEH 13] MEHDIPOUR I., MOGHADAM A.E., MEHDIPOUR C., "Application of an electrostatically actuated cantilevered carbon nanotube with an attached mass as a bio-mass sensor", *Current Applied Physics*, vol. 13, pp. e1463–e1469, 2013.

[MEI 97] MEIROVITCH L., *Principles and Techniques of Vibrations*, Prentice-Hall International, Inc., New Jersey, 1997.

[MEY 07] MEYER J. C., GEIM A. K., KATSNELSON M. I. *et al.*, "The structure of suspended graphene sheets", *Nature*, vol. 446, pp. 60–63, 2007.

[MIL 00] MILLER R.E., SHENOY V.B., "Size-dependent elastic properties of nanosized structural elements", *Nanotechnology*, vol. 11, no. S0957, pp. 139–147, 2000.

[MOH 08] MOHANTY N., BERRY V., "Graphene-based single-bacterium resolution biodevice and DNA transistor: interfacing graphene derivatives with nanoscale and microscale biocomponents", *Nano Letters*, vol. 8, no. 12, pp. 4469–4476, 2008.

[MOO 11] MOON J., GASKILL D., "Graphene: its fundamentals to future application", *IEEE Transactions on Microwave Theory and Techniques*, vol. 59, pp. 2702–2708, 2011.

[MOR 08] MORADIAN R., MOHAMMADI Y., GHOBADI N., "Investigation of gas sensing properties of armchair graphene nanoribbons", *Journal of Physics-Condensed Matter*, vol. 20, no. 42, 2008.

[MUK 06] MUKHERJEE S., ALURU N.R., "Applications in micro- and nano electromechanical systems", *Engg. Anal. Boundary Ele.*, vol. 30, p. 909, 2006.

[MUR 01] MURMU T., ADHIKARI S., "Nonlocal effects in the longitudinal vibration of double-nanorod systems", *Physica E*, vol. 43, pp. 415–422, 2010.

[MUR 09a] MURMU T., PRADHAN S.C., "Vibration analysis of nano-single-layered grapheme sheets embedded in elastic medium based on nonlocal elasticity theory", *Journal of Applied Physics*, vol. 105, no. 1, 2009.

[MUR 09b] MURMU T., PRADHAN S.C., "Buckling of biaxially compressed orthotropic plates at small scale", *Mechanics Research Communications*, vol. 36, no. 8, pp. 933–938, 2009.

[MUR 09c] MURMU T., PRADHAN S.C., "Small-scale effect on the free in-plane vibration of nanoplates by non-local continuum model", *Physica E-Low-Dimensional Systems & Nanostructures*, vol. 41, pp. 1628–1633, 2009.

[MUR 09d] MURMU T., PRADHAN S.C., "Vibration analysis of nanoplates under axial prestressed conditions via nonlocal elasticity", *Journal of Applied Physics*, vol. 106, p. 104301, 2009.

[MUR 09e] MURMU T., PRADHAN S.C., "Buckling analysis of a single-walled carbon nanotube embedded in an elastic medium based on nonlocal elasticity and Timoshenko beam", *Physica E*, vol. 41, p. 1232, 2009.

[MUR 09f] MURMU T., PRADHAN S.C., "Themo-mechanical vibration of a single-walled carbon nanotube embedded in an elastic medium based on nonlocal elasticity theory", *Comput. Mater. Sci.*, vol. 46, p. 854, 2009.

[MUR 10] MURMU T., ADHIKARI S., "Nonlocal effects in the longitudinal vibration of double-nanorod systems", *Physica E: Low-dimensional Systems and Nanostructures*, vol. 43, no. 1, pp. 415–422, 2010.

[MUR 10a] MURMU T., ADHIKARI S., "Nonlocal effects in the longitudinal vibration of double-nanorod systems", *Physica E: Low-dimensional Systems and Nanostructures*, vol. 43, no. 1, pp. 415–422, 2010.

[MUR 10b] MURMU T., PRADHAN S.C., "Thermal effect in the stability of embedded carbon nanotubes", *Comput. Mater. Sci.*, vol. 47, p. 721, 2010.

[MUR 10c] MURMU T., ADHIKARI S., "Nonlocal transverse vibration of double-nanobeam-systems", *Journal of Applied Physics*, vol. 108, p. 083514, 2010.

[MUR 11a] MURMU T., ADHIKARI S., "Nonlocal vibration of carbon nanotubes with attached buckyballs at tip", *Mechanics Research Communications*, vol. 38, no. 1, pp. 62–67, 2011.

[MUR 11b] MURMU T., ADHIKARI S., "Nonlocal vibration of bonded double-nanoplatesystems", *Composites Part B: Engineering*, vol. 42, no. 7, pp. 1901–1911, 2011.

[MUR 11c] MURMU T., ADHIKARI S., WANG C.Y., "Torsional vibration of carbon nanotube–buckyball systems based on nonlocal elasticity theory", *Physica E: Low-Dimensional Systems and Nanostructures*, vol. 43, no. 6, pp. 1276–1280, 2011.

[MUR 11d] MURMU T., ADHIKARI S., "Axial instability of double-nanobeam-systems", *Physics Letters A,* vol. 375, pp. 601–608, 2011.

[MUR 11e] MURMU T., SIENZ J., ADHIKARI S. *et al.*, "Nonlocal buckling behavior of bonded double-nanoplate-systems", *Journal of Applied Physics,* vol. 110, no. 8, pp. 084316–084316, 2011.

[MUR 12a] MURMU T., ADHIKARI S., "Nonlocal elasticity based vibration of initially prestressed coupled nanobeam systems", *European Journal of Mechanics – A/Solids*, vol. 34, no. 1, pp. 52–62, 2012.

[MUR 12b] MURMU T., ADHIKARI S., "Sensors and Actuators A", *Physical,* vol. 173, pp. 41–48, 2012.

[MUR 12c] MURMU T., ADHIKARI S., MCCARTHY M., "Vibration response of doublewalled carbon nanotubes subjected to an externally applied longitudinal magnetic field: a nonlocal elasticity approach", *Journal of Sound and Vibration*, vol. 331, no. 23, pp. 5069–5086, 2012.

[MUR 13] MURMU T., SIENZ J., ADHIKARI S. *et al.*, "Nonlocal buckling of double-nanoplate-systems under biaxial compression", *Composites Part B: Engineering*, vol. 44, no. 1, pp. 84–94, 2013.

[NAN 11] NARENDAR S. "Terahertz wave propagation in uniform nanorods: a nonlocal continuum mechanics formulation including the effect of lateral inertia", *Physica E: Low-dimensional Systems and Nanostructures*, vol. 43, no. 4, pp. 1015–1020, 2011.

[NAR 10] NARENDAR S., GOPALAKRISHNAN S., "Non local scale effects on ultrasonic wave characteristics nanorods", *Physica E-Low-Dimensional Systems & Nanostructures*, vol. 42, no. 5, pp. 1601–1604, 2010.

[NAR 11] NARENDAR S., GOPALAKRISHNAN S., "Axial wave propagation in coupled nanorod system with nonlocal small scale effects", *Composites Part B: Engineering*, vol. 42, no. 7, pp. 2013–2023, 2011.

[NAR 11a] NARENDAR S., GOPALAKRISHNAN S., "Axial wave propagation in coupled nanorod system with nonlocal small scale effects", *Composites Part B-Engineering*, vol. 42, no. 2013, 2011.

[NAR 11b] NARENDAR S., GOPALAKRISHNAN S., "Spectral finite element formulation for nanorods via nonlocal continuum mechanics", *Journal of Applied Mechanics-Transactions of the Asme*, vol. 78, 2011.

[NAR 11c] NARENDAR S., GOPALAKRISHNAN S., "Critical buckling temperature of single-walled carbon nanotubes embedded in a one-parameter elastic medium based on non-local continuum mechanics", *Physica E*, vol. 43, p. 1185, 2011.

[NAR 11d] NARENDAR S., MAHAPATRA R., GOPALAKRISHNAN S., "Prediction of non-local scaling parameter for armchair and zigzag single-walled carbon nontubes based on molecular structural mechanics, nonlocal elasticity and wave preparation", *Int. J. Engg. Sci.*, vol. 49, p. 509522, 2011.

[NIU 09] NIU J.C., LIM C.W., LEUNG A.Y.T., "Third-order non-local beam theories for the analysis of symmetrical nanobeams", *Proc. Inst. Mech. Eng., Part C: J. Mech. Eng. Sci.*, vol. 223, p. 2451, 2009.

[OBR 13a] O'BRIEN N.P., MCCARTHY M.A., CURTIN W.A., "Improved inter-tube coupling in CNT bundles through carbon ion irradiation", *Carbon,* vol. 51, no. 0, pp. 173–184, 2013.

[OBR 13b] O'BRIEN N.P., MCCARTHY M.A., CURTIN W.A., "A theoretical quantification of the possible improvement in the mechanical properties of carbon nanotube bundles by carbon ion irradiation", *Carbon,* vol. 53, no. 0, pp. 346–356, 2013.

[ODE 02] ODEGARD G.M., GATES T.S., NICHOLSON L.M. *et al.*, "Equivalent-continuum modeling of nano-structured materials", *Composites Science and Technology*, vol. 62, no. 14, pp. 1869–1880, 2002.

[OLS 07] OLSSON P.A.T., MELIN S., PERSSON C., "Atomistic simulations of tensile and bending properties of single-crystal bcc iron nanobeams", *Physical Review B*, vol. 76, no. 224112, 2007.

[ONI 00] ONISZCZUK Z., "Free transverse vibrations of an elastically connected rectangular simply supported double-plate complex system", *Journal of Sound and Vibration*, vol. 236, no. 4, pp. 595–608, 2000.

[PAC 08] PACILE D., MEYER J.C., GIRIT C.O. *et al.*, "The two-dimensional phase of boron nitride: few-atomic-layer sheets and suspended membranes", *Applied Physics Letters*, vol. 92, no. 13, pp. 133107–133107, 2008.

[PAR 02] PARK W., KIM D., JUNG S. *et al.*, "Metalorganic vapor-phase epitaxial growth of vertically well-aligned ZnO nanorods", *Applied Physics Letters*, vol. 80, no. 22, pp. 4232–4234, 2002.

[PAR 06] PARK S.K., GAO S.L., "Bernoulli–Euler beam model based on a modified couple stress theory", *Journal of Micromechanics and Microengineering*, vol. 16, pp. 2355–2359, 2006.

[PAS 08] PASTEWKA L., POU P., PEREZ R. *et al.*, "Describing bond-breaking processes by reactive potentials: importance of an environment-dependent interaction range", *Physical Review B*, vol. 78, no. 16, 2008.

[PAZ 80] PAZ M., *Structural Dynamics: Theory and Computation*, 2nd edition, Van Nostrand, Reinhold, 1980.

[PED 03] PEDDIESON J., BUCHANAN G., MCNITT R., "Application of nonlocal continuum models to nanotechnology", *International Journal of Engineering Science*, vol. 41, no. 305, 2003.

[PEN 06] PENG H.B., CHANG C.W., ALONI S. *et al.*, "Ultrahigh frequency nanotube resonators", *Physical Review Letters*, vol. 97, no. 8, p. 087203, 2006.

[PET 98] PETYT M., *Introduction to Finite Element Vibration Analysis*, Cambridge University Press, Cambridge, UK, 1998.

[PHA 10] PHADIKAR J.K., PRADHAN S.C., "Variational formulation and finite element analysis for nonlocal elastic nanobeams and nanoplates", *Computational Materials Science*, vol. 49, no. 3, pp. 492–499, 2010.

[PIS 03] PISANO A., FUSCHI P., "Closed form solution for a nonlocal elastic bar in tension", *International Journal of Solids and Structures*, vol. 40, no. 1, pp. 13–23, 2003.

[PIS 09] PISANO A.A., SOFI A., FUSCHI P., "Nonlocal integral elasticity: 2D finite element based solutions", *International Journal of Solids and Structures*, vol. 46, no. 21, pp. 3836–3849, 2009.

[PLI 95] PLIMPTON S., "Fast parallel algorithms for short-range molecular-dynamics", *Journal of Computational Physics,* vol. 117, no. 1, pp. 1–19, 1995.

[POU 12] POURESMAEELI S., FAZELZADEH S.A., GHAVANLOO E., "Exact solution for nonlocal vibration of double-orthotropic nanoplates embedded in elastic medium", *Composites Part B: Engineering*, vol. 43, no. 8, pp. 3384–3390, 2012.

[PRA 09a] PRADHAN S.C., "Buckling of single layer graphene sheet based on nonlocal elasticity and higher order shear deformation theory", *Physics Letters A*, vol. 373, no. 45, pp. 4182–4188, 2009.

[PRA 09b] PRADHAN S.C., PHADIKAR J.K., *Physics Letters A,* vol. 373, p. 1062, 2009.

[PRA 09c] PRADHAN S.C., PHADIKAR J.K., "Nonlocal elasticity theory for vibration of nanoplates", *Journal of Sound and Vibration*, vol. 325, no. 1, pp. 206–223, 2009.

[PRA 09d] PRADHAN S.C., PHADIKAR J.K., "Small-scale effect on vibration of embedded multilayered graphene sheets based on nonlocal continuum models", *Physics Letters A*, vol. 373, no. 11, pp. 1062–1069, 2009.

[PRA 09e] PRADHAN S.C., MURMU T., "Small scale effect on the buckling of single-layered graphene sheets under bi-axial compression via nonlocal continuum mechnanics", *Computational Materials Science,* vol. 47, p. 268, 2009.

[PRA 09f] PRADHAN S.C., MURMU T., "Vibration analysis of a single-walled carbon nonotubes embedded in an elastic medium based on non-local elasticity and Timoshenko beam theory", *J. Appl. Phys.*, vol. 105, p. 124306, 2009.

[PRA 10] PRADHAN S.C., MURMU T., "Small scale effect on the buckling analysis of single-layered graphene sheet embedded in an elastic medium based on non-local plate theory", *Physica E-Low-Dimensional Systems & Nanostructures,* vol. 42, no. 5, pp. 1293–1301, 2010.

[PRA 12] PRADHAN S.C., "Nonlocal finite element analysis and small scale effects of CNTs with Timoshenko beam theory", *Finite Elements in Analysis and Design*, vol. 50, no. 8, 2012.

[PUM 10] PUMERA M., AMBROSI A., BONANNI A. *et al.*, "Graphene for electrochemical sensing and biosensing", *Trac-Trends in Analytical Chemistry*, vol. 29, no. 9, pp. 954–965, 2010.

[QAZ 08] QAZI M., KOLEY G., "NO_2 detection using microcantilever based potentiometry", *Sensors*, vol. 8, no. 11, pp. 7144–7156, 2008.

[RAP 92] RAPPE A. K., CASEWIT C. J., COLWELL K. S. *et al.*, "UFF, a full periodic-table force-field for molecular mechanics and molecular-dynamics simulations", *Journal of the American Chemical Society*, vol. 114, no. 25, pp. 10024–10035, 1992.

[RAS 53] RAŠKOVIĆ D., "On some characteristics of the frequency equation of torsional vibrations of light shafts with several disks", *Publications de l'Institut Mathématique V*, available at: http://elib.mi.sanu.ac.rs/files/journals/publ/11/16.pdf, no. 05, pp. 155–164, 1953.

[RAY 77] RAYLEIGH L., *Theory of Sound* (two volumes), 2nd edition, Dover Publications, New York, 1945 re-issue, 1877.

[RED 02] REDDY J.N., *Energy Principles and Variational Methods in Applied Mechanics,* John Wiley & Sons, 2002.

[RED 04] REDDY J.N., *Mechanics of Laminated Composite Plates and Shells: Theory and Analysis/JN Reddy,* CRC Press, 2004.

[RED 07a] REDDY J.N., "Nonlocal theories for bending, buckling and vibration of beams", *International Journal of Engineering Science*, vol. 45, nos. 2–8, pp. 288–307, 2007.

[RED 07b] RED'KIN A.N., MAKOVEI Z.I., GRUZINTSEV A.N. *et al.*, "Vapor phase synthesis of aligned ZnO nanorod arrays from elements", *Inorganis Materials*, vol. 43, no. 3, pp. 253–257, 2007.

[RED 08] REDDY J.N., PANG S.D., "Non-local continuum theories of beams for the analysis of carbon nanotubes", *J. Appd. Phys.*, vol. 103, p. 023511, 2008.

[ROS 08] ROSALES L., PACHECO M., BARTICEVIC Z. *et al.*, "Transport properties of graphene nanoribbons with side-attached organic molecules", *Nanotechnology*, vol. 19, no. 6, 2008.

[RUU 94] RUUD J.A., JERVIS T.R., SPAEPAN F., Nanoindentation of Ag/Ni multilayered thin films, *Journal of Applied Physics,* vol. 75, no. 10, pp. 4969–4974, 1994.

[SAM 11] SAMAEI A.T., ABBASION S., MIRSAYAR M.M., "Buckling analysis of a singlelayer graphene sheet embedded in an elastic medium based on nonlocal Mindlin plate theory", *Mechanics Research Communications*, vol. 38, no. 7, pp. 481–485, 2011.

[SAN 09] SANTOS H., CHICO L., BREY L., "Carbon nanoelectronics: unzipping tubes into graphene ribbons", *Physical Review Letters*, vol. 103, no. 8, 2009.

[SAP 06] SAPMAZ S., JARILLO-HERRERO P., BLANTER Y. *et al.*, "Tunneling in suspended carbon nanotubes assisted by longitudinal phonons", *Physical Review Letters*, vol. 96, no. 2, 2006.

[SAP 06] SAPMAZ S., JARILLO-HERRERO P., BLANTER Y.M. *et al.*, "Tunneling in suspended carbon nanotubes assisted by longitudinal phonons", *Physical Review Letters*, vol. 96, no. 2, 2006.

[SCA 08] SCARPA F., ADHIKARI S., "A mechanical equivalence for the Poisson's ratio and thickness of C-C bonds in single wall carbon nanotubes", *Journal of Physics D: Applied Physics*, vol. 41, no. 085306, pp. 1–5, 2008.

[SCA 10] SCARPA F., ADHIKARI S., GIL A. J. *et al.*, "The bending of single layer grapheme sheets: lattice versus continuum approach", *Nanotechnology*, vol. 20, no. 12, p. 085405, 2010.

[SHA 04] SHAO Y., JIN Y., DONG S., "Synthesis of gold nanoplates by asparate reduction of gold chloride", *ChemCommun (Camb).*, vol. 9, p. 1104, 2004.

[SHA 10] SHAO Y., WANG J., WU H. *et al.*, "Graphene based electrochemical sensors and biosensors: a review", *Electroanalysis*, vol. 22, no. 10, pp. 1027–1036, 2010.

[SHE 08] SHENOY V. B., REDDY C. D., RAMASUBRAMANIAM A. *et al.*, "Edge-stressinduced warping of graphene sheets and nanoribbons", *Physical Review Letters*, vol. 101, no. 245501, pp. 1–4, 2008.

[SHE 10] SHEN L., SHEN H., ZHANG C., "Nonlocal plate model for nonlinear vibration of single layer graphene sheets in thermal environments", *Computational Materials Science,* vol. 48, pp. 680–685, 2010.

[SHI 11] SHI J., NI Q., LEI X. *et al.*, "Nonlocal elasticity theory for the buckling of double-layer graphene nanoribbons based on a continuum model", *Computational Materials Science,* vol. 50, pp. 3085–3090, 2011.

[SIM 10] SIMSEK M., "Vibration analysis of a single-walled carbon nanotube under action of a moving harmonic load based on nonlocal plasticity theory", *Physica E*, vol. 43, p. 182, 2010.

[SIM 11a] SIMSEK M., "Forced vibration of an embedded single-walled carbon nanotube traversed by a moving load using nonlocal Timoshenko beam theory", *Steel Comp. Struct.*, vol. 11, p. 59, 2011.

[SIM 11b] SIMSEK M., "Nonlocal effects in the forced vibration of an elastically connected double carbon nanotube system under a moving nanoparticle", *Computational Materials Science*, vol. 50, p. 2112, 2011.

[SIM 12] ŞIMŞEK M., "Nonlocal effects in the free longitudinal vibration of axially functionally graded tapered nanorods", *Computational Materials Science*, vol. 61, pp. 257–265, 2012.

[SOE 04] SOEDEL W., *Vibrations of Shells and Plates*, 3rd edition, Marcel Dekker Inc, New York, 2004.

[STO 13] STOJANOVIĆ V., KOZIĆ P., JANEVSKI G., "Exact closed-form solutions for the natural frequencies and stability of elastically connected multiple beam system using Timoshenko and high-order shear deformation theory", *Journal of Sound Vibration,* vol. 332, pp. 563–576, 2013.

[SUD 03] SUDAK L.J., Column buckling of multiwalled carbon nanotubes using nonlocal continuum mechanics, *Journal of Applied Physics*, vol. 94, pp. 7281–7287, 2003.

[SUN 03] SUN C.T., ZHANG H., "Size-dependent elastic moduli of platelike nanomaterials", *Journal of Applied Physics*, vol. 93, no. 2, pp. 1212–1218, 2003.

[TAN 08] TANG C., MENG L., SUN L. *et al.*, "Molecular dynamics study of ripples in graphene nanoribbons on 6H-SiC(0001): temperature and size effects", *Journal of Applied Physics*, vol. 104, no. 1, 2008.

[TAN 09] TANG C., GUO W., CHEN C., "Molecular dynamics simulation of tensile elongation of carbon nanotubes: temperature and size effects", *Physical Review B,* vol. 79, 2009.

[TIM 40] TIMOSHENKO S., *Theory of Plates and Shells*, McGraw-Hill Inc, New York, 1940.

[TON 09] TONG F. M., WANG C. Y., ADHIKARI S., "Axial buckling of multiwall carbon nanotubes with heterogeneous boundary conditions", *Journal of Applied Physics*, vol. 105, pp. 094325:1-7, May 2009.

[TOU 09a] TOUNSI A., HEIRECHE H., ADDA BEDIA E.A., "Comment on 'Free transverse vibration of the fluid-conveying single-walled carbon nanotube using nonlocal elasticity theory' (J. Appl. Phys. 103, 024302 (2008))", *J. Appl. Phys.*, vol. 105, p. 126105, 2009.

[TOU 09b] TOUNSI A., HEIRECHE H., BENZAIR A. *et al.*, "Comment on 'Vibration analysis of fluid-conveying double-walled carbon nanotubes based on nonlocal elastic theory'", *J. Phys.: Condens. Matter,* vol. 21, p. 448001, 2009.

[TSE 05] TSERPES K. I., PAPANIKOS P., "Finite Element modelling of single-walled carbon nanotubes", *Composites B*, vol. 36, p. 468, 2005.

[TSI 09] TSIATAS G.C., "A new Kirchhoff plate model based on a modified couple stress theory", *International Journal of Solids and Structures*, vol. 46, pp. 2757–2764, 2009.

[VER 07] VERNEREY F., LIU W.K., MORAN B., "Multi-scale micromorphic theory for hierarchical materials", *Journal of the Mechanics and Physics of Solids*, vol. 55, pp. 2603–2651, 2007.

[WAN 01] WANG Z.L., GAO R.P., PAN Z.W. *et al.*, "Nano-scale mechanics of nanotubes, nanowires, and nanobelts", *Advanced Engineering Materials*, vol. 3, no. 9, p. 657, 2001.

[WAN 03a] WANG R.M., XING Y.J., XU J. *et al.*, "Fabrication and microstructure analysis on zinc oxide nanotubes", *New J. Phys.*, vol. 5, p. 115, 2003.

[WAN 03b] WANG S.G., ZHANG Q., WANG R. *et al.*, "A novel multi-walled carbon nanotube-based biosensor for glucose detection", *Biochemical and Biophysical Research Communications,* vol. 311, pp. 572–576, 2003.

[WAN 05a] WANG J., "Carbon-nanotube based electrochemical biosensors: a review", *Electroanalysis*, vol. 17, no. 1, pp. 7–14, 2005.

[WAN 05b] WANG L.F., HU H.Y., "Flexural wave propagation in single-walled carbon nanotubes", *Physical Review B*, vol. 71, no. 195412, 2005.

[WAN 06] WANG C.M., ZHANG Y.Y., RAMESH S.S. *et al.*, "Buckling analysis of micro- and nano-rods/tubes based on nonlocal Timoshenko beam theory", *J. Phys. D*, vol. 39, p. 3904, 2006.

[WAN 07] WANG Q., WANG C.M., The constitutive relation and small scale parameter of nonlocal continuum mechanics for modelling carbon nanotubes, *Nanotechnology*, vol. 18, no. 075702, 2007.

[WAN 08a] WANG Q., HAN Q.K., WEN B.C., "Estimate of Material Property of Carbon Nanotubes via Nonlocal Elasticity", *Advanced Theoretical Applied Mechanics*, vol. 1, pp.1–10, 2008.

[WAN 08b] WANG C. M., DUAN W. H., "Free vibration of nanorings/arches based on nonlocal elasticity", *Journal of Applied Physics*, vol. 104, no. 1, 2008.

[WAN 08c] WANG C.M., KITIPORNCHAI S., LIM C.W. *et al.*, "Beam bending solutions based on nonlocal Timoshenko beam theory", *J. Eng. Mech.*, vol. 134, p. 475, 2008.

[WAN 09] WANG L., "Vibration and instability analysis of tubular nano- and micro-beams conveying fluid using nonlocal elastic theory", *Physica E,* vol. 41, p. 1835, 2009.

[WAN 10] WANG Y.Z., LI F.M., KISHIMOTO K., "Scale effects on the longitudinal wave propagation in nanoplates", *Physica E,* vol. 42, p. 1356, 2010.

[WAR 09] WARNER J., SCHAFFEL F., RUMMELI M. *et al.*, "Examining the edges of multilayer graphene sheets", *Chemistry of Materials*, vol. 21, pp. 2418–2421, 2009.

[WAT 10] WATER W., JHAO R.Y., JI L.W. *et al.* "Love wave ultraviolet photodetector using ZnO nanorods synthesized on 90°-rotated ST-cut (42° 45′) quartz", *Sensors and Actuators A: Physical*, vol. 161, no. 1, pp. 6–11, 2010.

[WEI 08] WEI X.L., LIU Y., CHEN Q.*et al.*, "The very-low shear modulus of multi-walled carbon nanotubes determined simultaneously with the axial young's modulus via *in situ* experiments", *Advanced Functional Materials*, vol. 18, no. 10, pp. 1555–1562, 2008.

[WEN 03] WEN J., LAO J., WANG D. *et al.*., "Self-assembly of semiconducting oxide nanowires, nanorods, and nanoribbons", *Chemical Physics Letters*, vol. 372, no. 5–6, pp. 717–722, 2003.

[WON 97] WONG E.W., SHEEHAN, P.E., LIEBER C.M, "Nanobeam mechanics: elasticity, strength, and toughness of nanorods and nanotubes", *Science*, vol. 277, no. 5334, pp. 1971–1975, 1997.

[XIA 03] XIA Y., YANG P., SUN Y.*et al.*, "One-dimensional nanostructures: synthesis, characterization, and applications", *Advanced Materials*, vol. 15, no. 5, pp. 353–389, 2003.

[XIA 06] XIAO S., HOU W., *Fullerenes Nanotubes and Carbon Nanostructures,* vol. 14, pp. 9–16, 2006.

[YAN 08] YANG J., JIA X., KITIPORNCHAI S., "Pull-in instability of nano-switches using nonlocal elasticity theory", *Journal of Physics D-Applied Physics*, vol. 41, no. 1, 2008.

[YAN 10a] YANG J., KE L.L., KITIPORNCHAI S., "Nonlinear free vibration of single-walled carbon nanotubes using nonlocal Timoshenko beam theory", *Physica E*, vol. 42, p. 1727, 2010.

[YAN 10b] YAN Y., WANG W.Q., ZHANG L.X., "Nonlocal effect on axially compressed backling of triple-walled carbon nanotubes under temperature field", *Appl. Math. Model.*, vol. 34, p. 3422, 2010.

[YOO 11] YOON J.W., HWANG H.J., "Molecular dynamics modeling and simulations of a single-walledcarbon-nanotube-resonator encapsulating a finite nanoparticle", *Computational Materials Science* vol. 50, pp. 2741–2744, 2011.

[YU 08] YU S. S., ZHENG W. T., JIANG Q., "Oxidation of graphene nanoribbon by molecular oxygen", *IEEE Transactions on Nanotechnology*, vol. 7, no. 5, pp. 628–635, 2008.

[ZHA 04] ZHANG H.Y., GU X., ZHANG X.H. *et al.*, "Gong", *Phys. Lett. A.,* pp. 331–332, 2004.

[ZHA 08] ZHANG Y.Q., LU Y., WANG S.L. *et. al.*, "Vibration and buckling of a double-beam system under compressive axial loading", *J. Sound Vib.*, vol. 318, pp. 341–352, 2008.

[ZHA 09a] ZHAO H., MIN K., ALURU N., *Nano Letters,* vol. 9, p. 3012, 2009.

[ZHA 09b] ZHAO Y.H., GUO Y.Z., WEI Q. *et al.*, "Influence of specimen dimensions and strain measurement methods on tensile stress-strain curves", *Materials Science and Engineering A-Structural Material Properties Microstructure and Processing*, vol. 525, nos. 1–2, pp. 68–77, 2009.

[ZHA 10] ZHANG Y.Y., WANG C.M., CHALLAMEL N., "Bending, buckling, and vibration of micro/nanobeams by hybrid nonlocal beam model", *Journal of Engineering Mechanics-ASCE*, vol. 136, p. 562, 2010.

[ZHU 04] ZHU F.Q., FAN D., ZHU X. *et al.*, "Ultrahigh density arrays of ferromagnetic nanorings on macroscopic areas", *Advanced Materials*, vol. 16, nos. 23–24, pp. 2155–2159, 2004.

[ZHU 10] ZHU Y., MURALI S., CAI W. *et al.*, "Graphene and graphene oxide: synthesis, properties, and applications", *Advanced Materials*, vol. 22, no. 35, pp. 3906–3924, 2010.

[ZIE 10] ZIENERT A., SCHUSTER J., STREITER R. *et al.*, "Transport in carbon nanotubes: contact models and size effects", *Physica Status Solidi (B)–Basic Solid State Physics,* vol. 247, pp. 3002–3005, 2010.

[ZIE 91] ZIENKIEWICZ O.C., TAYLOR R.L., *The Finite Element Method*, 4th edition, McGraw-Hill, London, 1991.

Index

A, B, C

D, E, F

I, L, M, N

P, R S

T, U, V, W, Y, Z

Other titles from

ISTE

in

Mechanical Engineering and Solid Mechanics

2015

SAB Karam, LEBÉE Arthur
Homogenization of Heterogeneous Thin and Thick Plates

2014

ATANACKOVIC M. Teodor, PILIPOVIC Stevan, STANKOVIC Bogoljub, ZORICA Dusan
Fractional Calculus with Applications in Mechanics: Vibrations and Diffusion Processes

ATANACKOVIC M. Teodor, PILIPOVIC Stevan, STANKOVIC Bogoljub, ZORICA Dusan
Fractional Calculus with Applications in Mechanics: Wave Propagation, Impact and Variational Principles

CIBLAC Thierry, MOREL Jean-Claude
Sustainable Masonry: Stability and Behavior of Structures

ILANKO Sinniah, MONTERRUBIO Luis E., MOCHIDA Yusuke
The Rayleigh–Ritz Method for Structural Analysis

LALANNE Christian
Mechanical Vibration and Shock Analysis – 5-volume series – 3rd edition
Sinusoidal Vibration – volume 1
Mechanical Shock – volume 2
Random Vibration – volume 3
Fatigue Damage – volume 4
Specification Development – volume 5

LEMAIRE Maurice
Uncertainty and Mechanics

2013

ADHIKARI Sondipon
Structural Dynamic Analysis with Generalized Damping Models: Identification

ADHIKARI Sondipon
Structural Dynamic Analysis with Generalized Damping Models: Analysis

BAILLY Patrice
Materials and Structures under Shock and Impact

BASTIEN Jérôme, BERNARDIN Frédéric, LAMARQUE Claude-Henri
Non-smooth Deterministic or Stochastic Discrete Dynamical Systems: Applications to Models with Friction or Impact

EL HAMI Abdelkhalak, BOUCHAIB Radi
Uncertainty and Optimization in Structural Mechanics

KIRILLOV Oleg N., PELINOVSKY Dmitry E.
Nonlinear Physical Systems: Spectral Analysis, Stability and Bifurcations

LUONGO Angelo, ZULLI Daniele
Mathematical Models of Beams and Cables

SALENÇON Jean
Yield Design

2012

DAVIM J. Paulo
Mechanical Engineering Education

DUPEUX Michel, BRACCINI Muriel
Mechanics of Solid Interfaces

ELISHAKOFF Isaac *et al.*
Carbon Nanotubes and Nanosensors: Vibration, Buckling and Ballistic Impact

GRÉDIAC Michel, HILD François
Full-Field Measurements and Identification in Solid Mechanics

GROUS Ammar
Fracture Mechanics – 3-volume series
Analysis of Reliability and Quality Control – volume 1
Applied Reliability – volume 2
Applied Quality Control – volume 3

RECHO Naman
Fracture Mechanics and Crack Growth

2011

KRYSINSKI Tomasz, MALBURET François
Mechanical Instability

SOUSTELLE Michel
An Introduction to Chemical Kinetics

2010

BREITKOPF Piotr, FILOMENO COELHO Rajan
Multidisciplinary Design Optimization in Computational Mechanics

DAVIM J. Paulo
Biotribolgy

PAULTRE Patrick
Dynamics of Structures

SOUSTELLE Michel
Handbook of Heterogenous Kinetics

2009

BERLIOZ Alain, TROMPETTE Philippe
Solid Mechanics using the Finite Element Method

LEMAIRE Maurice
Structural Reliability

2007

GIRARD Alain, ROY Nicolas
Structural Dynamics in Industry

GUINEBRETIÈRE René
X-ray Diffraction by Polycrystalline Materials

KRYSINSKI Tomasz, MALBURET François
Mechanical Vibrations

SIH George C. *et al.*
Particle and Continuum Aspects of Mesomechanics

Printed and bound by CPI Group (UK) Ltd, Croydon, CR0 4YY

13/09/2026

14961076-0007